Springer Series in Advanced Microelectronics

Volume 50

The Springer Series in Advanced Microelectronics provides systematic information on all the topics relevant for the design, processing, and manufacturing of microelectronic devices. The books, each prepared by leading researchers or engineers in their fields, cover the basic and advanced aspects of topics such as wafer processing, materials, device design, device technologies, circuit design, VLSI implementation, and subsystem technology. The series forms a bridge between physics and engineering and the volumes will appeal to practicing engineers as well as research scientists.

More information about this series at http://www.springer.com/series/4076

Bernhard Goll · Horst Zimmermann

Comparators in Nanometer CMOS Technology

Springer

Bernhard Goll
Horst Zimmermann
Institute of Electrodynamics, Microwave
and Circuit Engineering (EMCE)
TU Wien
Vienna
Austria

ISSN 1437-0387 ISSN 2197-6643 (electronic)
ISBN 978-3-662-44481-8 ISBN 978-3-662-44482-5 (eBook)
DOI 10.1007/978-3-662-44482-5

Library of Congress Control Number: 2014946766

Springer Heidelberg New York Dordrecht London

Printed on acid-free paper

Springer is part of Springer Science+Business Media (www.springer.com)

Preface

This book describes circuit engineering efforts for comparators. Comparators are key components for analog-digital converters and consequently for modern wireless and mobile communication. Electronic circuits for latest mass applications like WIFI or WLAN modems, mobile phones, and smart phones are realized as so-called systems on chip (SoC). Such SoCs are realized in nanometer CMOS technology and contain a lot of digital circuits for digital signal processing but they also contain analog circuits which often form the key devices for good overall performance. The importance of comparators nowadays is often underestimated. They are, however, important to obtain high-performance analog-digital converters and wireless receivers and transmitters. The improvement of analog-digital converters is the final issue in bringing the digital signal processing as close as possible to the antenna. Therefore, comparators need to be improved also. Sense-amplifier type comparators are also important for SRAMs and DRAMs as well as some optical receivers.

In fact, progress in CMOS technology and circuit design allows the revolution in modern wireless and mobile communication. UMTS, HSPA+, and LTE as well as Ultra Wide Band (UWB) and Software Defined Radio were important keywords in research of the last years for SoCs. All these applications require high-samplerate analog-digital converters, which rely on high-performance comparators. For application in mobile devices, low-voltage operation and low power consumption are important issues of comparators.

To make all this possible for a mass market, analog circuit design in nanometer CMOS technology is an important key factor. This book concentrates on one subtopic of analog circuit design, i.e., comparators. This is especially justified since in the literature very often only analog-digital converters are described without characterizing the properties of the comparators being implemented in detail. Starting from the basics of comparators and the poor transistor characteristics in nanometer CMOS, seven high-performance comparators developed by the authors in 120 nm and 65 nm CMOS are described extensively. These comparators cover sample rates from 500 MHz to 7 GHz and supply voltages from 0.5 to 1.5 V. Their power consumption is in the milliwatt range down to 18 μW. Detailed descriptions of measurement methods for the characterization of advanced comparators are introduced in addition.

Although comparators are being used since several decades in many integrated circuits, it was still possible to develop new topologies or to modify and expand known comparator topologies to improve their performance. This book introduces newest results of development of comparators in deep-sub-micron and nanometer CMOS and describes methods for comparator design dealing successfully with the nanometer hell of physics. Numerous detailed circuit diagrams and plots of measured results allow a fast comprehension.

The authors would like to thank their colleagues at the Institute of Electrodynamics, Microwave and Circuit Engineering at Vienna University of Technology for fruitful discussions and valuable support, especially Kerstin Schneider-Hornstein, Franz Schlögl, and Robert Kolm. Furthermore, special thanks are directed to A. Bertl, F. Kuttner, C. Sandner, L. Dörrer, M. Haas, and R. Petschacher from Infineon Technologies Austria AG in Villach for their technical and financial support as well as the opportunity to use the design environment.

Vienna, Austria

Bernhard Goll
Horst Zimmermann

Contents

Symbols

C_{ox}	Gate oxide capacitance per gate area $\left[\frac{\text{F}}{m^2}\right]$
C_x	Capacitance [F] (x stands for an index, which depends on the context)
D_x	Diode (x stands for an index, which depends on the context)
e	Elementary charge, 1.60218×10^{-19}As
f_x	Frequency [Hz] (x stands for an index, which depends on the context)
f_t	Transit frequency [Hz]
GB	Gain-bandwidth product
g_x	Transconductance of a transistor or small-signal conductance (x stands for an index, which depends on the context)
g_{DS}	Output conductance [S]
g_m	Effective small-signal transconductance of the latch of a comparator [S]
$high$	Logical high
I_{dsat}	Saturation current [A]
$\overline{i_{sn}^2}$	Mean-square of current due to shot noise [A^2]
$\overline{i_{tn}^2}$	Mean-square of current due to thermal noise [A^2]
I_x	Current [A] (x stands for an index, which depends on the context)
i_x	Small-signal current [A] (x stands for an index, which depends on the context)
k	Boltzmann constant, $1.38065 \times 10^{-23} \frac{\text{Ws}}{\text{K}}$
L_x	Gate length [µm] (x stands for an index, which denotes a transistor)
low	Logical low
M_{meta}	Average number of metastability errors per second $\left[\frac{1}{s}\right]$
N_x	n-MOS transistor (x is a number)
P[X]	Probability that an event X occurs
V_{ST}, V_{STx}	Switching threshold of an inverter [V] (x is the number of the inverter)
P_x	Power [dBm] (x stands for an index, which depends on the context)
Px	p-MOS transistor (x is a number)
Q_x	Charge [As] (x stands for an index, which depends on the context)

$Q(\zeta)$	Q-function
R_x	Resistor [Ω] (x stands for an index, which depends on the context)
r_x	Small-signal resistance [Ω] (x stands for an index, which depends on the context)
t_x	Time [s] (x stands for an index, which depends on the context)
u_x	Small-signal voltage [V] (x stands for an index, which depends on the context)
V_x	Voltage [V] (x stands for an index, which depends on the context)
V_{tx}	Threshold voltage [V] (x stands for an index, which denotes a transistor)
V_A	Early voltage [V]
V_{DD}	Supply voltage [V]
V_{SS}	Ground/reference potential (0V)
V_{GD}	Gate-drain voltage
V_{DS}	Drain-source voltage
ΔV_0	Initial voltage difference, which causes a latch to regenerate
$\overline{v^2{}_{fn}}$	Mean square of voltage due to flicker (1/f) noise [V^2]
$\overline{v^2{}_{tn}}$	Mean square of voltage due to thermal noise [V^2]
W_x	Gate width [μm] (x stands for an index, which denotes a transistor)
β_x	Transconductance parameter of a MOS transistor $\left[\frac{A}{V^2}\right]$ (x stands for an index, which denotes a transistor)
ε_r	Relative permittivity (depends on material)
ε_0	Vacuum permittivity $8.85419 \times 10^{-12}\,\frac{\mathrm{As}}{\mathrm{Vm}}$
ϑ	Temperature in Kelvin [K]
μ_n, μ_p	Mobility of an electron (index n) and a hole (index p) $\left[\frac{\mathrm{cm}^2}{\mathrm{Vs}}\right]$
μ_{OS}	Mean offset voltage [V]
μ_X	Mean of X (unit depends on X)
σ_N	Standard deviation of noise, which is assumed to be a stationary stochastic process with a mean free Gaussian pdf [V]
σ_{OS}	Standard deviation of the offset voltage [V]
σ_X	Standard deviation of X (unit depends on X)

Acronyms

AC	Alternating Current
ADC	Analog-to-Digital Converter
AND	Logical AND
BER	Bit Error Ratio
BiCMOS	Bipolar and CMOS
BPG	Bit Pattern Generator
BPR	Bit Pattern Receiver
CGRAM	Character Generator RAM
CGROM	Character Generator ROM
CML	Current Mode Logic
CMOS	Complementary Metal-Oxide-Semiconductor
COM	Communication Port
DAC	Digital-to-Analog Converter
DC	Direct Current
DEMUX	Demultiplexer
DMM	Digital Multimeter
DRAM	Dynamic Random Access Memory
DSP	Digital Signal Processor
FET	Field Effect Transistor
GUI	Graphical User Interface
HF	High Frequency
HSPA+	High Speed Packet Access +
I2C	Inter-IC bus
IC	Integrated Circuit
IOS	Input Offset Storage
LCD	Liquid Crystal Display
LSB	Least Significant Bit
LP	Low-Power
LTE	Long Term Evolution
MC	Microcontroller

MCML	MOS Current Mode Logic
MOS	Metal-Oxide-Semiconductor
MUX	Multiplexer
NOR	Not OR
NRZ	Non-Return-to-Zero
OOS	Output Offset Storage
OpAmp	Operational Amplifier
OR	Logical OR
pdf	Probability density function
PC	Personal Computer
PDN	Pull-Down Network
PLL	Phase Locked Loop
PRBS	Pseudo-Random-Bit-Sequence
PUN	Pull-Up Network
RAM	Random Access Memory
RF	Radio Frequency
ROM	Read Only Memory
SDR	Software Defined Radio
SMA	Sub-Miniature A
SMB	System Management Bus
SNM	Static Noise Margin
SNR	Signal-to-Noise Ratio
SoC	Systems on a Chip
SPI	Serial Peripheral Interface
SRAM	Static Random Access Memory
STI	Shallow Trench Isolation
UART	Universal Asynchronous Receiver Transmitter
UDSM	Ultra-Deep Submicron
UMTS	Universal Mobile Telecommunications System
USB	Universal Serial Bus
UWB	Ultra Wide Band
VGA	Variable Gain Amplifier
VNA	Vector Network Analyzer
WLAN	Wireless Local Area Network
WPAN	Wireless Personal Area Networks
XNOR	Exclusive NOR
XOR	Exclusive OR

Chapter 1
Introduction

In this introduction the motivation for investigating deep-sub-micron and nanometer CMOS comparators in detail is described. Furthermore, comparators are classified and different types of analog continuous-time and clocked comparators are introduced. In addition digital comparators are shortly addressed.

1.1 Motivation

The further progress of technology is in most cases coupled with the increasing demand of handling more operations in a shorter time. As a tendency more functional blocks are implemented into the digital domain to have the possibility of easier and more individual implementation and adaptation with software. At the link between the analog and digital domain, analog-to-digital converters (ADCs) translate analog signals into digital values, which are discrete in time and amplitude. The translation from the digital into the analog domain is done by digital-to-analog converters (DACs). A fast computation in the digital domain also forces among other things the need of fast ADCs. For example in the future fast ADCs will be used in Ultra-WideBand (UWB) communication systems and in Software Defined Radio (SDR):

- The UWB technology [1] will cover wireless data transmission in the frequency band of 3.1–10.6 GHz. Data transmissions of higher than 100 MBit/s in distances of below 10 m are intended for Wireless Personal Area Networks (WPANs). In an UWB receiver ADCs may be placed in the base-band for a further decoding of the received signal with e.g. a Digital Signal Processor (DSP). So functions like decoding, correction sequences or controlling of e.g. Variable Gain Amplifiers (VGAs) or Phase Locked Loops (PLLs) in the receiver are implemented in an easier way with software.
- The concept of a SDR transceiver [2] is to place the ADC and DAC as close as possible to the antenna (see Fig. 1.1). This makes the transceiver to be reconfigured

B. Goll and H. Zimmermann, *Comparators in Nanometer CMOS Technology*,
Springer Series in Advanced Microelectronics 50, DOI 10.1007/978-3-662-44482-5_1

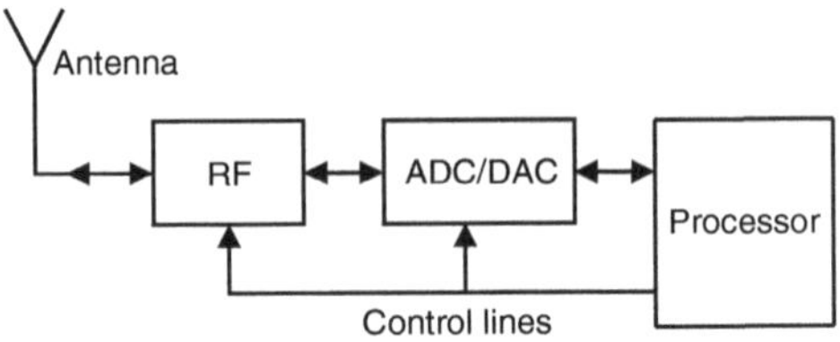

Fig. 1.1 Concept of an ideal SDR transceiver

very easily with the help of changing or extending the software. Various wireless standards may be handled with only a moderate amount of hardware components, which as a consequence reduces costs. It should be noticed, that in today's SDR transceivers there are many trade-offs, because the ideal implementation is very optimistic and would consume a considerable amount of power with state-of-the-art technology.

To save costs of production steps and to minimize the final product, several functional circuit blocks are combined into one chip and form a System on a Chip (SoC). So it is common to e.g. implement digital and analog blocks into one chip with additional ADCs or DACs. A key element in an ADC is the comparator. It compares an input voltage or signal with a reference voltage or another signal and has as a result a logical stage, which indicates whether the voltage at one input is lower or higher than at the other input at a distinct time point. In most cases in fast ADCs, e.g. a flash ADC or a folding converter system [3, 4], many parallel clocked regenerative comparators are implemented, which work in parallel. They use positive feedback to force a fast decision during a half clock period. The rest of the clock is generally used to reset the decision of the comparator.

Other applications of clocked regenerative comparator circuits are their usages in an analog rank-order extractor [5] or as a voltage sense amplifier in a SRAM [6–8] or for data regeneration in e.g. a fast demultiplexer [9, 10].

This work focuses on clocked regenerative comparators for an intended implementation in an ADC, but also other applications may be considered. The demands to the comparator are among many others e.g. a low power consumption, a high sampling clock, mostly a small area consumption on the chip and a sufficient sensitivity and offset.

1.2 Classification of Comparators

The principle electrical function of a comparator [11, 12] is depicted in Fig. 1.2. A comparator compares the input value VA_p with VA_n and gives as result a logical value at the output which determines, whether the VA_p is higher or lower then VA_n. In Fig. 1.2 the logical value is represented by an non-inverted output OUT and an inverted output $\overline{OUT}$. In another point of view it can be said that a comparator detects the sign of the difference $VA_p - VA_n$. The following sections give a rough general overview of different types of comparators, which may be implemented in CMOS circuit design.

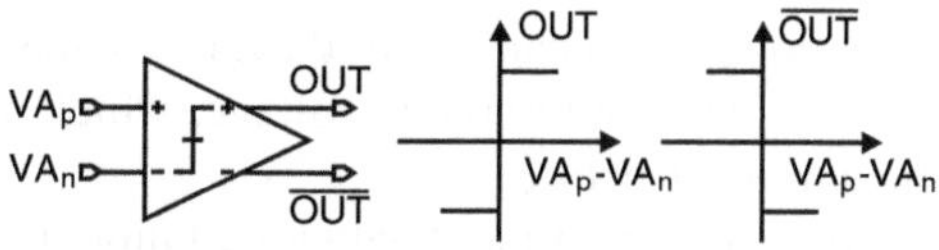

Fig. 1.2 Principle electrical function of a comparator

1.2.1 Comparators for Comparing Analog Values

Classified to the different types of analog input values, in CMOS circuit design typically comparators may compare voltages, currents or charges. As a result the comparator gives a logical value, e.g. the voltage levels of the supply voltage and the ground level for logical *high* and *low*, which indicates, whether one input value is higher or lower than the other one. Equality of both input values are mostly not treated, because the probability of appearing is extremely minimized. For this it is common, that a comparator is designed to have as high gain as possible to detect even small differences at the inputs. Other possibilities for treating equality are to introduce a hysteresis or adding sub-circuits for indication of equal input values.

Comparison of inputs can also be done by the comparator at distinct time points. Mostly these comparators need a clock or trigger signal. So it can be distinguished in a rough way between clocked or continuous-time comparators. Clocked comparators have to do their decision in a limited time slot. To minimize probability of appearance of errors due to equality of input values, in most cases positive feedback is used to enhance gain. In the literature sometimes clocked comparators are called dynamic comparators. This is due to the similarity to the function of dynamic logic [13], which needs during a clock period a pre-charge phase and a following evaluation phase, where a pull-up network (PUN) and/or a pull-down network (PDN) defines the logic state. Figure 1.3 shows the basic concepts of dynamic and complementary static logic styles. Complementary static CMOS logic, where the simplest implementation is the static CMOS inverter, consists of only a PUN and a PDN without a clock.

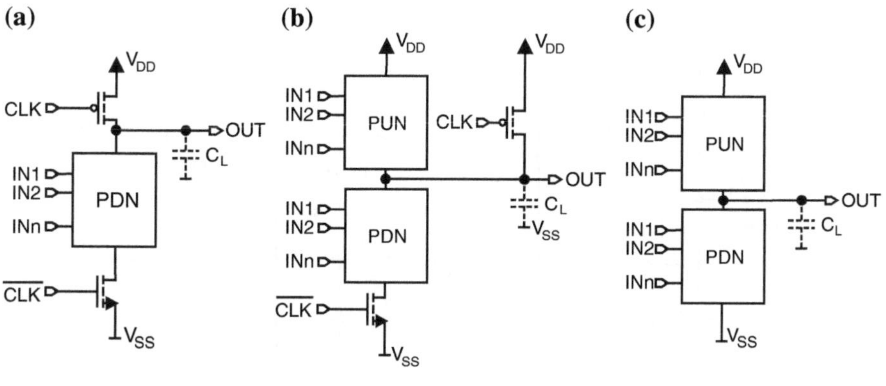

Fig. 1.3 Difference between static and dynamic logic. **a** Dynamic logic gate with pull-down network. **b** Dynamic complementary logic gate. **c** Static complementary logic gate

A comparator can also be designed to have only one input value, which is compared with an internal reference voltage, which may be defined by the operating point of the circuit structure.

To compare an input value with more than one reference values, multilevel comparators are built. They are used in e.g. multilevel logic, where more than two states exist in opposite to *high* and *low* of conventional logic or e.g. in a simple 1.5-bit ADC [14].

Figures 1.4, 1.5 and 1.6 show some examples of different types of CMOS comparators from the literature. A simple realization of a non-clocked voltage comparator is the **uncompensated OpAmp (operational amplifier) comparator** [15, 16], which is depicted in Fig. 1.4a. The input differential amplifier consists of input transistors $P1$ and $P2$, a current source transistor $P0$ and an active current-mirror load $N0$ and $N1$ for differential to single ended conversion. The output voltage OUT is generated by additional amplification with an active-loaded, common-source amplifier (transistors

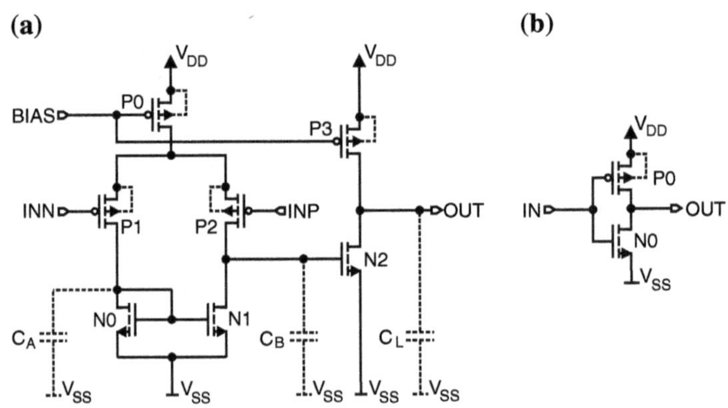

Fig. 1.4 Continuous-time CMOS voltage comparators. **a** OpAmp comparator [15, 16]. **b** Simple inverter as comparator [17]

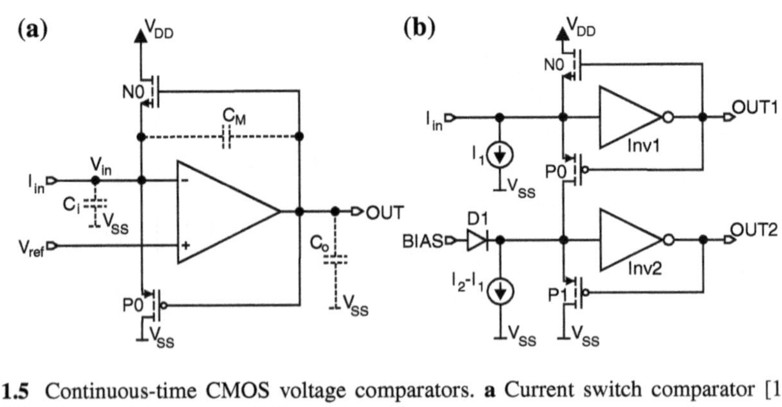

Fig. 1.5 Continuous-time CMOS voltage comparators. **a** Current switch comparator [19, 20]. **b** Multilevel current comparator [14]

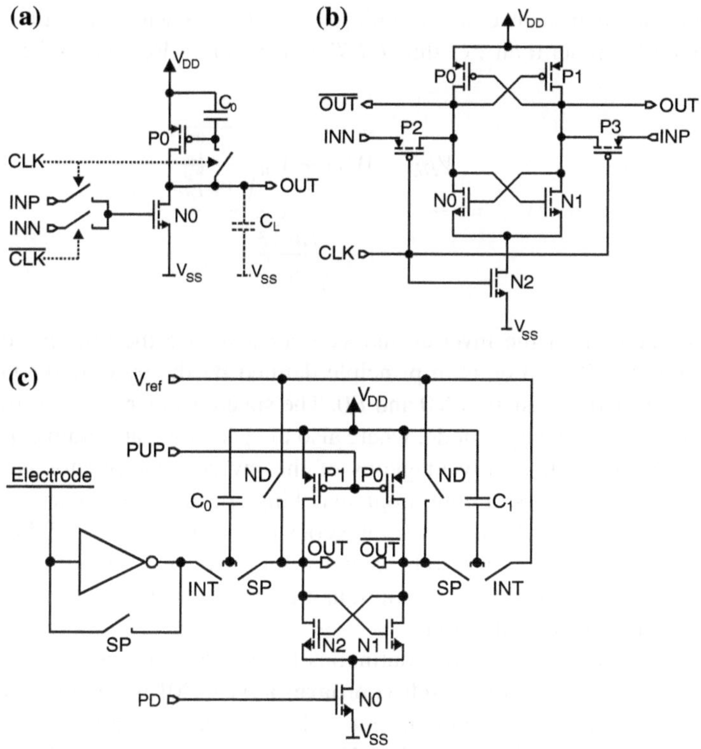

Fig. 1.6 Clocked CMOS comparators. **a** Simple dynamic CMOS voltage comparator [21]. **b** Dynamic CMOS latch [15]. **c** CMOS charge comparator [22]

$N2$ and $P3$). Here a typical compensation (e.g. Miller compensation) of the OpAmp is omitted to somewhat speed up the comparator and because to a comparator usually no circuitry for feedback is attached in opposite to an OpAmp. This comparator structure is still inherent slow, because during switching the internal parasitic capacitors C_A, C_B and the load capacitance C_L at the output node OUT have to be charged. Furthermore the minimum input voltage difference $INP-INN$, which causes a valid logical level at the output is defined by the amplification of the OpAmp comparator. In most Deep-Sub-Micron processes this amplification is typically low (in the area of 40 dB) due to different degrading effects of short channel MOS transistors. In opposite if these channels are designed longer, the parasitic capacitances increase due to larger gate-source or gate-drain capacitances thus reducing speed.

Figure 1.4b shows a **simple inverter used as comparator**. Such comparators were implemented in a flash ADC in [17]. Here the switching threshold V_{ST} of the inverter is the reference voltage to be compared, which is defined as the input voltage IN, that causes an equal output voltage OUT [18]. V_{ST} can be calculated in the ideal case with 1.1, where V_{tn} and V_{tp} are the threshold voltages, W_n and W_p the gate widths and L_n and L_p the gate lengths of transistors $N0$ and $P0$ respectively.

The mobility of electrons is denoted with μ_n and of holes with μ_p. If the voltage at IN is sufficiently higher than V_{ST} than OUT will switch to V_{SS} and in the other case to V_{DD}.

$$V_{ST} = \frac{V_{DD} - |V_{tp}| + V_{tn}\sqrt{\frac{\mu_n \frac{W_n}{L_n}}{\mu_p \frac{W_p}{L_p}}}}{1 + \sqrt{\frac{\mu_n \frac{W_n}{L_n}}{\mu_p \frac{W_p}{L_p}}}} \tag{1.1}$$

The amplification of the inverter and as a consequence the minimal detectable voltage difference $IN - V_{ST}$ is in principle defined by the transconductances and output resistances of transistors $N0$ and $P0$. The speed is determined mostly by the load capacitance of the output node, where also the parasitic gate-drain capacitances of $N0$ and $P0$ are added. Advantageous of the inverter comparator is the small amount of used transistors, a quite high switching speed and in most cases only a dynamic current consumption. Disadvantageous is, that V_{ST} suffers on the influence of mismatch, that the layout has to be changed for different V_{STH} and that there are several cases of static current consumption, e.g. if IN is between $V_{DD} - |V_{tp}|$ and $V_{DD} + |V_{tn}|$ for a rough estimation.

Figure 1.5 shows examples of continuous-time current comparators, where in Fig. 1.5a a so called **current switch comparator** [19, 20] is depicted. This comparator detects, whether the input current I_{in} is positive or negative. If as an initial condition is assumed, that $V_{in} = OUT = V_{ref}$ and transistors $N0$ and $P0$ are switched off, then a small positive input current I_{in}, which flows into the input node, will raise V_{in} slightly. This causes the output voltage OUT to decrease more than V_{in} raises due to the amplifier. Transistor $P0$ turns on and holds as a consequence V_{in} at level V_{ref}. The input node becomes a virtual ground. In the opposite case of a negative current I_{in} transistor $N0$ holds the input node at V_{ref}. OUT reaches a logical level and indicates, if I_{in} is positive or negative. If an ideal amplifier is assumed and the influence of the parasitic capacitance C_M, which consist of gate-source capacitances of $N0$ and $P0$, is neglected, a simple expression for $\Delta OUT(t)$ can be found with $\Delta V_{in} = (I_{in}/C_{in})t$ (see 1.2). GB is the gain-bandwidth of the amplifier.

$$\Delta OUT(t) = -\frac{GB}{2}\frac{I_{in}}{C_{in}}t^2 \tag{1.2}$$

Because of the parasitic capacitor C_M degrades the comparator's behavior, in [20] a current steering comparator was introduced, where instead of Fig. 1.5a the drain of $N0$ and $P0$ is connected to the output node and each gate is biased with an extra bias voltage.

To give an example of an **multilevel current comparator**, the comparator of Fig. 1.5a can be extended to that one of Fig. 1.5b [14]. Inverters (see Fig. 1.4b) are used instead of amplifiers, where V_{ref} from Fig. 1.5b is replaced by the inverter switching thresholds. Diode $D1$ with bias voltage $BIAS$ is only added to speed up the circuit minimizing the voltage swing, where *BIAS* is chosen low enough to keep the node at the input of inverter Inv2 below the switching thresholds of Inv1 and Inv2. Following it will be assumed, that the current differences are sufficiently large so that the outputs of Inv1 and Inv2 switches between valid logical voltage levels like V_{DD} or V_{SS}. In the case that I_{in} is smaller than I_1 and I_1 smaller than I_2, $P0$ is off and $N0$ turns on to compensate $I_1 - I_{in}$ thus $OUT1$ and $OUT2$ are at V_{DD}. If I_{in} is between I_1 and I_2, $P0$ turns on and $OUT1$ switches to V_{SS}. The current difference $I_{in} - I_1$ flows to the input node of inverter Inv2 and is compared with the current difference $I_2 - I_1$. So $OUT2$ is V_{DD} if $I_{in} < I_2$ and switches to V_{SS}, if $I_{in} > I_2$. The application of such a multilevel comparator may be a simple 1.5-bit ADC.

Examples of different **clocked comparators** are given in Fig. 1.6. A **simple dynamic comparator** is depicted in Fig. 1.6a [21]. If *CLK* is *high*, pin *INP* is connected to the gate of transistor $N0$. Due to the fact, that transistor $P0$ is connected as MOS-diode and a capacitive load is assumed at node OUT, the appearing output voltage level is stored in capacitance C_0 (pre-charge phase). When *CLK* changes to *low*, *INP* is disconnected from the gate of $N0$, OUT is disconnected from the gate of $P0$ and instead *INN* is connected to the gate of $N0$. So in this phase the gate-source voltage of $P0$, caused by the current through $P0$ and $N0$ from the previous clock phase is stored in C_0. This current is now compared with the drain current, caused by *INN* (evaluation phase). If *INP* is larger than *INN*, the output node is charged by the difference current and the voltage level at OUT raises. In the opposite case node OUT is discharged and the voltage at OUT decreases. The advantage of this type of comparator is, that in principle the problem of transistor mismatch is avoided. Otherwise care has to be taken to parasitic charge injection of switches and the parasitic gate-drain capacitance of $P0$, which degrades the gain during evaluation. In [21] this comparator was built in a 0.9 μm CMOS process with a power supply of 1.5–3 V. At 3 V an offset of 1 mV was achieved.

The fastest types of comparators use **positive feedback**. In Fig. 1.6b a comparator is built with a **dynamic CMOS latch** [15]. The principle circuit block consists of two inverters $N0$-$P0$ and $N1$-$P1$, which are cross-coupled to a latch to achieve positive feedback. When *CLK* is *low*, transistors $P2$ and $P3$ are switched on and transistor $N2$ is switched off thus disconnecting current flow to V_{SS}. The output nodes OUT and $\overline{OUT}$ are pre-charged with *INP* and *INN*. When *CLK* changes to *high*, the latch is connected to V_{SS} and nodes *INP* and *INN* are disconnected. Due to the initial imbalance at the output nodes from the previous clock phase and the positive feedback, the latch pulls OUT to V_{DD} and $\overline{OUT}$ to V_{SS}, if *INP* was higher than *INN* and vice versa in the case *INP* is smaller than *INN*. The positive feedback of the latch is also called the regeneration of the latch. Due to the fast switching and the inherent high gain, caused by positive feedback, similar types of comparators are widely used in fast flash ADCs.

A charge comparator, which also has two n-MOS transistors $N1$ and $N2$ connected to positive feedback, is depicted in Fig. 1.6c [22]. The goal of this design was to measure the charge, which accumulates during a distinct time period to the electrode. The function of the comparator is divided into three phases. In the first phase, switches *ND* and *INT* are on, switch *SP* is off and transistors $P0$, $P1$ (*PUP*) and $N0$ (*PD*) are off. The charge, which is collected by the electrode causes a small change in the electrode voltage, which is amplified by the inverter and finally stored in capacitor C_0. A reference voltage V_{ref} is connected to capacitor C_1 and nodes OUT and $\overline{OUT}$ are also pre-charged to V_{ref}. In the second phase *SP* is switched on and *INT* is released, while the other switches keep their position. The output of the inverter and the electrode is reset to the inverter's switching threshold. Node OUT is charged by capacitor C_0 and node $\overline{OUT}$ is charged from reference capacitor C_1. In the third phase switches *ND* and *SP* are off and switch *INT* is turned on. *PUP* changes to *low* and *PD* switches to *high* thus turning load transistors $P0$ and $P1$ and transistor $N0$ on. The latch regenerates due to positive feedback depending on the charge difference at nodes OUT and $\overline{OUT}$. In this phase capacitor C_0 begins to store the output voltage of the inverter, which depends on the accumulated charge, for next comparison. Capacitor C_1 is connected to V_{ref}. After a sufficient charge accumulation time the first phase is initiated once again.

This book concerns on **clocked regenerative comparators**, which compares two voltage levels and are designed in Deep-Sub-Micron processes. Such comparators are widely used in integrated flash ADCs.

1.2.2 Comparators for Comparing Digital Values

Comparators, which compare digital, binary values, are implemented in most cases in microprocessors. Figure 1.7 shows an example of an implementation of a comparator, which compares two 4-bit values *A* and *B* [23]. The XOR gate determines at what bits *A* differs from *B* and places there a "1". Equality of *A* and *B* can be detected by using a NOR gate at the 4-bit result after the XOR gate. Only if a value 0000*b* appears after the XOR gate, the NOR gate indicates a equality by having a "1" as result. With the priority encoder the most significant "1" (= most significant unequal bit of *A* and *B*) is detected from the value after the XOR gate. The 4-bit result, which contains the

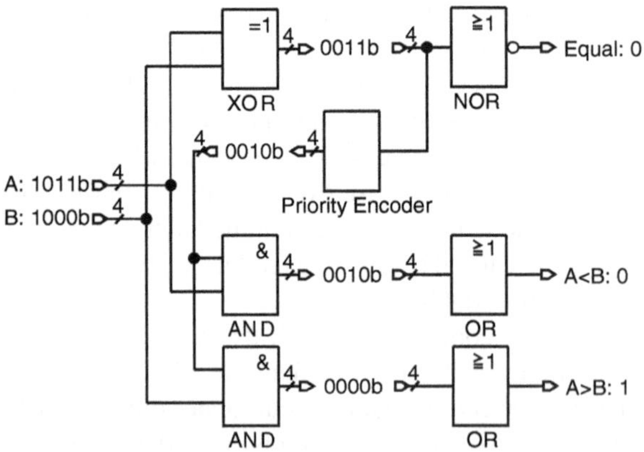

Fig. 1.7 Block diagram of a 4-bit comparator with a priority encoder [23]

most significant bit is used to detect from which value A or B it is originated. For this the AND- and OR gates are used. As a result a "1" at $A > B$ and a "0" at $A < B$ indicate that A is bigger than B and vice versa.

Comparators for comparing digital binary values are beyond the scope of this book.

References

1. T.K.K. Tsang, M.N. El-Gamal: Ultra-Wideband (UWB) Communication Systems: An Overview, in IEEE Northeast Workshop on Circuits and Systems, June 2005, pp. 381–386
2. J. Mitola, The software radio architecture. IEEE Commun. Mag. **33**(5), 26–38 (1995)
3. R. van de Plasche, *CMOS Integrated Analog-to-Digital and Digital-to-Analog Converters* (Kluwer Academic Publishers, Boston, 2003)
4. R. Patzelt, H. Schweinzer, *Elektrische Messtechnik* (Springer, New York, 1996)
5. Y.-C. Hung, B.-D. Liu, 1-V CMOS comparator for programmable analog rank-order extractor. IEEE Trans. Circuits Syst. I **50**(5), 673–677 (2003)
6. B. Wicht, *Current Sense Amplifiers for Embedded SRAM in High-Performance System-on-a-Chip Designs* (Springer, Heidelberg, 2003)
7. B. Wicht, T. Nirschl, D. Schmitt-Landsiedel, Yield and speed optimization of a latch-type voltage sense amplifier. IEEE J. Solid-State Circuits **39**(7), 1148–1158 (2004)
8. B. Wicht, T. Nirschl, D. Schmitt-Landsiedel: A Yield-Optimized Latch-Type SRAM Sense Amplifier. IEEE European Solid-State Circuits Conference, September 2003, pp. 409–412
9. Y. Okaniwa, H. Tamura, M. Kibune, D. Yamazaki, T.-S. Cheung, J. Ogawa, N. Tzartzanis, W.W. Walker, T. Kuroda, A 40-Gb/s CMOS clocked comparator with bandwidth modulation technique. IEEE J. Solid-State Circuits **40**(8), 1680–1685 (2005)
10. Y. Okaniwa, H. Tamura, M. Kibune, D. Yamazaki, T.-S. Cheung, J. Ogawa, N. Tzartzanis, W.W. Walker, T. Kuroda: A 0.11 μm CMOS Clocked Comparator for High-Speed Serial Communications. IEEE Symposium on VLSI Circuits, June 2004, pp. 198–201
11. F. Maloberti, *Analog Design for CMOS VLSI Systems* (Kluwer Academic Publishers, Boston, 2001)
12. P.E. Allen, D.R. Holberg, *CMOS Analog Circuit Design* (Oxford University Press, New York, 2002)
13. J.M. Rabaey, A. Chandrakasan, B. Nicolić, *Digital Integrated Circuits* (Prentice Hall, New York, 2003)
14. H. Gustat, Fast CMOS multilevel current comparator. IET Electron. Lett. **29**(7), 592–593 (1993)
15. R. Gregorian, *Introduction to CMOS Op-Amps and Comparators (A Wiley-Interscience Publication* (Wiley, New York, 1999)
16. P.R. Gray, P.J. Hurst, S.H. Lews, R.G. Meyer, *Analysis and Design of Analog Integrated Circuits* (Wiley, New York, 2001)
17. J. Yoo, D. Lee, K. Choi, A. Tangel: Future-Ready Ultrafast 8Bit CMOS ADC for System-on-Chip Applications. IEEE 14th Annual IEEE International ASIC/SOC Conference, September 2001, pp. 455–459
18. A.S. Sedra, K.C. Smith, *Microelectronic Circuits* (Oxford University Press, New York, 2004)
19. H. Trff, Novel approach to high speed cmos current comparator. IET. Electron. Lett. **28**(3), 310–312 (1992)
20. R. del Río-Fernández, G. Liñán-Cembrano, R. Domínguez-Castro, A. Rodríguez-Vázquez: A Mismatch-Insensitive High-Accuracy High-Speed Continuous-Time Current Comparator in Low Voltage CMOS, 2nd IEEE-CAS Region 8 Workshop on Analog and Mixed IC Design, September 1997, pp. 111–116

21. K. Azadet, A.G. Dickinson, D.A. Inglis, A mismatch free cmos dynamic comparator. IEEE Int. Symp. Circuits Syst. **3**, 2116–2119 (1995)
22. J.V. Hatfield, P.J. Hicks: A Sensitive Integrated CMOS Charge Comparator, IET Colloquium on Current Mode Analogue Circuits, February 1989, pp. 4/1–4/6
23. C.-H. Huang, J.-S. Wang, High-performance and power-efficient CMOS comparators. IEEE. J. Solid-State Circuits **38**(2), 254–262 (2003)

Chapter 2
Fundamentals of Clocked, Regenerative Comparators

The fastest ADC structure is the flash ADC, where in principle for a resolution of N bits $2^N - 1$ comparators are arranged in parallel for fast analog to digital conversion. Due to the large amount of placed comparators, the demands to the design of one comparator are mainly low power and area consumption, fast switching capabilities and a sufficiently low standard deviation of the input referred offset voltage to meet an intended resolution. Hence to meet the requirements of resolution, also pre-amplifiers may be placed in front of the comparator or offset compensation techniques are added.

The most popular comparators for flash ADCs are from the clocked and regenerative type. They use transistors, which are connected in positive feedback to force a fast decision with a small power and area consumption. Clocked, regenerative comparators are also used in e.g. data receivers, repeaters, or in memories as sense amplifiers.

2.1 The Static Latch

The fundamental circuit of most clocked, regenerative comparators is the static complementary latch [1], which consists of two cross-coupled static inverters to provide positive feedback (see Fig. 2.1). The first inverter consists of transistors $N1$, $P1$ and the second inverter is built with $N2$, $P2$, where both inverters equals each other in electric functionality and size. The characteristic of the static latch, which is built out of the transfer characteristic of both inverters, is depicted in Fig. 2.1b. It consists of one metastable- and two stable points. The metastable point V_M is the condition that the voltages of both nodes u_1 and u_2 are equal. If a small disturbance $\Delta V_0 = u_2(t = 0) - u_1(t = 0)$ is introduced, positive feedback starts. Depending on the polarity of ΔV_0, the latch switches to the stable point $u_1 = V_{DD}$ and $u_2 = V_{SS}$ in the case of a positive sign (see Fig. 2.1d). In the other case it switches to the stable point $u_1 = V_{SS}$ and $u_2 = V_{DD}$. The influence of a positive sign of ΔV_0 and as follows positive feedback is illustrated in Fig. 2.1b with dashed arrows in the characteristic. The small signal equivalent circuit of the static latch is depicted

B. Goll and H. Zimmermann, *Comparators in Nanometer CMOS Technology*,
Springer Series in Advanced Microelectronics 50, DOI 10.1007/978-3-662-44482-5_2

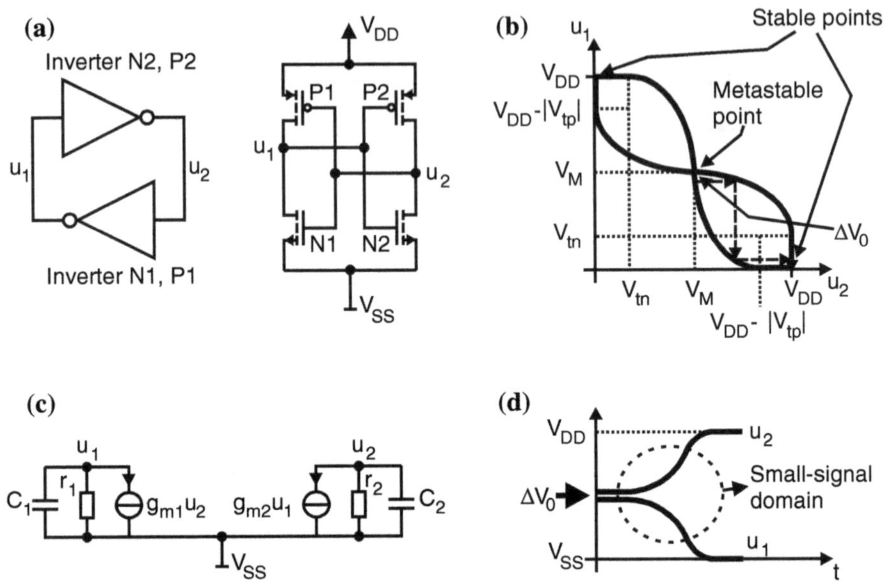

Fig. 2.1 The static latch [1]. **a** Cross-coupled inverters. **b** Characteristics. **c** Small-signal equivalent circuit. **d** Voltages versus time

in Fig. 2.1c. Because of both inverters are equal, it can be assumed, that also both inverter's transconductances, which are the sums of the appropriate transistor pair, are equal ($g_m = g_{m1} = g_{m2} = g_{mn1} + g_{mp1} = g_{mn2} + g_{mp2}$) and also both output resistances are equal ($r = r_1 = r_2 = r_{n1}||r_{p1} = r_{n2}||r_{p2}$). If furthermore it is assumed, that the capacitive loads at nodes u_1 and u_2 are the same ($C = C_1 = C_2$), then for the output difference $\Delta u(t) = u_2(t) - u_1(t)$ 2.1 can be written.

$$\frac{d}{dt}\Delta u(t) = \Delta u(t)\frac{1}{C}\left(g_m - \frac{1}{r}\right) \tag{2.1}$$

The result of differential equation 2.1 is an exponential raising of $\Delta u(t)$, where $\Delta u(t = 0) = \Delta V_0$ is the initial voltage difference (see 2.2). The exponential raising of the output voltage difference is limited by the supply rails of the latch.

$$\Delta u(t) = \Delta V_0 e^{\frac{g_m - \frac{1}{r}}{C}t} \tag{2.2}$$

To reach a distinct voltage difference $\Delta u(t_d) = \Delta V_{end}$ a delay time t_d has to be wait, which can be calculated out of 2.2 with the result in 2.2.

$$t_d = \frac{C}{g_m - \frac{1}{r}} ln\left(\frac{\Delta V_{end}}{\Delta V_0}\right), \quad g_m r > 1 \tag{2.3}$$

It can be seen, that the delay time t_d lasts longer, if C is larger and g_m, r and ΔV_0 are smaller.

In reality the characteristics of each inverter differs from the other due to mismatch. So if a comparator is in the reset phase, nodes u_1 and u_2 are connected together with a switch. The working point, which occurs due to shortening is in between the, now different, switching thresholds (see 1.1) of the inverters. So in the evaluation phase as a rough estimation, the initial voltage difference has to overcome the difference of switching thresholds. This can be seen with a low level calculation of 2.4 and 2.5, where in the small signal equivalent circuit of Fig. 2.1c the resistances are neglected, $C = C_1 = C_2$ ($\Delta C = 0$) is assumed and the switching thresholds V_{ST1} and V_{ST2} of inverters $N1$, $P1$ and $N2$, $P2$ are introduced.

$$(u_1(t) - V_{ST2})\left(g_m + \frac{\Delta g_m}{2}\right) + \left(C + \frac{\Delta C}{2}\right)\frac{du_2(t)}{dt} = 0 \tag{2.4}$$

$$(u_2(t) - V_{ST1})\left(g_m - \frac{\Delta g_m}{2}\right) + \left(C - \frac{\Delta C}{2}\right)\frac{du_1(t)}{dt} = 0 \tag{2.5}$$

With $\Delta u(t) = u_2(t) - u_1(t)$, $\overline{U} = 0.5(u_1(t) + u_2(t))$, $\Delta V_{ST} = V_{ST2} - V_{ST1}$, $\overline{V_{ST}} = 0.5(V_{ST2} + V_{ST1})$ and $\Delta C = 0$ the result of these equations is depicted in 2.6. $\overline{U}$ and $\overline{V_{ST}}$ are assumed to be constant.

$$\Delta u(t) = \left[\frac{\Delta g_m}{g_m}\left(\overline{U} - \overline{V_{ST}}\right) - \Delta V_{ST}\right] + \left[\Delta V_0 - \frac{\Delta g_m}{g_m}\left(\overline{U} - \overline{V_{ST}}\right) + \Delta V_{ST}\right] e^{\frac{g_m}{C}t} \tag{2.6}$$

In this simple model it can be seen, that the initial voltage difference ΔV_0 has to overcome $(\Delta g_m/g_m)(\overline{U} - \overline{V_{ST}}) - \Delta V_{ST}$, which can be interpreted as offset.

Another principle schematics of latches are shown in Fig. 2.2, where instead of the complementary static latch of Fig. 2.1 an active load is connected to only one cross-coupled transistor pair. Such latches are treated in a similar way. The difference

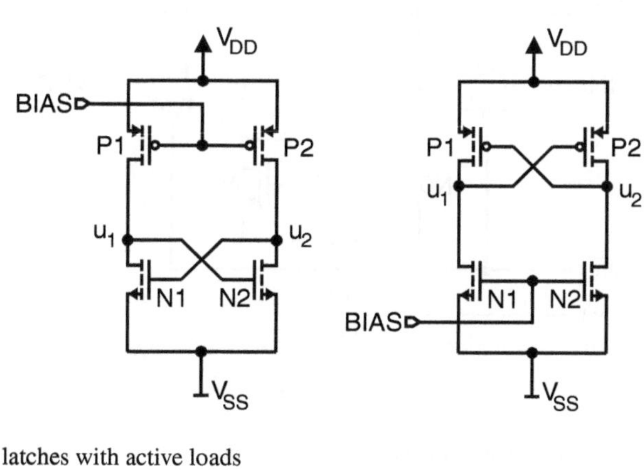

Fig. 2.2 Static latches with active loads

to the complementary latch is, that there is no switching between both supply rails. The voltage swing is defined by the current through the active load. The advantage is, that these latches have a better performance at lower supply voltages.

2.2 Basic Clocked, Regenerative Comparator Circuits

In the literature, there are several variants to implement clocked regenerative comparators. In this chapter the basic structures of the most popular circuits are described.

The latched comparator in Fig. 2.3 [2] is a direct consequence of Sect. 2.1. Typically offset compensated pre-amplifiers are added in front. There are several ways for the timing of the switches. In [2] three phases are implemented. In the first phase $\overline{SW}$ is *high*, *EN* is *low* and $\overline{RES}$ is *low*. So the latch is disconnected from the inputs and the power supply rails and both output nodes are shortened by transistor *P*3. In the next phase $\overline{RES}$ switches to *high* thus turning transistor *P*3 off. At the same time $\overline{SW}$ changes to *low* and the inputs are connected to the latch thus pre-charging the output nodes with an initial voltage difference. In the last phase the input nodes are disconnected from the latch, transistor *P*3 keeps off and *EN* changes to *high* to activate positive feedback of the latch. Depending on the initial voltage difference the latch switches at node *OUT* to *high* and at $\overline{OUT}$ to *low* or vice versa and keeps the state. For a new comparison the first phase is started again. Another possibility is to use only two phases, where the reset is released delayed [3]. So in the first phase

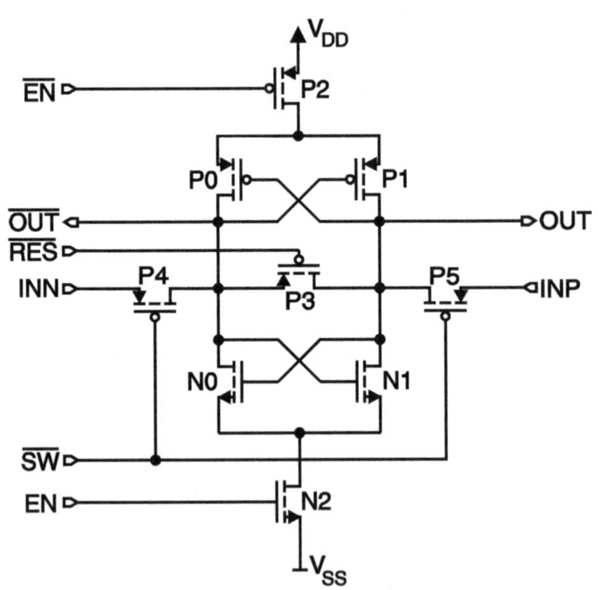

Fig. 2.3 Clocked regenerative dynamic latched comparator

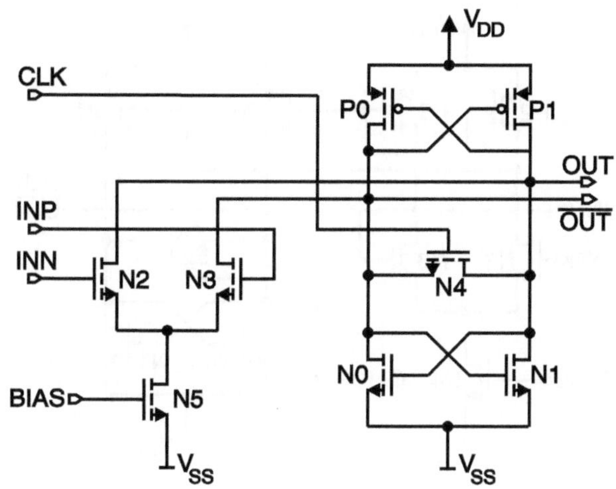

Fig. 2.4 Clocked regenerative comparator using a static latch with input differential pair

$\overline{SW}$ is *high*, *EN* is *low* and $\overline{RES}$ is *low* thus disconnecting the latch from the input voltages and power supply and shorting both output nodes. In the second phase $\overline{SW}$ is *low*, *EN* and $\overline{RES}$ changes to high for positive feedback. To overcome influence of charge injection of the switches and to give the pre-amplifier time for pre-charging, signal $\overline{RES}$ is delayed by about 40 ps.

Figure 2.4 shows an extension of the circuit in Fig. 2.3. Transistors $N2$, $N3$ and $N5$ have been added and build a differential input amplifier pair [4–6]. This has the advantage, that the comparator has high-impedance inputs due to the gates of transistors $N2$, $N3$. During reset phase transistor $N4$ shorts the output node OUT with $\overline{OUT}$ to force the initial condition $\Delta u(t=0) = OUT(t=0) - \overline{OUT}(t=0) = 0$. When transistor $N4$ is turned off, comparison phase starts and in principle the small signal equivalent circuit of Fig. 2.1c is valid but with the difference, that now also a current difference $g_{mn2}\Delta V_{in}$ is introduced to the output nodes. The transconductances of transistors $N2$ and $N3$, $g_{mn2} = g_{mn3}$ are assumed to be equal and $\Delta V_{in} = INP - INN$ is the input voltage difference, which is assumed to be constant. With assuming, that $g_m = g_{mn0} + g_{mp0} = g_{mn1} + g_{mp1}$ is the transconductance and $r = r_{n0}||r_{p0} = r_{n1}||r_{p1}$ is the overall output resistance of the whole latch and C_L is the load capacitance of each node OUT and $\overline{OUT}$, 2.7 can be written ($\Delta u(t) = OUT(t) - \overline{OUT}(t)$).

$$C_L \frac{d}{dt} \Delta u(t) + \Delta u(t) \left(\frac{1}{r} - g_m \right) = g_{mn2} \Delta V_{in} \tag{2.7}$$

The result of 2.7 is depicted in 2.8 and is similar to that of 2.2.

$$\Delta u(t) = g_{mn2} \Delta V_{in} \left(\frac{1}{\frac{1}{r} - g_m} + \frac{1}{g_m - \frac{1}{r}} e^{\frac{g_m - \frac{1}{r}}{C_L} t} \right) \tag{2.8}$$

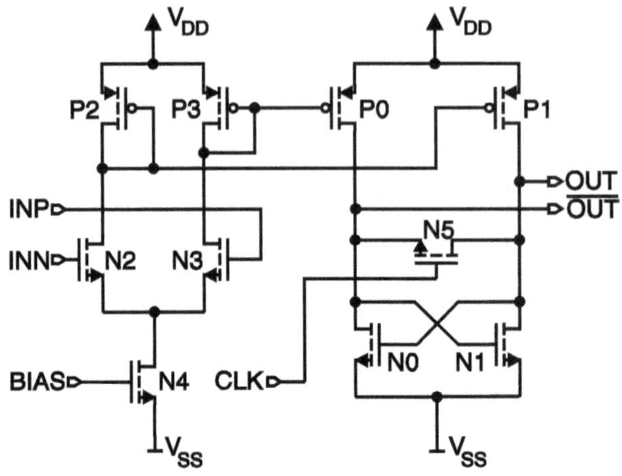

Fig. 2.5 Clocked regenerative comparator with current mirrors

In opposite to the dynamic latched comparator of Fig. 2.3 this comparator switches not between both rails, because of if e.g. OUT goes *high*, then transistor $P1$ is on, but also through transistor $N2$ a current flows due to *INN*. So in principle the level V_{DD} cannot be reached by an output node and it has to be considered, if a valid logic level is guaranteed.

In Fig. 2.5 a comparator is shown, where a current difference, which is generated by the input differential pair $N2$ and $N3$, is mirrored via transistors $P0$, $P1$, $P2$ and $P3$ to the final latch $N0$, $N1$ [7–9]. If *CLK* is *high*, then the latch is in reset and transistor $N6$ shorts both output nodes to have the initial condition $\Delta u(t = 0) \approx 0$. If the pre-amplifier is assumed to be fast enough, then the current difference, which is introduced to nodes OUT and $\overline{OUT}$ of the latch is depicted in 2.9, where ΔV_{in} is assumed to be constant.

$$\Delta I_{in} = g_{mp0} \frac{g_{mn2}}{g_{mp2}} \Delta V_{in} \tag{2.9}$$

Here it is also assumed, that the transconductances of transistors $N2$ and $N3$ ($g_{mn2} = g_{mn3}$), of transistors $P2$ and $P3$ ($g_{mp2} = g_{mp3}$), of transistors $P0$ and $P1$ ($g_{mp0} = g_{mp1}$) and the transconductances of the transistors $N0$ and $N1$ ($g_{mn0} = g_{mn1}$), are equal. So the output voltage difference $\Delta u(t)$ is calculated in a similar way as above and the result is depicted in 2.10, where $r = r_{n0}||r_{p0} = r_{n1}||r_{p1}$ is the overall output resistance of the latch and C_L is the load capacitance of each output node.

$$\Delta u(t) = g_{mp0} \frac{g_{mn2}}{g_{mp2}} \Delta V_{in} \left(\frac{1}{\frac{1}{r} - g_{mn0}} + \frac{1}{g_{mn0} - \frac{1}{r}} e^{\frac{g_{mn0} - \frac{1}{r}}{C_L} t} \right) \tag{2.10}$$

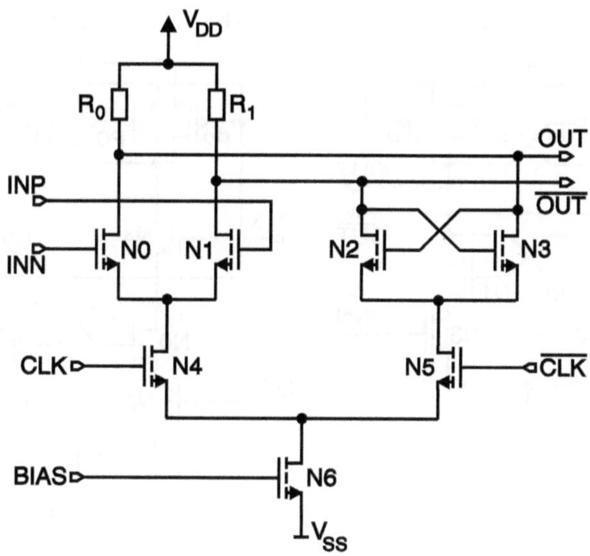

Fig. 2.6 Clocked regenerative CML latched comparator

The advantage of this comparator is, that distortions, which are fed back to the input and which are caused by fast switching of the latch, are low, because they are coupled back via serial parasitic capacitances and because the output resistance of the pre-amplifier is lower due to the diode connected loads $P2$ and $P3$. An output node can't reach in principle the voltage level V_{SS} because due to the comparator structure, there is always a current flow through transistors $P0$ or $P1$, if $\overline{OUT}$ or OUT are low respectively.

A Current-Mode-Logic (CML) or more exactly a MOS-Current-Mode-Logic (MCML) latch [10, 11] is drawn in Fig. 2.6 and can be also used as comparator. The overall function of the comparator consists of two phases in a clock period. If CLK is at voltage level V_{DD}, the current I_{n6} , which is determined by the bias voltage $BIAS$ and transistor $N6$, flows through the differential amplifier $N0$, $N1$, R_0, R_1, because transistor $N4$ is switched on. Due to the fact, that transistor $N5$ is off, no positive feedback of the latch $N2$, $N3$ occur. If a sufficient fast input differential amplifier is assumed and the influence of transistors $N2$ and $N3$ are neglected, then the initial voltage difference ΔV_0 for the next clock phase can be calculated with 2.11, where it is assumed, that $\Delta V_{in} = INP - INN$ is constant and the transconductances of $N0$ and $N1$ ($g_{mn0} = g_{mn1}$) and resistors $R_0 = R_1$ are equal.

$$\Delta V_0 \approx -g_{mn0} R_0 \Delta V_{in} \tag{2.11}$$

When CLK changes to *low* transistor $N4$ is cut off and $N5$ is switched on thus activating positive feedback of the latch $N2, N3$. 2.12 shows in principle the behavior of the output voltage difference $\Delta u(t) = OUT(t) - \overline{OUT}(t)$, when the latch is

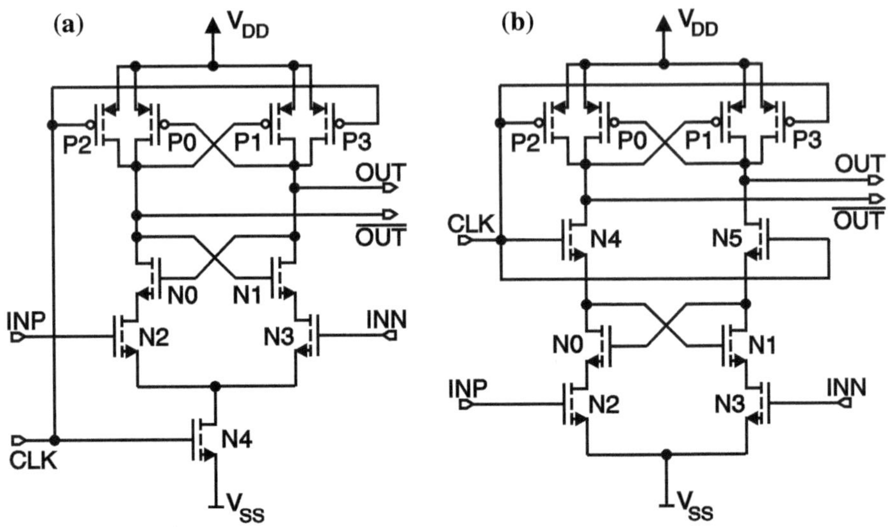

Fig. 2.7 Clocked regenerative comparators. **a** Comparator with input transistor pair below the latch. **b** Comparator of Fig. 2.7a with changed arrangement of switches

activated. The transconductances of $N2$ and $N3$ ($g_{mn2} = g_{mn3}$) and resistors $R_0 = R_1$ are assumed to be equal. C_L is the load capacitance of an output node.

$$\Delta u(t) \approx -g_{mn0} R_0 \Delta V_{in} e^{\frac{g_{mn2} - \frac{1}{R_0}}{C_L} t} \tag{2.12}$$

CML-latched comparators are in principle fast, because of their lower than rail-to-rail output voltage swing. Figure 2.7a shows a comparator, where the input transistor pair $N2$, $N3$ is placed below the latch $N0$, $P0$, $N1$, $P1$ [12–14]. The operation of the comparator is divided into two clock-phases. When CLK is at voltage level V_{SS}, transistor $N4$ is cut off and transistors $P2$ and $P3$ pull both output nodes to voltage level V_{DD} to reset the comparator. If CLK changes to level V_{DD} the initial condition is, that both output nodes are at level V_{DD}, transistors $P0$, $P1$, $P2$ and $P3$ are off and transistors $N0$ and $N1$ are on. Transistors $N2$ and $N3$ start to discharge both output nodes with a current difference, depicted in 2.13, where $\Delta V_{in} = INP - INN$ is the constant assumed input voltage difference, $g_{mn2} = g_{mn3}$ are the equally assumed transconductances of transistors $N2$ and $N3$, $\beta_{n2} = \beta_{n3}$ are their transconductance parameters and I_{n4} is the tail current through switch transistor $N4$.

$$\Delta I_{n4} = g_{mn2} \Delta V_{in} = \sqrt{2\beta_{n2} I_{n4}} \tag{2.13}$$

If one output node reaches the voltage $V_{DD} - |V_{tp0}|$, where $V_{tp0} = V_{tp1}$ are the equally assumed threshold voltages of transistors $P0$ and $P1$, then positive feedback of cross-coupled inverters $N0$, $P0$ and $N1$, $P1$ begins and the latch regenerates to $OUT = V_{DD}$ and $\overline{OUT} = V_{SS}$ in the case $INP > INN$ and vice versa. The time duration t_0 from the time point, discharging of output nodes starts till the time point, where the output node, which is faster discharged, reaches $V_{DD} - |V_{tp0}|$ can be estimated with 2.14, where C_L is the overall load capacitance of an output node.

$$t_0 \approx \frac{2C_L|V_{tp0}|}{I_{n4}} \tag{2.14}$$

The output voltage difference $\Delta u(t = t_0) = OUT(t = t_0) - \overline{OUT}(t = t_0)$ (see 2.15) at time t_0 determines the initial voltage difference of positive feedback of the latch.

$$\Delta u(t = t_0) \approx 2|V_{tp0}|\frac{g_{mn2}\Delta V_{in}}{I_{n4}} \tag{2.15}$$

With this initial voltage difference the latch regenerates similar as depicted above in 2.2 (see 2.16).

$$\Delta u(t \geq t_0) = 2|V_{tp0}|\frac{g_{mn2}\Delta V_{in}}{I_{n4}} e^{\frac{g_m - \frac{1}{r}}{C_L}t} \tag{2.16}$$

In 2.16 g_m is the overall transconductance and r is the overall output resistance of the latch $P0, P1, N0, N1$. In principle the comparator structure of Fig. 2.7a has a rail-to-rail output voltage swing and consumes only dynamic power. A similar structure, is depicted in Fig. 2.7b [15, 16], where instead during reset, transistors $N0$ to $N3$ are cut off by transistors $N4$ and $N5$. In principle this structure is treated in a similar way as the comparator of Fig. 2.7a.

It should be mentioned, that there are various ways to implement reset switches into a comparator. A possibility is, that a reset switch only shorts both output nodes together to force a metastable working point. Here the disadvantage is, that this metastable point is somewhere in the middle between V_{DD} and V_{SS} and defines therefore in most cases no valid logical voltage level. Another possibility to implement a reset is to pull both output nodes to V_{DD} or V_{SS}, which defines a valid logical voltage level. A disadvantage may be, that an additional discharging or pre-charging of output nodes before the latch regenerates is necessary, which might need time. Otherwise a larger initial voltage difference before positive feedback of the latch starts may occur and may compensate the loss of time.

Finally, instead of a single input difference transistor pair, two such pairs may be implemented to have the possibility to compare one difference voltage with another.

2.3 Analog MOS Switches

Analog MOS switches are important for the functionality of clocked, regenerative CMOS comparators, because the circuits in most cases consist of switches for "Reset" and switches for "Enable". Furthermore analog MOS switches can be used to hold dynamically the decision of the comparator during the reset phase in a following stage.

2.3.1 Analog Switch Implementation

Figure 2.8 shows different variants of implementing analog MOS switches [17–19]. It is assumed, that circuit 1 works as a driver and has a low output impedance and that circuit 2 has a capacitive input impedance like an input gate of a following stage. Depending on the input voltage range, a single n-MOS transistor $N0$ or a single p-MOS transistor $P0$ may be sufficient for a switch. In the assumed case of Fig. 2.8 the gate is clocked with signal CLK and the input voltage range is in between $V_{SS} \leq V_i \leq V_{DD}$. Transistor $N0$ is switched off, if its gate voltage $V_{GN0} \leq V_{tn0}$ (V_{GN0} related to V_{SS}) and conducts if $V_{GN0} > V_{i,max} + V_{tn0}$, where $V_{i,max}$ is the maximum input voltage and V_{tn0} is the threshold voltage of $N0$. The on-resistance R_{N0} of $N0$ is calculated in 2.17 with the assumption, that the transistor operates in the linear region and that the drain-source voltage V_{DS} is very small.

$$R_{N0} \approx \frac{V_{DS}}{I_D} = \frac{1}{\beta_{n0}\left(V_{GS} - V_{tn0}\right)} = \frac{1}{\beta_{n0}\left(V_{GN0} - V_i - V_{tn0}\right)} \tag{2.17}$$

I_D denotes the drain current and β_{n0} the transconductance parameter. In the case of a single p-MOS transistor (see Fig. 2.8, $V_{GP0} = \overline{CLK}$, V_{GP0} related to V_{SS}), switch $P0$ is off, if the gate voltage $V_{GP0} \geq V_{DD} - |V_{tp0}|$ and conducts if $V_{i,min} - |V_{tp0}| > V_{GP0}$,

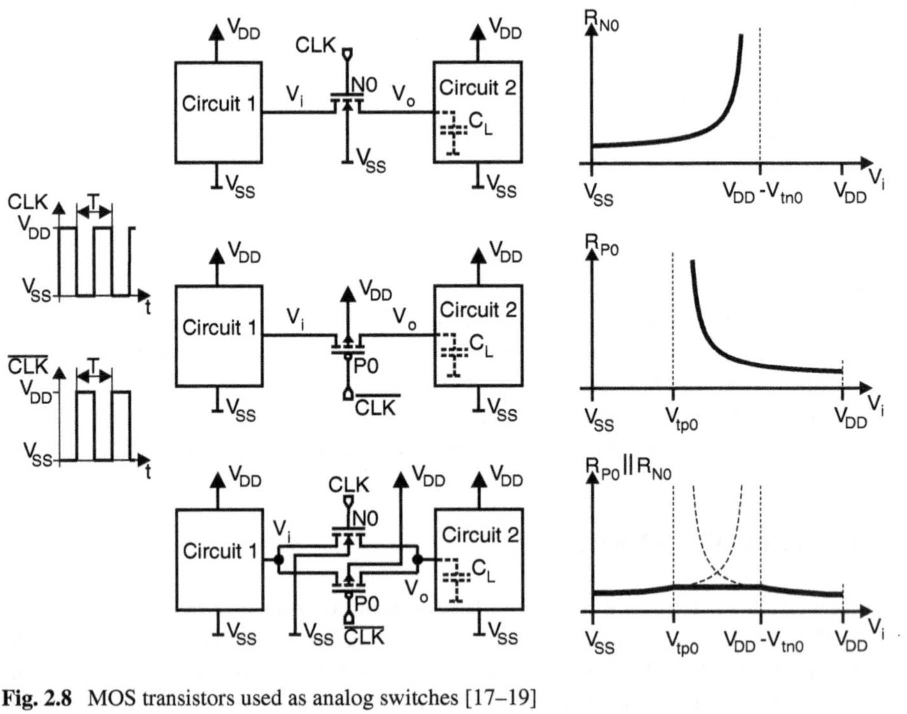

Fig. 2.8 MOS transistors used as analog switches [17–19]

where $V_{i,min}$ is the minimum input voltage and V_{tp0} is the threshold voltage of $P0$. With the same assumptions as for $N0$ the on-resistance R_{P0} can be calculated with 2.18, where β_{p0} is the transconductance parameter.

$$R_{P0} \approx \frac{V_{DS}}{I_D} = \frac{1}{\beta_{p0}\left(V_{SG} - |V_{tp0}|\right)} = \frac{1}{\beta_{p0}\left(V_i - V_{GP0} - |V_{tp0}|\right)} \tag{2.18}$$

If transistors $N0$ and $P0$ are switched off ($V_{GN0} = V_{SS}$ and $V_{GP0} = V_{DD}$ in case of Fig. 2.8), $R_{N0} \rightarrow \infty$ and $R_{P0} \rightarrow \infty$ can be assumed, if subthreshold and leakage currents are neglected. It depends on the application whether subthreshold- and leakage currents have to be considered. The problem of a single n-MOS or p-MOS switch is, that the on-resistance strongly depends on the input voltage V_i and is therefore not constant during switching. To extend the input voltage range to $V_{SS} \leq V_i \leq V_{DD}$ a n-MOS switch can be connected in parallel to a p-MOS switch as depicted in Fig. 2.8, where the different gates are clocked with the complementary signals CLK and $\overline{CLK}$. The resulting on-resistance can be calculated with $R_{P0}||R_{N0}$. Such a complementary switch is called transmission gate.

2.3.2 Charge Injection and Clock Feedthrough

In principle there are two important reasons, why an additional error voltage appears at the capacitively loaded output node when switching MOS switches. These reasons are charge injection and clock feedthrough [17–19], which are shown in principle in Fig. 2.9.

- Charge injection occurs, if the MOS switch is turned off, a part $Q_i \approx Q_{ch}/2$ of the channel charge Q_{ch} is absorbed by the generally low-impedance load of the input and the other part $Q_i \approx Q_{ch}/2$ is stored at the input capacitance C_L of the following circuit and causes there an error-voltage step.
- Clock feedthrough is the capacitive coupling from the gate via the overlap capacitances to source (via C_{GSov}) or to drain (via C_{GDov}) caused by clock edges.

When the switch changes from "off" to "on" all the charge injection will be absorbed in principle by the low-impedance input [19]. In the case of a transition from "on" to "off", it depends on how long the MOS switch is turned on during the time duration of the ramp of a clock-edge. Here some amount of the charge, injected or coupled to the capacitive output will be absorbed by the low-impedance input via the switch. For a more detailed analysis of charge injection also slow and fast transition time of clock edges in comparison to the channel time-constant have to be considered [18]. A first-order analysis of charge injection and clock feedthrough is given in [19]. At a fast clock-edge transition from "on" to "off" the resulting error voltage $\Delta V_E \approx Q_i/(2C_L) + Q_{GDov}/C_L$ can be assumed. Figure 2.10 shows two simple variants to compensate charge injection and clock feedthrough. In Fig. 2.10a a dummy MOS switch, where source and drain are connected with each other and

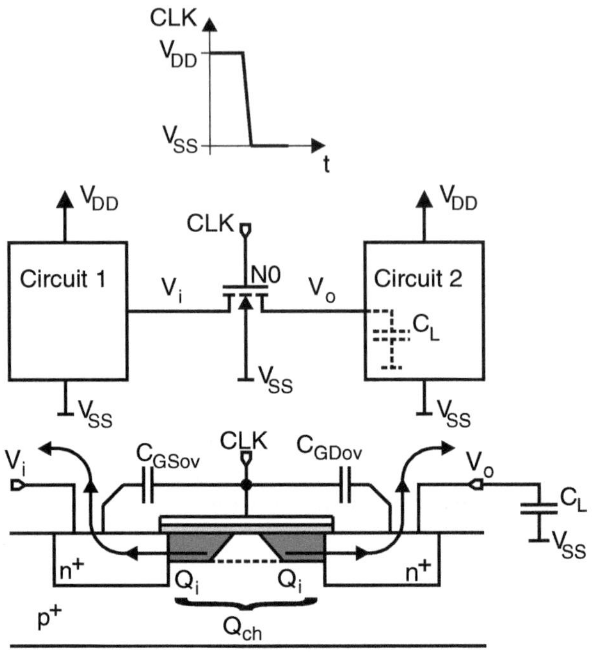

Fig. 2.9 Charge injection and clock feedthrough

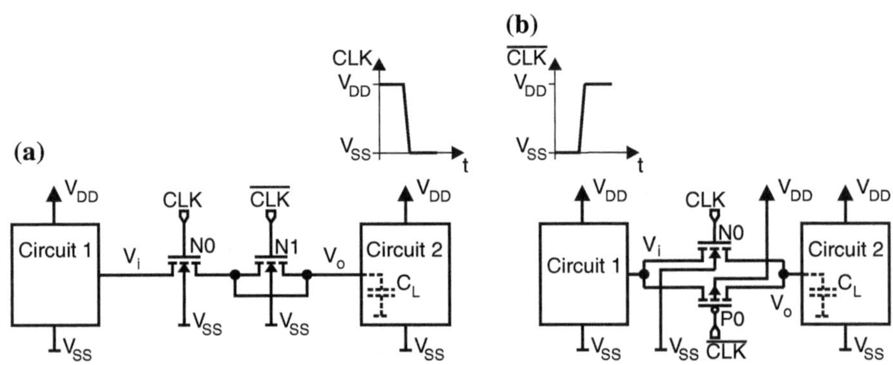

Fig. 2.10 Compensation of charge injection and clock feedthrough. **a** Dummy switch. **b** Complementary switches

$W_{N1}L_{N1} \approx W_{N0}L_{N0}/2$ (WL = gate area) is clocked with the complementary clock signal to take away the injected charge from transistor $N0$ for a first order. Also with a transmission gate, which consists of two complementary transistors (see Fig. 2.10b), charge injection and clock feedthrough can be reduced to some extend, because a part of the unwanted error voltage is canceled due to complementary clock signals and channel loads of transistors involved. In principle for complementary switches best compensation occurs at $W_{N0}L_{N0} = W_{P0}L_{P0}$.

2.4 Characterization of Comparators

There are several characteristic values to describe a clocked regenerative comparator, which are treated in this chapter.

2.4.1 Noise, Offset and Hysteresis

In Fig. 2.11 a summary of the principle influence of an offset and a hysteresis to a comparator, where the decision is affected by noise, is depicted [17, 20, 21]. The voltages *INP* and *INN* are applied to the inputs, where in comparison to the period of the clock *CLK*, *INP* a slow varying triangle signal. *INN* is the reference voltage.

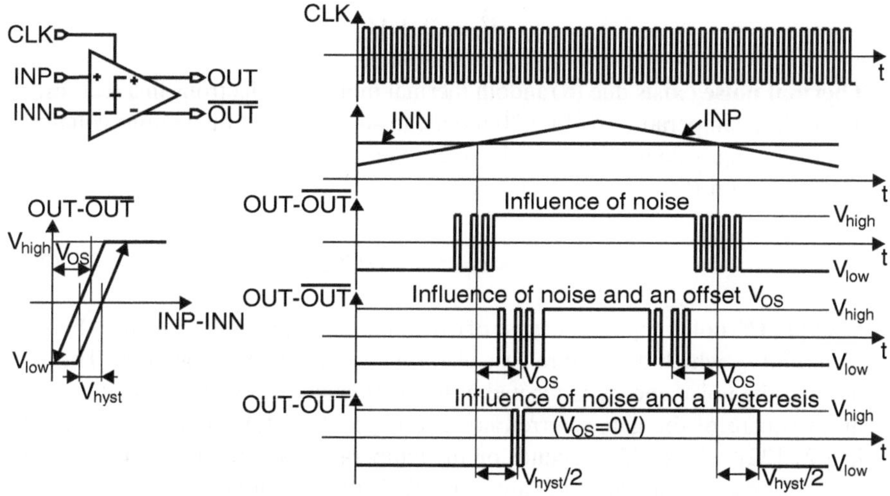

Fig. 2.11 Offset and hysteresis of a comparator, which is affected by noise: It is assumed that noise dominates and no metastability error occurs

2.4.1.1 Noise

The influence of the noise can be seen in Fig. 2.11 as a uncertainty in the decision near the cross-over points of *INP* and *INN*, where the noise causes the comparator to random decisions. It is assumed, that the comparator has time for a valid decision so that no metastabilty error occurs. The higher the input voltage difference *INP*−*INN*, the lower is the chance for a wrong decision. In most cases the noise is assumed to be a stationary random stochastic process, where the probability density function (pdf) of the amplitude of the noise is assumed to be mean free and Gaussian. There are different noise sources, which can be combined to an overall input referred noise source.

- The **noise of the sources of the signals**, which are applied to the input of the comparator disturb the decision. This might be uncertainty in amplitude or in time (jitter). Furthermore noise via the supply-voltage and ground-lines, which occur mostly due to switching of neighboring circuit components, may also cause a wrong decision of the comparator.
- In principle there are three sources of noise in CMOS devices [20]. **Shot noise** is the always superimposed random variation of a current caused by fluctuation of charge carriers (electrons or holes). The mean square $\overline{i_{sn}^2}$ of this variation is given by 2.19, where $e = 1.602 \times 10^{-19} C$ is the elementary charge, I is the average current and Δf is the bandwidth.

$$\overline{i_{sn}^2} = 2eI\Delta f \tag{2.19}$$

  Thermal noise exists due to random thermal motion of electrons in a real resistor at absolute temperature ϑ. In 2.20 the mean square of the fluctuating voltage $\overline{v_{tn}}$ and of the equivalent current $\overline{v_{tn}}$ is given. For a MOS transistor in saturation the approximation $R \approx \frac{3}{2g_m}$ can be used.

$$\overline{v_{tn}^2} = \overline{i_{tn}^2} R^2 = 4k\vartheta R\Delta f \tag{2.20}$$

  Flicker (1/f) noise appears because of extra electron states at the boundary between Si and SiO_2, which trap and release electrons in a relatively slow time. Therefore most of the noise energy is located at lower frequency. An approximation of the mean square of the gate referred noise voltage of a MOS transistor is depicted in 2.21 [22], where KF_F depends on the temperature and the fabrication process, C_{ox} is the gate capacitance per area and *WL* is the gate area.

$$\overline{v_{fn}^2} = \frac{KF_F}{C_{ox}^2 WL} \frac{\Delta f}{f} \tag{2.21}$$

- **Kickback noise** [23] is caused by the comparator itself, when at the output nodes a fast transition due to positive feedback in the decision phase or due to pulling the

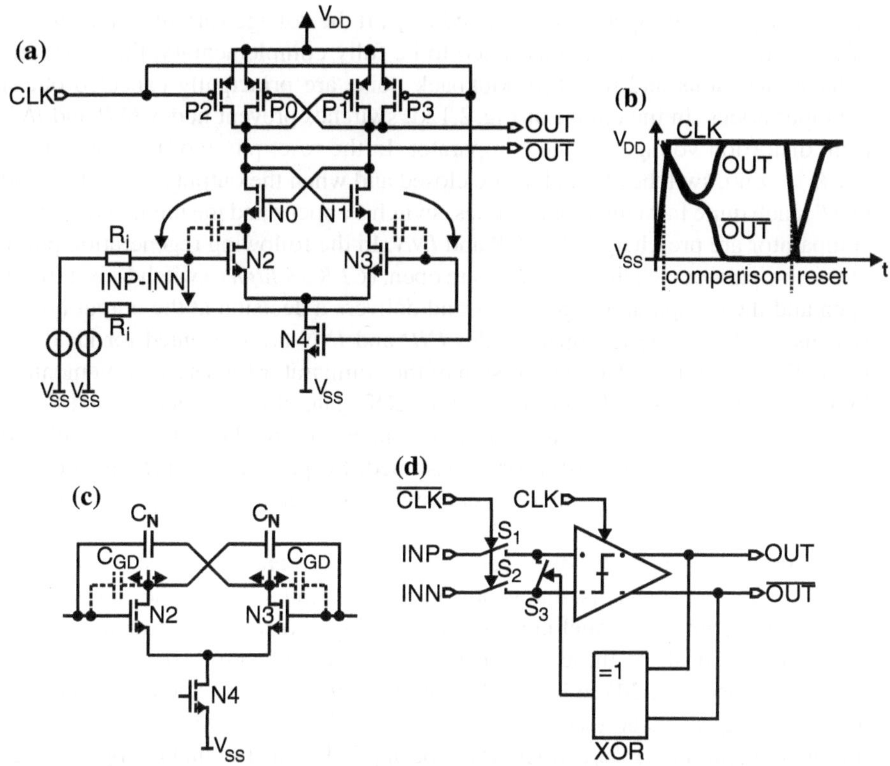

Fig. 2.12 Kickback noise [23]. **a** Kickback noise: Coupled back disturbances. **b** Typical waveform diagram. **c** Kickback noise neutralization technique. **d** Kickback noise reduction technique

output nodes to a defined voltage level in the reset phase occurs, which is coupled back via parasitic capacitances to input nodes (see Fig. 2.12a, b).

Another source of coupled back disturbances is caused by the fast ramps of the clock. In most cases this noise can be neglected, because it results mostly in a common-mode disturbance at the inputs if also equal loads are considered at each input and the influence of mismatch is low. Also at the time, when the output nodes regenerate or reset, the output nodes cause a considerable voltage at the input nodes, the comparator has already done most of the decision and is therefore more immune against noise, if also a small enough source resistance R_i (see Fig. 2.12a) is considered. The real problem of kickback noise occurs, when e.g. more comparators are placed in parallel in e.g. a flash ADC, where the reference and input voltages are affected from several comparators with e.g. different decision times (the higher the input voltage difference is the shorter is in common the decision time) or e.g. the comparators decide at different times due to clock delays. Figure 2.12c, d shows in principle techniques against kickback noise [23]. In the neutralization technique two capacitors C_N are added, which have the same size as the parasitic capacitances, named C_{GD}. If the voltage variation at the drains of transistors $N2$ and $N3$ are assumed to be fully complementary, then parasitic voltage variations and therefore kickback noise are principally canceled out at the input nodes. In the circuit of Fig. 2.12d, switches prevent nodes *INP* and *INN* from distortion voltages of the comparator. In the reset phase of the comparator ($CLK = low$), switches S_1 and S_2 are closed and when the output nodes OUT and $\overline{OUT}$ reach quite fast equal logic values, switch S_3 opens and the input nodes of the comparator are pre-charged by *INP* and *INN*. In the following regeneration phase of the comparator switches S_1 and S_2 are open, ($CLK = high$), switch S_3 is initially open and the comparator regenerates and delivers a decision at the output nodes. Because of S_1 and S_2 are open, nodes *INP* and *INN* are prevented for kickback noise. Due to the fact, that the decision of the comparator forces a complementary logic value at nodes OUT and $\overline{OUT}$, the XOR gate closes switch S_3 and therefore parasitic charge injections from the comparator are shorted and canceled if an ideal complementary distortion is assumed. So parasitic charge injections at the beginning of the reset phase, when switches S_1 and S_2 are opened are also eliminated.

- **Static noise** [24] is a DC disturbance, e.g. a disturbance of the working point caused by leakage currents, that is present in logic gates. This noise can be either series-voltage noise, parallel-current noise, voltage noise at the ground or power supply line and all of these can also appear in combination. In static random access memory (SRAM) cells, which consists in principal of static latched circuits to store a logical value (see Fig. 2.13a), the **static noise margin (SNM)** is the minimal magnitude ΔV_{ser}, which has to be applied both at u_1 and in opposite sign at u_2 to change the logic state, if we only consider series-voltage noise. However, parallel-current noise and voltage noise at the ground- or power supply line are treated in a similar way. It can be stated, that ΔV_{ser} represents a border to which static noise does not affect the stored logic state of the static latch. In the case of Fig. 2.13b, c the static noise margin is represented by ΔV_{ser}, which causes in the case depicted in Fig. 2.13c that the metastable point B equals the stable point A. If a higher DC noise voltage than ΔV_{ser} occurs, the logic state flips. Also for a comparator, which has already decided, it may be considered, that static noise may also affect the decision. But due to the fact, that a decision of the comparator in common only lasts for a half clock cycle and the clock period is small in the case of a fast comparator, the influence of DC disturbances is not considered, because if a long-time disturbance larger than the SNM occurs, the comparator always decides in one direction, because only one stable point is left. In e.g. SRAMs which have to hold a logical value for a comparably long time, the concept of a static noise makes sense. A calculation of SNM can be found in [25]. The influence of device fluctuations of the process to SNM was analyzed in [26].

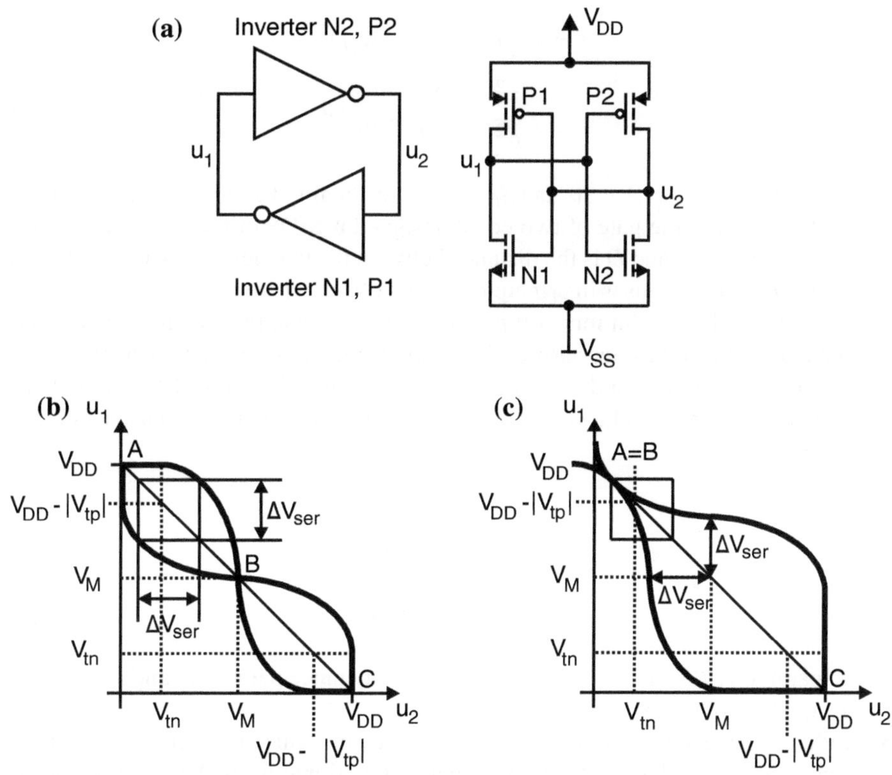

Fig. 2.13 Static noise [24]. **a** The static latch. **b** Characteristics of static latch. **c** Shift due to static noise margin ($u_1 - \Delta V_{ser}, u_2 + \Delta V_{ser}$)

2.4.1.2 Offset

The offset V_{OS} of a comparator is defined as the input voltage difference, which has to be applied to obtain the crossing point between logic level *low* and *high*. In Fig. 2.11 it can be seen, that the switching point of the comparator is shifted away from the cross-point of *INP* and *INN* by the offset V_{OS}, if the delay time of the comparator is small in comparison to the triangle. The offset is caused by the reasons stated next.

- The mismatch of different electronic parts, e.g. resistors, capacitances ([27]) and transistors in the circuit results as a consequence in a deviation from an ideal symmetrical circuit structure. The mismatch between two transistors [28, 29], which is for p-MOS and n-MOS transistors mainly the mismatch of their threshold voltages V_{tn} and V_{tp} and their current factors β_p and β_n, can be described with 2.22 and 2.23. These laws for the standard deviations for the transconductance parameters and threshold voltages for two equally designed n-MOS or two equally designed p-MOS transistors are based on a Gaussian probability distribution.

$$\sigma^2(V_t) \approx \frac{A_{V_t}^2}{WL} + S_{V_t}^2 D^2 \tag{2.22}$$

$$\frac{\sigma_\beta^2}{\beta^2} \approx \frac{A_\beta^2}{WL} + S_\beta^2 D^2 \tag{2.23}$$

The parameters $A_{\Delta V_t}$, A_β, S_{Vt} and S_β are proportionality factors. W is the width and L is the length of the gate of two equal designed n-MOS or two equally designed p-MOS transistors and D is the distance between the two devices. The influence of process tolerances causes in principle random offset, where the mean value is zero. Typically in a layout dummy components are added and the electronic parts to be matched are placed as near as possible and designed as big as possible to reduce influence of mismatch and avoid offset cancellation techniques. If additionally the distance D is very small a very simple model for mismatch can be used (see 2.24 and 2.25).

$$\sigma(V_t) \approx \frac{A_{V_t}}{\sqrt{WL}} \tag{2.24}$$

$$\sigma_\beta \approx \frac{A_B}{\sqrt{WL}} \tag{2.25}$$

The parameter $\beta = \mu C_{ox} W/L$ is the transconductance parameter and A_B (2.25) is not equal to A_β (2.23).

- Gradients of e.g. temperature or stress, which are introduced by neighboring circuit blocks have an influence to the symmetric functionality of the comparator and cause typically systematic offsets (the mean is not zero). The offset caused by gradients can be minimized with an intelligent placement of components in the layout.
- Systematic offsets are also introduced by asymmetry due to the used circuit structure.
- The offset also may change in some cases between different clock frequencies (dynamic effects), e.g. if a mismatch in the reset switches is present.

In principle there are two main ways to compensate the offset of a clocked regenerative comparator. The first is the **static offset cancellation** (trimming), where additional transistors are added to the comparator to introduce an additional current to compensate mismatch. This current is adjusted only one time in a separate offset cancellation run or with a bias voltage or current applied from outside the chip and stays constant when the comparator does its assigned task [31]. Another attempt of static offset cancellation is to program the floating gate of p-MOS input transistors of a comparator in a separate offset cancellation run with the help of hot-electron injection by raising the supply voltage [32].

The more important offset cancellation technique in ADCs is the **dynamic offset cancellation**. Typically in ADCs one or more pre-amplifiers are placed in front of the comparator/latch. In [30] two main dynamic offset cancellation techniques are described, input offset storage (IOS) and output offset storage (OOS), which are

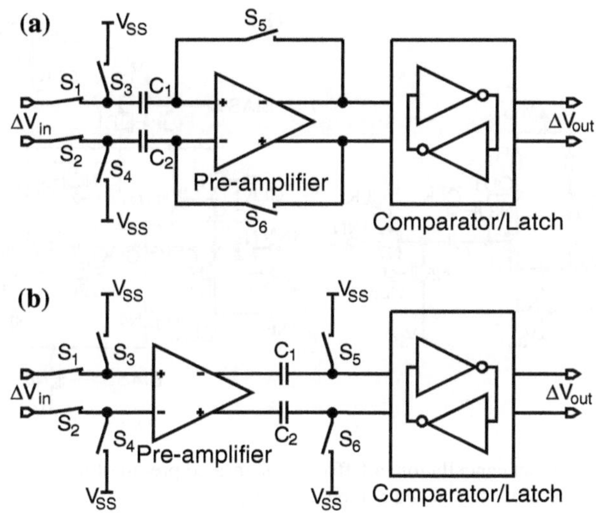

Fig. 2.14 Dynamic offset cancellation techniques [30]. **a** Input offset storage (IOS). **b** Output offset storage (OOS)

depicted in Fig. 2.14. If IOS is used the cancellation is performed when switches S_3, S_4, S_5 and S_6 are closed, S_1 and S_2 are open. The unity-gain loop around the pre-amplifier is closed and the offset is stored in the input capacitances C_1 and C_2 as depicted in Fig. 2.14a. For comparison phase of the comparator switches S_3, S_4, S_5 and S_6 are opened and S_1 and S_2 are closed. Then the input referred offset V_{OS} after calibration is stated in 2.26.

$$V_{OS} = \frac{V_{OSA}}{1 + A_0} + \frac{\Delta Q}{C} + \frac{V_{OSL}}{A_0} \tag{2.26}$$

V_{OSA} and A_0 are the input offset and the gain of the pre-amplifier, V_{OSL} is the offset of the latch and ΔQ is the mismatch of charge injection of switches S_5 and S_6 onto capacitance $C = C_1 = C_2$.

If OOS is used the offset cancellation takes place, when switches S_3, S_4, S_5 and S_6 are closed and S_1, S_2 are open. Then for the comparison phase (S_3, S_4, S_5 and S_6 are open and S_1, S_2 are closed) an input referred offset occur, which is stated in 2.27.

$$V_{OS} = \frac{\Delta Q}{A_0 C} + \frac{V_{OSL}}{A_0} \tag{2.27}$$

The advantage of OOS compared to IOS is, that in principle the input referred offset can be smaller as it can be seen if 2.27 is compared with 2.26. Another disadvantage of IOS is that seen from the input nodes, a higher capacitance is present. The problem of OOS is, that it might saturate at the output nodes, because a quite higher voltage difference is stored in the capacitances. In principle OOS is generally preferable in

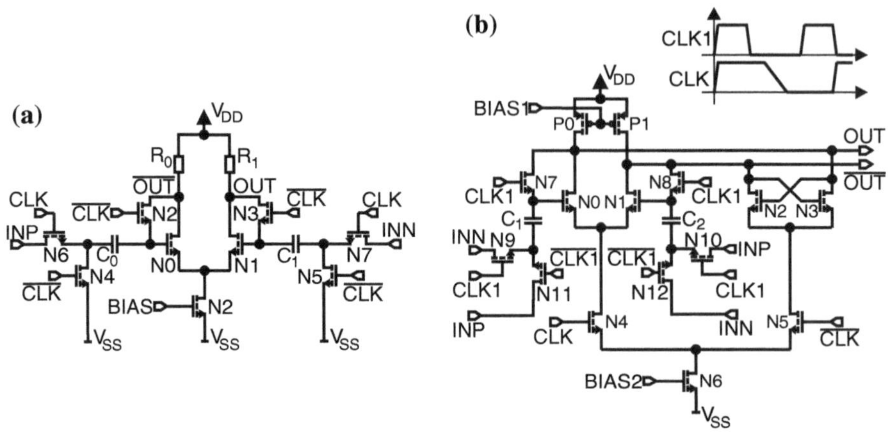

Fig. 2.15 Dynamic offset cancellation. **a** Offset compensated pre-amplifier [9, 20]. **b** Offset compensated pre-amplifier of CML latched comparator [20]

flash ADC stages. An example of an offset compensated pre-amplifier is depicted in Fig. 2.15a [9, 20], where offset storing takes place when $CLK = low$ and amplification when $CLK = high$. Care has to be taken that the voltages at the gates of the input transistors $N0$ and $N1$ do not exceed the level for hot carrier injection or oxide breakdown. So to avoid this, instead of V_{SS} at the sources of $N0$ and $N1$ an appropriate bias voltage could be connected. In Fig. 2.15b [20] an CML latched comparator with offset compensated pre-amplifier is depicted, where additionally the clock cycles of signals CLK and $CLK1$ are drawn schematically for clarity. In [33] dynamic offset cancellation takes place at one input of a multi input floating gate n-MOS transistor. The potential at the floating gate is adjusted via a separate input to cancel the offset during offset cancellation phase of the clock cycle.

Other options for offset compensation are introducing a mismatch of the load capacitances at the inverted and non-inverted outputs of a clocked regenerative comparator [34]. There the amount of the capaciotors can be controlled digitally with p-MOS varactors. In [35] the offset is callibrated with the help of the back-gate effect of the p-MOS transistors at the input. and in [36] the p-MOS transistors at the input are floating gate transistors, where each potential at the floating gates is controlled with hot carrier injection and tunneling. The different potentials at the floating gates effects the threshold voltage of the input transistors and thus the offset can be compensated. An analysis of static and dynamic offsets on a comparator structure similar to that in Fig. 2.7b can be found in [37]. A dynamic offset is caused by the mismatch of the parasitic load and coupling capacitances of the different circuit nodes.

2.4.1.3 Hysteresis

When a comparator has a hysteresis [17, 20], the comparison threshold for input signals is different in the case that the output changes from *low* to *high* and the case of a change from *high* to *low*. The distance between these comparison thresholds is characterized in Fig. 2.11 by the hysteresis voltage V_{hyst}. The advantage of an hysteresis is, that the comparator is more immune against noise, because noise has to overcome the voltage V_{hyst} to flip back the decision of the comparator. In an ADC a hysteresis might limit the performance, because it can cause different codes depending on the sign of the input signal derivative. A typical source of creating a hysteresis in a clocked regenerative comparator is, if the reset switches are dimensioned too small. So at the end of the reset phase a considerable voltage difference exists still between the output nodes, which has to be overcome by the input voltage in the following comparison phase and a hysteresis is established.

2.4.2 Sensitivity, Metastability Error and Bit Error

The sensitivity [17, 20], sometimes called minimal input resolution, of a clocked regenerative comparator is a measure for which minimal input voltage difference (or another value, e.g. current or charge) a proper operation is guaranteed. Typically it depends on several parameters, like e.g. clock frequency, input common-mode level or supply voltage. In a simple flash ADC with N bits, $2^N - 1$ comparators in parallel are necessary. So every comparator has to be designed for a sensitivity of 1/2 least significant bit (LSB), which has for a reference voltage V_{ref} the value $0.5\,V_{ref}/(2^N - 1)$. In reality no exact value for the sensitivity can be measured, because there is a considerable influence of noise. If at an ideal offset- and hysteresis free comparator the influence of noise is assumed to be not present, the sensitivity is defined by the minimal input voltage difference, at which the comparator in a distinct time (in most cases the half clock period) still has a correct decision, which can be interpreted by following logic gates. For an input voltage difference, which is smaller than the sensitivity the output voltage difference of the comparator is too small to be interpreted by a following logic gate and a metastability error occur. If noise is present, than there is additionally the possibility, that the comparator might give a valid, but wrong decision.

2.4.2.1 Metastability Error

A treatise of metastability errors and a therefore occurring failure rate at a flip-flop used as synchronizer of an incoming asynchronous data stream was done in [38] (see Fig. 2.16). Such a consideration may also be important for flash ADCs, because there might also appear a metastability error for the comparator, when the analog input voltage, which is in general distributed uniformly in the voltage interval of a least-significant bit, occurs near the reference voltage of the appropriate comparator. For the following considerations it is assumed that no noise influences the function of the comparator with the static latch. Metastability error means, that e.g. a comparator cannot deliver a valid decision in a distinct time, which should be smaller than the

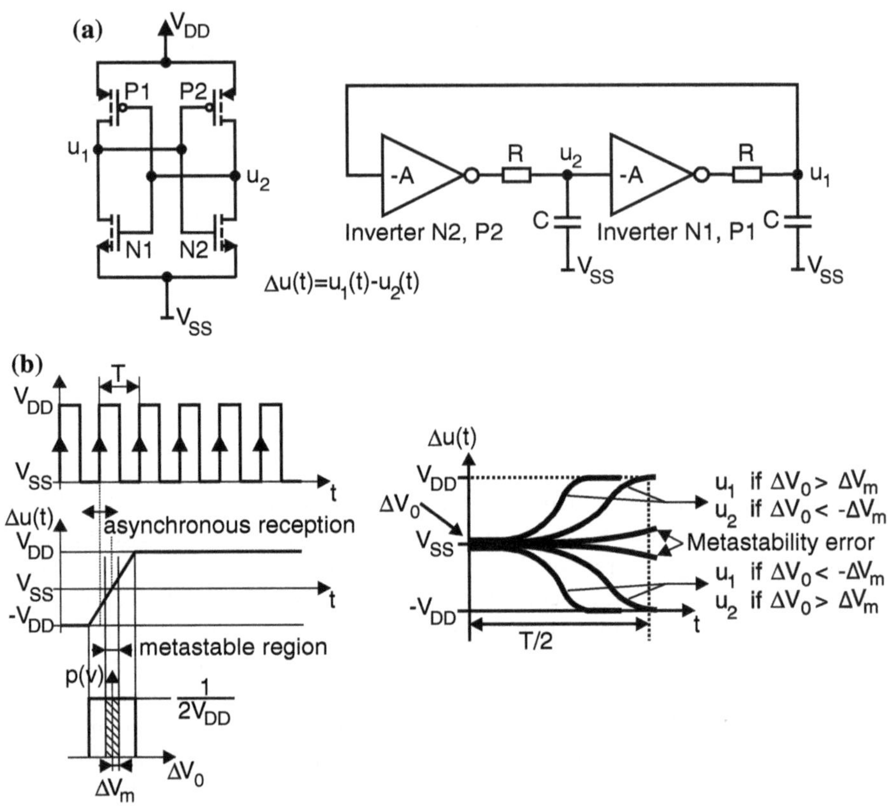

Fig. 2.16 Occurrence of metastability errors [38]. **a** Model of the static latch. **b** A metastability error occurs, if the input voltage difference $|\Delta V_0| \leq \Delta V_m$

time slot for decision, given by the clock. The comparator remains therefore in a metastable state during this time slot and a following logic gate cannot recognize a valid decision. Using the model of Fig. 2.16a of the static latch (switches for comparison and reset phase are not drawn) and neglecting the exponential function with the negative expression, the output voltage difference at e.g. the half clock period $T/2$, $\Delta u(t = T/2) = u_2(t = T/2) - u_1(t = T/2)$, can be calculated similar as in Sect. 2.1 with the initial condition $\Delta V_0 = u_2(t = 0) - u_1(t = 0)$ (see 2.28).

$$\Delta u(t = T/2) = \Delta V_0 e^{\frac{A-1}{RC}\frac{T}{2}} \rightarrow \frac{\Delta V_m}{\Delta u(t = T/2) = V_{DD}} = e^{-\frac{A-1}{RC}\frac{T}{2}} \tag{2.28}$$

With assuming $V_{SS} = 0V$, the probability of sampling a voltage $|\Delta V_0| \leq \Delta V_m$ (a metastable error occur) is $P(|\Delta V_0| \leq \Delta V_m) = \Delta V_m / V_{DD}$ (see Fig. 2.16b). So an expression for the average number of metastable states (errors) per second M_{meta} lasting longer a distinct time, here $T/2$ (50 % duty cycle assumed, T/2 for comparison and T/2 for reset), was calculated in [7, 38], adapted here for a clock $f_{clk} = 1/T$, is depicted in 2.29

$$M_{meta} = f_{clk} P(|\Delta V_0| \leq \Delta V_m) = f_{clk} e^{-\frac{A-1}{RC}\frac{T}{2}} = f_{clk} e^{-\frac{A-1}{RC}\frac{1}{2f_{clk}}} \tag{2.29}$$

So for a clocked regenerative comparator the chance that a metastability error occur is at a lower clock frequency smaller than at a higher frequency. With consideration that the input voltage difference of a comparator, which is compared, changes from comparison to comparison, but is uniformly distributed in a distinct voltage interval, [38] figured out, that the average rate of the occurence of a metastability error, is independent of circuit noise.

2.4.2.2 Wrong Decisions Caused by Noise

Different types of noise cause a clocked regenerative comparator to a, for a following logic gate, valid, but wrong decision, which is in opposite to a metastability error no remaining in a metastable state after a distinct time. The different noise sources were described in Sect. 2.4.1.1. For the following considerations a comparator with infinite amplification is assumed so that no metastability error can occur. To calculate the bit error rate (BER) the whole input-referred combined noise of the comparator is assumed to be induced by a stationary stochastic process [39], where the probability density function (pdf) $p_n(\zeta)$ does not change at different time points. The pdf itself is assumed to be a mean free ($\mu_N = 0$), Gaussian distribution with standard deviation σ_N. Then the probability $P[E]$ of a false decision for a synchronous data signal with amplitude ΔV_0 around V_{ref} is illustrated in Fig. 2.17. Here only false or correct decisions of the comparator due to the impact of noise are treated, because due to the synchronous input data, a fix ΔV_0, where no metastability error occurs due to the supposed infinite amplification of the comparator, is assumed. With the Gaussian normal distribution (see 2.30) and the Q-function (see 2.31) the probability $P[E]$, that a false decision occurs can be calculated by integration of the lined area in Fig. 2.17 (see 2.32) is:

$$p_n(\zeta) = \frac{1}{\sqrt{2\pi}\sigma_N} e^{-\frac{1}{2}\left(\frac{\zeta}{\sigma_N}\right)^2} \tag{2.30}$$

$$Q(\zeta) = \frac{1}{\sqrt{2\pi}} \int_{\zeta}^{\infty} e^{-\frac{\alpha^2}{2}} d\alpha \tag{2.31}$$

$$P[E] = 0.5Q\left(\frac{\Delta V_0}{\sigma_N}\right) + 0.5\left(1 - Q\left(\frac{-\Delta V_0}{\sigma_N}\right)\right) = Q\left(\frac{\Delta V_0}{\sigma_N}\right) \tag{2.32}$$

So a measured BER can be described sufficiently with the Q-function ($P[E]$), where it can be seen, that a higher ΔV_0 has a lower BER as a consequence.

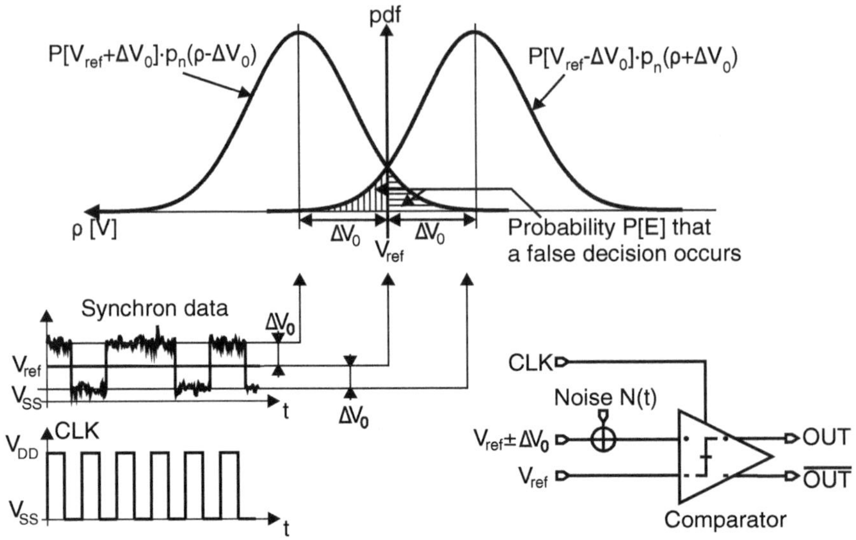

Fig. 2.17 Occurrence of bit errors due to influence of noise when a comparator is assumed, which has an ideal, infinite amplification so that no metastability error can appear [39]

2.4.2.3 Sensitivity and BER Measurements

For the comparators measured in this book bit error ratio (BER) measurements have been done to determine a sensitivity of the comparator. Due to the presence of noise a metastability error as explained in Sect. 2.4.2.1 might not occur like it would be in ideal, noise-less simulations. This noisy situation is shown in Fig. 2.18. The comparator metastability in the presence of noise is mathematically treated in [40]. BER measurements are done with a pseudo-random-bit-sequence (PRBS), which emulates a real bit sequence with the probability of 0.5 that a logical high (probability of 0.5 for a low) occurs. Therefore the mean voltage value of this bit stream is located in the middle between the voltage level for the logical high and low. If an offset-free comparator with infinite amplification is considered so that no metastability error occurs, then for large noise the BER→0.5, which means that half of the decisions are correct. Between two comparison phases the comparator has to be reseted. So there exists additional logic in the inverter chains, which holds the decision during reset. For BER measurements it is often easier to use one output of the comparator only as shown in Fig. 2.18. If now a comparator with finite amplification in a noiseless system is considered then a metastability error occurs when ΔV_0 is small enough. Typically the following logic gate of the chain of inverters has another switching threshold than the operating point of the latch in the comparator. So it happens that $COUT$ and $\overline{COUT}$ have the same logic voltage value when a metastability error occurs, but a valid logic level. So if one output $COUT$ is measured it has alwas the same logical voltage level in the case of a metastability error, which is theoretically a

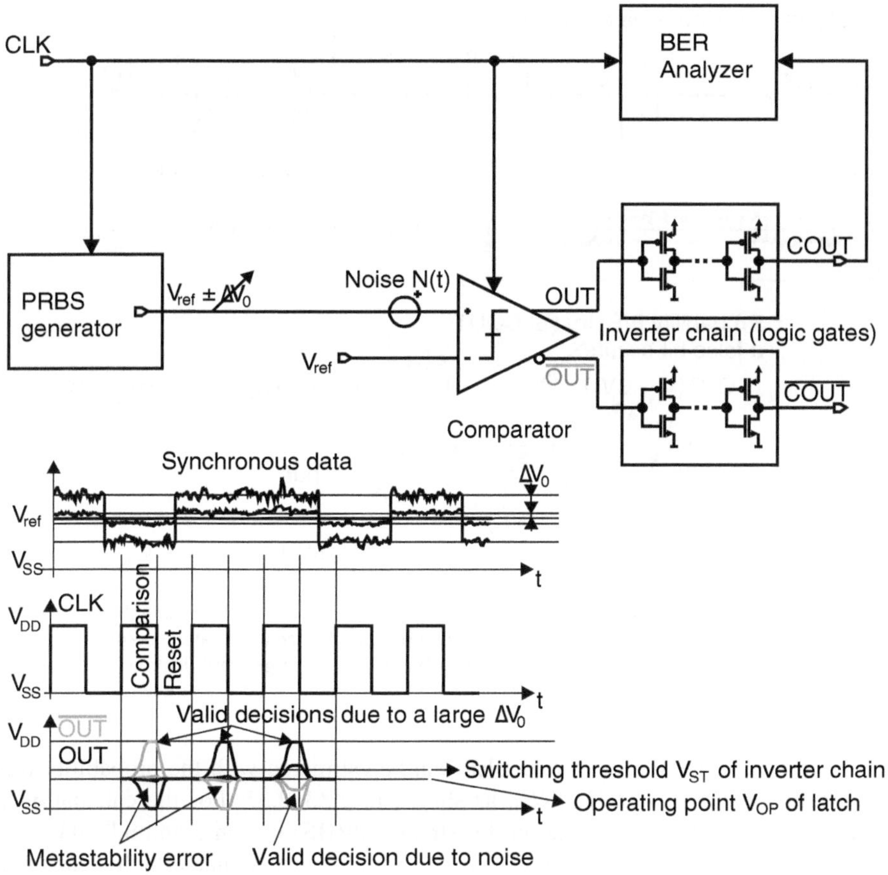

Fig. 2.18 Influence of noise on the decision of a comparator for a large and a small amplitude ΔV_0. Due to noise there is a chance for a valid decision for a small amplitude, but also a metastability error might occur for large amplitudes [40]. In other words due to noise there exist also valid decisions for small amplitudes

BER of 0.5, because half of the bits are transmitted correct. However a metastability error can be distinguished from a wrong decision due to noise when $COUT$ and $\overline{COUT}$ have the same logical level instead of an inverted and non-inverted output. In [41] a variance of noise which effects the output signal has been calculated on a similar comparator structure as shown in Fig. 2.7a in dividing the operation of the comparator during decision into two phases. In [40] a variance of noise which effects the output voltages and as a consequence the metastability error rate of a simple latch has been calculated, where it consists of a noise power at the beginning of comparison and a varianace which is generated during comparison. This variance grows exponentially in time, because the amplification of a latch grows exponentially as well in time. Inspired from the calculations in [40] and adapted to the situation in

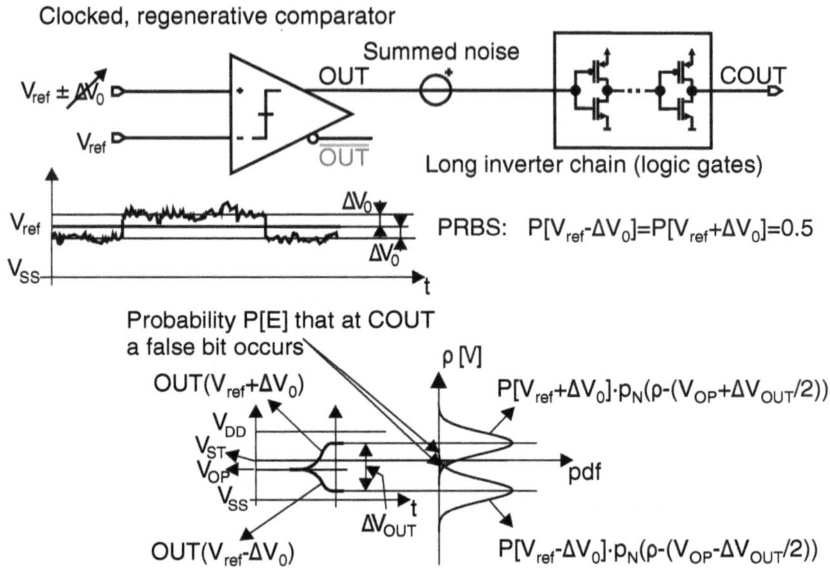

Fig. 2.19 BER measurements at one output of a comparator: V_{ST} is the switching threshold of the inverter chain, V_{OP} is the operating point of the comparator. The chance that the output *COUT* has no logical level (*OUT* very near to V_{ST}) is negligible due to the large amplification of the inverter chain

Figs. 2.18 and 2.19, the measured BER can be estimated with the probability that a false bit occurs at *COUT*, $P[E]$, as shown in 2.33–2.35. For the calculation it is assumed that the pseudo random bit stream (PRBS) has an amplitude ΔV_0 and that the probability of receiving a logical *high* or a *low* is equal, which means that $P[V_{ref} - \Delta V_0] = P[V_{ref} + \Delta V_0] = 0.5$.

$$p_N(\zeta) = \frac{1}{\sqrt{2\pi}\sigma_N} e^{-\frac{1}{2}\left(\frac{\zeta}{\sigma_N}\right)^2} \tag{2.33}$$

$$Q(\zeta) = \frac{1}{\sqrt{2\pi}} \int_{\zeta}^{\infty} e^{-\frac{\alpha^2}{2}} d\alpha \tag{2.34}$$

$$P[E] = 0.5\left(1 - Q\left(\frac{V_{ST} - (V_{OP} + \Delta V_{OUT}/2)}{\sigma_N}\right)\right) + 0.5Q\left(\frac{V_{ST} - (V_{OP} - \Delta V_{OUT}/2)}{\sigma_N}\right) \tag{2.35}$$

Variable σ_N^2 is the variance of the summed noise which effects the output voltage of the comparator. V_{ST} is the switching threshold of the inverter chain and V_{OP} is the operating point of the comparator. A metastability error ($OUT(V_{ref} + \Delta V_0) < V_{ST}$) of the comparator would result in a logical level at output *COUT*, because the amplification of the long inverter chain can be considered as large. Therfore also the chance that the output *COUT* has no logical level (*OUT* very near to V_{ST}) is negligible. In the design of the inverter chain care has to be taken that the switching times of the inverters are fast enough.

To compare comparators the sensitivity has been measured. The sensitivity is defined here as the amplitude ΔV_0 of a data stream at the input so that a BER of 10^{-9} occurs. This gives a measure of the combined influence of noise with the impact of metastability errors.

2.4.3 Delay Time, Overdrive Recovery Time

2.4.3.1 Delay Time

The delay time of a clocked regenerative comparator is the minimum time duration from beginning of the comparison phase (reset is released when the appropriate clock edge reaches typically 50 % of V_{DD}) till a valid logical voltage level is available at the output. Another definition is the minimum time duration from beginning of the comparison phase (reset is released when the appropriate clock edge reaches typically 50 % of V_{DD}) till one output node reaches the half supply voltage $V_{DD}/2$.

2.4.3.2 Overdrive Recovery Time

The overdrive recovery time [17] is the delay time, which is needed by the comparator, when the input voltage difference changes from a large value, where the gain stages of the comparator saturates, to a small value of opposite sign. In principle the overdrive recovery lasts longer than a typical delay time.

2.4.4 Power Consumption

A further important parameter to characterize a clocked regenerative comparator is the power consumption [17]. Typically it consists of a static and a dynamic part, which depends on the clock frequency and the capacitances, which have to be charged and discharged.

2.5 Measurement Techniques of Clocked, Regenerative Comparators from the Literature

The simplest method to measure a clocked regenerative comparator is to apply e.g. at the negative input a reference voltage and at the positive input a triangular wave with a ramp, which rises very slowly in comparison to the period of the clock, so

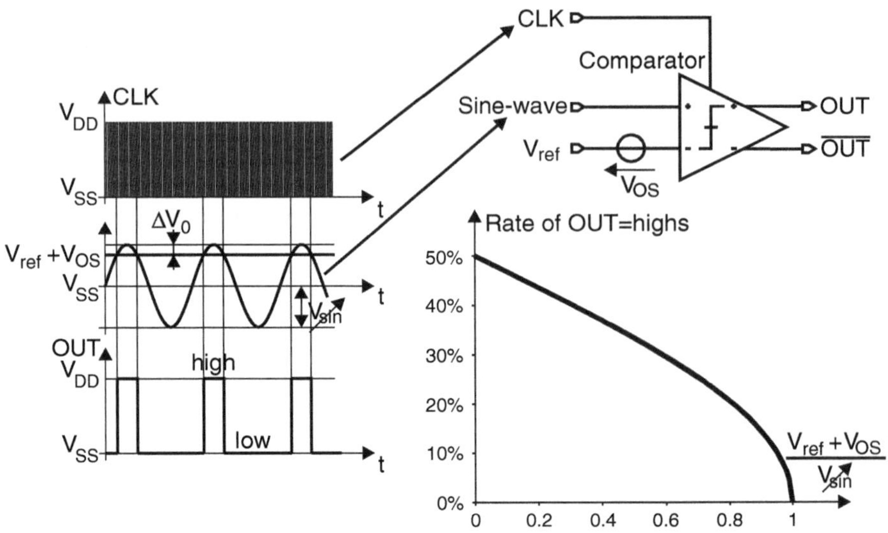

Fig. 2.20 Overdrive test [42]

that the delay time of the comparator can be neglected. Taking care of different delay times of cables then with an oscilloscope the offset and hysteresis of the comparator can be determined, as depicted in principle for different cases in Fig. 2.11.

In [21] the sensitivity and the maximum clock frequency of a clocked, regenerative comparator was measured by a so-called overdrive recovery test. At the negative input a precise, variable reference voltage was applied. At the positive input a synchronous square wave signal with a frequency of 1/8 of the clock frequency was applied. The reference voltage was adjusted so that the negative input voltage difference was a large value compared to the positive input voltage difference, which was in principle the minimum resolvable value to be measured at a distinct clock frequency.

Another test, which is called overdrive test, is shown in Fig. 2.20 and was described in [42]. At the negative input of the comparator a fixed reference voltage V_{ref} and at the positive input a sine wave with an adjustable amplitude V_{sin} is applied. The frequency of the sine wave is adjusted so that the clock frequency is an integer multiple factor of it (coherent sampling). If the sine wave's amplitude is larger than $V_{ref} + V_{OS}$, then *OUT* will switch to *high* and in the case, that it is smaller, *OUT* switches to *low*. The rate of *highs* is counted at different V_{ref} and so the offset voltage V_{OS} and the sensitivity of the comparator can be determined.

In Fig. 2.21a a clocked regenerative comparator is shown, which is measured with the help of a counter to count the random switching [43, 44], caused by the input referred noise of all noise sources, which is assumed to be a stationary stochastic process witch a Gaussian pdf of the amplitude. At the negative input of the comparator a fixed reference voltage V_{ref} and at the positive input an adjustable DC voltage V_{in} is applied. The latch after the comparator is added to store the decision during the reset

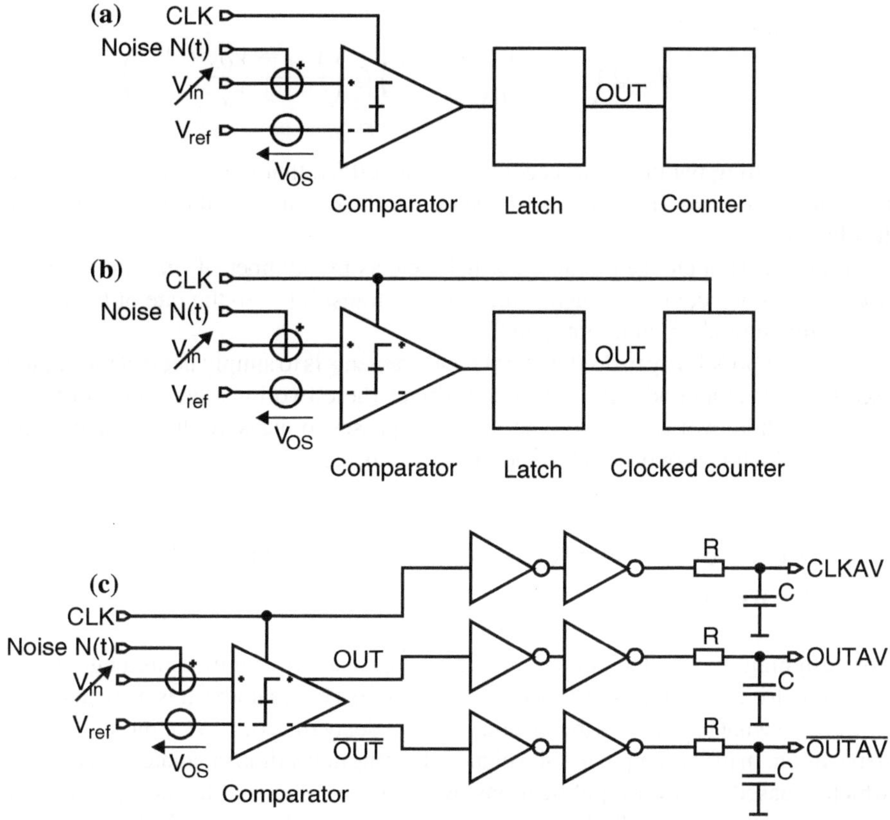

Fig. 2.21 Dynamic characterization of clocked regenerative comparators. **a** Dynamic characterization [43, 44]. **b** Dynamic characterization with a clocked counter. **c** Averaging the outputs of the comparator [32]

phase. The more V_{in} is adjusted to get nearer $V_{ref} + V_{OS}$, the more the comparator will switch at random from one clock period to the next. The maximum random switching frequency will occur, if V_{in} equals $V_{ref} + V_{OS}$. With this knowledge the offset and the standard deviation σ_N of the input referred noise can be measured. With the assumptions stated above a formula for the average measured random clock frequency $\overline{f_{OUT}}$ can be derived (see 2.37), with the possibility $P[0]$ that a *low* occurs (see 2.36), the possibility $P[1]$ that a *high* occurs (see 2.37) and the possibility $P[0 \rightarrow 1] \approx P[0]P[1]$ that a $0 \rightarrow 1$ transition occurs, which is counted by the counter.

$$P[0] = Q\left(\frac{V_{in} - V_{ref} - V_{OS}}{\sigma_N}\right) \tag{2.36}$$

$$P[1] = 1 - P[0] \tag{2.37}$$

$$
\begin{aligned}
\overline{f_{OUT}} &= mf_{clk}P[0 \rightarrow 1] \\
&= mf_{clk}Q\left(\frac{V_{in} - V_{ref} - V_{OS}}{\sigma_N}\right)Q\left(\frac{V_{ref} + V_{OS} - V_{in}}{\sigma_N}\right)
\end{aligned} \tag{2.38}
$$

In 2.38 fitting parameter m accounts for the effect, that the previous decision of the comparator may influence the present due to parasitic feedback (m = 1 for no feedback).

In Fig. 2.21b a clocked counter, which detects the number of *high*-decisions, is used instead of a counter, which detects $0 \rightarrow 1$ transitions. So the rate of *highs* $P[1]$ can be measured, which is stated in 2.37.

A further possibility to do statistical measurements is to simply neglect the counter (see Fig. 2.21c) and add after an output driver (here two inverters with rail-to-rail output) with known supply voltage, R-C low-passes to measure the average output voltage, which is proportional to the rate of *highs* [32].

2.6 A Low Level Consideration of the Delay Time Mismatch for Two Inverters

For designing of e.g. clock drivers, transfer stages, additional measurement circuits or output drivers, which are needed to characterize a comparator, it is also necessary to know on what parameters the mismatch of the delay time of two (or more) inverters depends on. In [45] a simple low-level model for the delay time mismatch is extracted, which is based on several publications, which have calculated the delay time of an inverter with finite rising/fall time of the input voltage, when the logical voltage level is changed. In this section another simplified calculation is presented to calculate the delay time of one inverter to see on what parameters the delay time depends. For simplification ideal MOS transistors are used. Figure 2.22 shows the inverter with its schematic representation of input and output signals. The appropriate delay times t_{Irf}

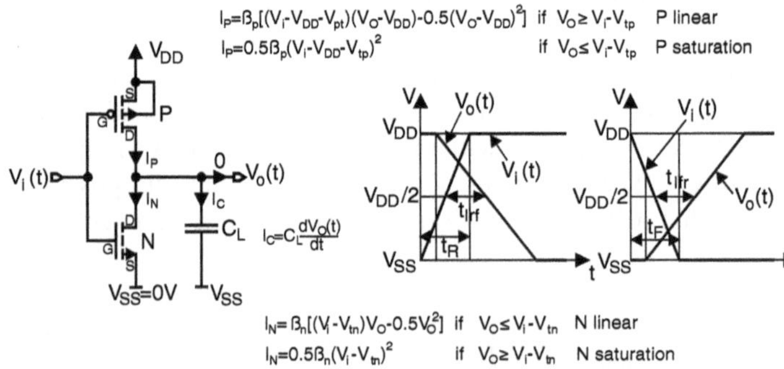

Fig. 2.22 Inverter and its schematic representation of its input voltage $V_i(t)$ and its output voltage $V_O(t)$

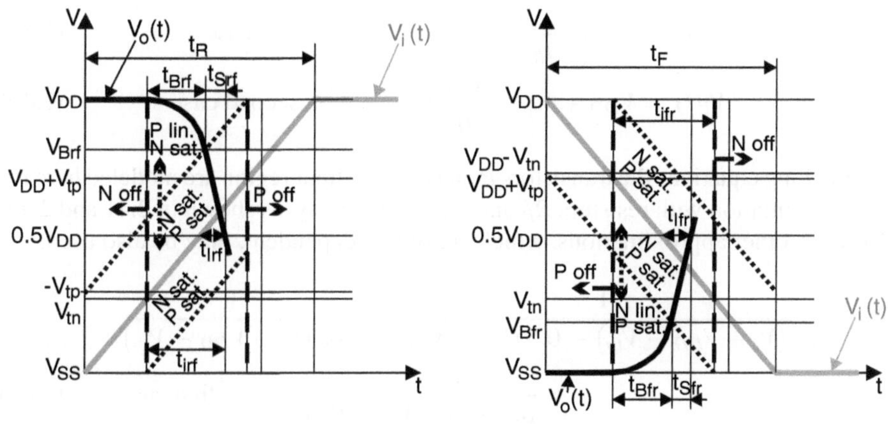

Fig. 2.23 Sketches of $V_O(t)$ in dependence on a linear rising $V_i(t)$ (*left side*, first case) and a linear falling $V_i(t)$ (*right side*, second case), which are used for calculations. Some inportatant regions for $V_O(t)$ of N and P are drawen

for a linear rising input voltage $V_i(t)$ and falling output voltage $V_O(t)$ as well as t_{Ifr} for a linear falling $V_i(t)$ and a rising $V_O(t)$ are sketched in Fig. 2.22. The delay time is definded here as the time difference between $V_i(t)$ has reached 50 % of V_{DD} and $V_O(t)$ has reached 50 % of V_{DD}. An overall delay time of the inverter can be defined by combining t_{Irf} and t_{Ifr} to $t_{Iabs} = (t_{Irf} + t_{Ifr})/2$. In Fig. 2.23 qualitative drawings of $V_O(t)$ in dependence of a linear rising $V_i(t)$ (left side, first case) and a linear falling $V_i(t)$ (right side, second case) are depicted. $V_i(t)$ rises from $V_{SS} = 0\,V$ to V_{DD} during time t_R in the first case and falls in the second case from V_{DD} to $V_{SS} = 0\,V$ during time t_F. When $V_i(t)$ reaches the threshold voltage $V_{tn} > 0$ of N in the first case, N starts to conduct and is in saturation while P is in linear mode. V_O falls while V_i is rising further and after a time t_{Brf}, a voltage $V_O(t_{Brf}) = V_{Brf}$ is reached, where P becomes saturated and works as well as N as a current source. In the second case ($V_i(t)$ is falling), *Vit* has to reach a voltage $V_{DD} + V_{tp}$ ($V_{tp} < 0$ is the threshold voltage of P), so that P starts to conduct in saturation while N is in linear mode. V_O is rising while V_i is falling further and after a time t_{Bfr}, a voltage $V_O(t_{Bfr}) = V_{Bfr}$ is reached, where N becomes saturated and works as well as P as a current source. For calculations it is assumed, that in the first case V_O starts in a region, where P works in linear mode and N is a current source, and reaches most likely $V_O = V_{DD}/2$ in the region, where N as well as P are saturated and work as current source. In the second case V_O starts in a region, where N works in linear mode and P as current source. $V_{DD}/2$ is reached by V_O most likely in a region, where both N and P work as current source.

If $t = 0$ is set to the time point, where N starts to conduct (N saturation, P linear) for the first case and where P starts to conduct (P saturation, N linear) for the second case, then the formulas for $V_i(t)$ can be written as shown in 2.39 and 2.40.

$$V_i(t) = V_{tn} + \frac{V_{DD}}{t_R}t \qquad \text{first case} \tag{2.39}$$

$$V_i(t) = V_{DD} + V_{tp} - \frac{V_{DD}}{t_F}t \qquad \text{second case} \tag{2.40}$$

To omit long equations the transistors P and N in the linar region are modeled here for an approximation with resistors R_P and R_N, respectively, as shown in 2.41 and 2.42. However, other approximations, which are time independent, may be also used.

$$R_P = \frac{1}{\beta_p\left(V_i - V_{DD} - V_{tp}\right) - 0.5\left(V_O - V_{DD}\right)} \approx \frac{1}{\beta_p\left(V_i - V_{DD} - V_{tp}\right)}$$
$$\approx \frac{1}{\beta_p\left(V_{DD} + V_{tp} - V_{tn}\right)} \qquad \text{first case} \tag{2.41}$$

$$R_N = \frac{1}{\beta_n\left(V_i - V_{tn}\right) - 0.5V_O} \approx \frac{1}{\beta_n\left(V_i - V_{tn}\right)}$$
$$\approx \frac{1}{\beta_n\left(V_{DD} + V_{tp} - V_{tn}\right)} \qquad \text{second case} \tag{2.42}$$

The differential equations for calculating the time dependence of $V_O(t)$ at the beginning for the first case (P linear, N saturated) and for the second case (N linear, P saturated) are found by applying the current law to the output node of the inverter, as it can be seen in 2.43 and 2.44.

$$\frac{V_{DD} - V_O(t)}{R_P} = \frac{\beta_n}{2}\left(V_i(t) - V_{tn}\right)^2 + C_L\frac{dV_O(t)}{dt} \quad V_O(0) = V_{DD} \qquad \text{first case} \tag{2.43}$$

$$\frac{\beta_P}{2}\left(V_i(t) - V_{DD} - V_{tp}\right)^2 = \frac{V_O(t)}{R_N} + C_L\frac{dV_O(t)}{dt} \quad V_O(0) = 0V \qquad \text{second case} \tag{2.44}$$

Inserting 2.39 into 2.43 for the first case and 2.40 into 2.44 for the second case results into 2.45 and 2.46.

$$\frac{dV_O(t)}{dt} + \frac{1}{R_PC_L}V_O(t) = \frac{V_{DD}}{R_PC_L} - \frac{\beta_n}{2C_L}\left(\frac{V_{DD}}{t_R}t\right)^2 \quad V_O(0) = V_{DD} \quad \text{first case} \tag{2.45}$$

$$\frac{dV_O(t)}{dt} + \frac{1}{R_NC_L}V_O(t) = \frac{\beta_P}{2C_L}\left(\frac{V_{DD}}{t_F}t\right)^2 \quad V_O(0) = 0V \qquad \text{second case} \tag{2.46}$$

Solving these equations leads to the exact results in 2.47 ($\tau_P = R_PC_L$) and 2.48 ($\tau_N = R_NC_L$).

$$V_O(t) = V_{DD} - \frac{\beta_n V_{DD}^2 R_P}{2t_R^2}\left(t^2 - 2\tau_P t + 2\tau_P^2\left(1 - e^{\frac{-t}{\tau_P}}\right)\right) \qquad \text{first case} \tag{2.47}$$

$$V_O(t) = \frac{\beta_p V_{DD}^2 R_N}{2t_F^2}\left(t^2 - 2\tau_N t + 2\tau_N^2\left(1 - e^{\frac{-t}{\tau_N}}\right)\right) \qquad \text{second case} \tag{2.48}$$

With $1 - e^{-t/\tau} \approx t/\tau$ 2.47 can be simplified to 2.49 for the first case and 2.48 can be simplified to 2.50 for the second case.

$$V_O(t) \approx V_{DD} - \frac{\beta_n V_{DD}^2 R_P}{2t_R^2} t^2 \qquad \text{first case} \tag{2.49}$$

$$V_O(t) \approx \frac{\beta_p V_{DD}^2 R_N}{2t_F^2} t^2 \qquad \text{second case} \tag{2.50}$$

These approximated results for $V_O(t)$ are used to determine the borders $V_O(t_{Brf}) = V_{Brf}$ for the first case and $V_O(t_{Bfr}) = V_{Bfr}$ when P as well as N starts working in saturation as current source. Here the conditions 2.51 for the first case and 2.52 for the second case are needed.

$$V_{Brf} = V_{tn} + \frac{V_{DD}}{t_R} t_{Brf} - V_{tp} = V_{DD} - \frac{\beta_n V_{DD}^2 R_P}{2t_R^2} t_{Brf}^2 \qquad \text{first case} \tag{2.51}$$

$$V_{Bfr} = V_{DD} + V_{tp} - \frac{V_{DD}}{t_F} t_{Bfr} - V_{tn} = \frac{\beta_p V_{DD}^2 R_N}{2t_F^2} t_{Bfr}^2 \qquad \text{second case} \tag{2.52}$$

The results of 2.51 and 2.52 can be found in 2.53 and 2.54.

$$\begin{aligned} t_{Brf} &= \frac{t_R}{\beta_n V_{DD} R_P}\left(\sqrt{1 + 2\beta_n R_P\left(V_{DD} + V_{tp} - V_{tn}\right)} - 1\right) \\ V_{Brf} &= V_{tn} - V_{tp} + \frac{1}{\beta_n R_P}\left(\sqrt{1 + 2\beta_n R_P\left(V_{DD} + V_{tp} - V_{tn}\right)} - 1\right) \end{aligned} \qquad \text{first case} \tag{2.53}$$

$$\begin{aligned} t_{Bfr} &= \frac{t_F}{\beta_p V_{DD} R_N}\left(\sqrt{1 + 2\beta_p R_N\left(V_{DD} + V_{tp} - V_{tn}\right)} - 1\right) \\ V_{Bfr} &= V_{DD} + V_{tp} - V_{tn} - \frac{1}{\beta_p R_N}\left(\sqrt{2\beta_p R_N\left(V_{DD} + V_{tp} - V_{tn}\right)} - 1\right) \end{aligned} \qquad \text{second case} \tag{2.54}$$

When $V_O(t)$ reaches at time point t_{Brf} the voltage level V_{Brf} in the first case or at t_{Bfr} the voltage level V_{Bfr} in the second case P and N are saturated and work as current sources. By the way $V_{DD}/2$ can be also reached when P is turned off and N is in linear or saturated mode in the first case or N is turned off and P is in linear and saturated mode for the second case. For simplicity the most likely case is assumed here, that the output voltage $V_O(t)$ reaches $V_{DD}/2$ in the region where P and N are current sources. Then an additional differential equation has to be solved, which is depicted in 2.55, where the boundary conditions are stated in 2.56 for the first case and 2.57 for the second case. The new time point $t = 0$ is now the former time point $t = t_{Brf}$ for the first case and $t = t_{Bfr}$ for the second case.

$$C_L \frac{dV_O(t)}{dt} = \frac{\beta_p}{2}\left(V_i - V_{DD} - V_{tp}\right)^2 - \frac{\beta_n}{2}\left(V_i - V_{tn}\right)^2 \tag{2.55}$$

$$V_O(t=0) = V_{Brf} \qquad V_i(t) = V_{tn} + \frac{V_{DD}}{t_R}\left(t + t_{Brf}\right) \qquad \text{first case} \tag{2.56}$$

$$V_O(t=0) = V_{Bfr} \quad V_i(t) = V_{DD} + V_{tp} - \frac{V_{DD}}{t_f}\left(t + t_{Bfr}\right) \qquad \text{second case} \tag{2.57}$$

The exact solutions of 2.55–2.57 are depicted for both cases in 2.58 and 2.59.

$$\begin{aligned} V_O(t) &= V_{Brf} + \frac{V_{DD}^2}{6C_L t_R^2}\left(\beta_p\left(\frac{V_{tn} - V_{tp} - V_{DD}}{V_{DD}} t_R + t_{Brf} + t\right)^3\right. \\ &\quad \left. -\beta_p\left(\frac{V_{tn} - V_{tp} - V_{DD}}{V_{DD}} t_R + t_{Brf}\right)^3 - \beta_n\left(\left(t + t_{Brf}\right)^3 - \left(t_{Brf}\right)^3\right)\right) \\ &\qquad \text{first case} \end{aligned} \tag{2.58}$$

$$\begin{aligned} V_O(t) &= V_{Bfr} + \frac{V_{DD}^2}{6C_L t_F^2}\left(\beta_p\left(\left(t + t_{Bfr}\right)^3 - t_{Bfr}^3\right)\right. \\ &\quad -\beta_n\left(\frac{V_{DD} + V_{tp} - V_{tn}}{V_{DD}} t_F + t_{Bfr} + t\right)^3 \\ &\quad \left. +\beta_n\left(\frac{V_{DD} + V_{tp} - V_{tn}}{V_{DD}} t_F + t_{Bfr}\right)^3\right) \qquad \text{second case} \end{aligned} \tag{2.59}$$

For simplification these equations can be linearized (Taylor row) and the results are shown in 2.60 and 2.61.

$$\begin{aligned} V_O(t) &\approx V_{Brf} + \frac{V_{DD}^2}{2C_L t_R^2}\left(\beta_p\left(\frac{V_{tn} - V_{tp} - V_{DD}}{V_{DD}} t_R + t_{Brf}\right)^2 - \beta_n t_{Brf}^2\right) t \\ &\qquad \text{first case} \end{aligned} \tag{2.60}$$

$$\begin{aligned} V_O(t) &\approx V_{Bfr} + \frac{V_{DD}^2}{2C_L t_F^2}\left(\beta_p t_{Bfr}^2 - \beta_n\left(\frac{V_{tp} + V_{DD} - V_{tn}}{V_{DD}} t_F + t_{Bfr}\right)^2\right) t \\ &\qquad \text{second case} \end{aligned} \tag{2.61}$$

With $V_O(t_{Srf}) = V_{DD}/2$ (first case) and $V_O(t_{Sfr}) = V_{DD}/2$ (second case) the time from the beginning of the region, where P and N are in saturation till $V_O(t)$ reaches 50 % of V_{DD}, can be calculated. The results are depicted in 2.62 for the first case and in 2.63 for the second case.

$$t_{Srf} = \frac{\frac{V_{DD}}{2} - V_{Brf}}{\frac{V_{DD}^2}{2C_L t_R^2}\left(\beta_p\left(\frac{V_{tn}-V_{tp}-V_{DD}}{V_{DD}}t_R + t_{Brf}\right)^2 - \beta_n t_{Brf}^2\right)} \quad \text{first case} \tag{2.62}$$

$$t_{Sfr} = \frac{\frac{V_{DD}}{2} - V_{Bfr}}{\frac{V_{DD}^2}{2C_L t_F^2}\left(\beta_p t_{Bfr}^2 - \beta_n\left(\frac{V_{tp}+V_{DD}-V_{tn}}{V_{DD}}t_F + t_{Bfr}\right)^2\right)} \quad \text{second case} \tag{2.63}$$

With the time durations to when the rising input voltage $V_i(t)$ reaches $V_{DD}/2$ (2.64) for the first case and to when the falling input voltage reaches $V_{DD}/2$ (2.66) for the second case, all times are given to calculate the delay times t_{Irf} (first case, 2.65) and t_{Ifr} (second case, 2.67).

$$t_{irf} = \left(\frac{V_{DD}}{2} - V_{tn}\right)\frac{t_R}{V_{DD}} \quad \text{first case} \tag{2.64}$$

$$t_{Irf} = t_{Brf} + t_{Srf} - t_{irf} \quad \text{first case} \tag{2.65}$$

$$t_{ifr} = \left(\frac{V_{DD}}{2} + V_{tp}\right)\frac{t_F}{V_{DD}} \quad \text{second case} \tag{2.66}$$

$$t_{Ifr} = t_{Bfr} + t_{Sfr} - t_{ifr} \quad \text{second case} \tag{2.67}$$

Finally the overall delay time t_{Iabs} can be calculated with $t_{Iabs} = (t_{Irf} + t_{Ifr})/2$.

The mismatch between two transistors can be described with 2.22 and 2.23 [28] (see Sect. 2.4.1). If the different influences are assumed to be uncorrelated, the influence of mismatch to the delay time difference Δt_{Iabs} between two inverters can be described with 2.68.

$$\sigma\left(\Delta t_{Iabs}\right)^2 \left(\frac{\partial t_{Iabs}}{\partial V_{tn}}\right)^2 \sigma\left(\Delta V_{tn}\right)^2 + \left(\frac{\partial t_{Iabs}}{\partial V_{tp}}\right)^2 \sigma\left(\Delta V_{tp}\right)^2 + \left(\frac{\partial t_{Iabs}}{\partial \beta_n}\right)^2 \sigma\left(\Delta\beta_n\right)^2 + \left(\frac{\partial t_{Iabs}}{\partial \beta_p}\right)^2 \sigma\left(\Delta\beta_p\right)^2 \tag{2.68}$$

As a result of this simplified model, it can be stated, that if the rise time t_R or fall time t_F of the input voltage V_i lasts longer, also the influence of mismatch to t_{Iabs} (Δt_{Iabs}) will be higher. Also if C_L is higher and β_n and β_p, which depend on the ratios of gate width to gate length, are designed small, the influence of mismatch on t_{Iabs} will be raised. On the other hand, if β_n and β_p are fixed and the gate area is enhanced, the dependence of t_{Iabs} on mismatch will be smaller. Also if V_{tn} or $|V_{tp}|$ are lower, the influence of mismatch is also raised to t_{Iabs}.

References

1. H. Klar, *Integrierte Digitale Schaltungen MOS/BICMOS* (Springer, Berlin, 1996)
2. V. Srinivas, S. Pavan, A. Lachhwani, N. Sasidhar, A distortion compensating flash anaolog-to-digital conversion technique. IEEE J. Solid-State Circuits **41**(9), 1959–1969 (2006)
3. B. Goll, H. Zimmermann, A Low Power 1.4 GSample/s Comparator for Flash-ADCs in 120 nm CMOS Technology, Austrochip 2004, pp. 39–42 (2005)
4. K. Uyttenhove, M. Steyaert, A CMOS 6-bit, 1 GHz ADC for IF sampling applications. IEEE MTT-S Int. Microwave Symp. Digest **3**, 2131–2134 (2001)
5. K. Cornelissens, P. Reynaert, M. Steyaert, A 0.18 μm CMOS switched capacitor voltage modulator. IEEE European Solid-State Circuits Conference, pp. 375–378 (2005)
6. P.M. Figueiredo, J.C. Vital, Low kickback noise techniques for CMOS latched comparators. IEEE Int. Symp. Circuits Syst. **1**, 537–540 (2004)
7. R. van de Plasche, *CMOS Integrated Analog-to-Digital and Digital-to-Analog Converters* (Kluwer Academic, Boston, 2003)
8. M. Frey, H.-A. Loeliger, On flash A/D-converters with low-precision comparators. IEEE International Symposium on Circuits and Systems, pp. 3926–3929 (2006)
9. C. Sandner, M. Clara, A. Santner, T. Hartnig, F. Kuttner, A 6bit, 1.2GSps low-power flash-ADC in 0.13μm digital CMOS, Design, Automation and Test in Europe Conference and Exhibition (DATE'05), Vol. 3, pp. 223–226 (2005)
10. H. Dang, M. Sawan, Y. Savaria, A novel approach for implementing ultra-high speed flash ADC using MCML circuits. IEEE Int. Symp. Circuits Syst. **6**, 6158–6161 (2005)
11. S. Sheikhaei, S. Mirabbasi, A. Ivanov, An encoder for a 5GS/s 4-bit flash ADC in 0.18μm CMOS. IEEE canadian conference on electrical and computer engineering, pp. 698–701 (2005)
12. T. Kobayashi, K. Nogami, T. Shirotori, Y. Fujimoto, A current-controlled latch sense amplifier and a static power-saving input buffer for low-power architecture. IEEE J. Solid-State Circuits **28**(4), 523–527 (1993)
13. P. Uthaichana, E. Leelarasmee, Low power CMOS dynamic latch comparators. IEEE Conference on Convergent Technologies for Asia-pacific Region (TENCON), vol.2, pp. 605–608 (2003)
14. B. Wicht, *Current Sense Amplifiers for Embedded SRAM in High-Performance System-on-a-Chip Designs* (Springer, Berlin, 2003)
15. L. Samid, P. Volz, Y. Manoli, A dynamic analysis of a latched CMOS comparator. IEEE international symposium on circuits and systems, vol. 1, pp. 181–184 (2004)
16. L. Sumanen, M. Waltari, V. Hakkarainen, K. Halonen, CMOS dynamic comparators for pipeline A/D converters. IEEE International Symposium on Circuits and Systems, vol. 1, pp. 157–160 (2002)
17. F. Maloberti, *Analog Design for CMOS VLSI Systems* (Kluwer Academic, Boston, 2001)
18. P.E. Allen, D.R. Holberg, *CMOS Analog Circuit Design* (Oxford University Press Inc, New York, 2002)
19. A. Baschirotto, *II. Analog Switches, Lecture Notes, Dipartimento di Ingegneria dell' Innovazione* (Universita' degli Studi di Lecce, Lecce, 2003)
20. R. Gregorian, *Introduction to CMOS Op-Amps and Comparators A Wiley-Interscience Publication* (Wiley, New York, 1999)
21. J.-T. Wu, B.A. Wooley, A 100-MHz pipelined CMOS comparator. IEEE J. Solid-State Circuits **23**(6), 1379–1385 (1988)
22. W.M.C. Sansen, *Analog Design Essentials* (Springer, Berlin, 2006)
23. P.M. Figueiredo, J.C. Vital, Kickback noise reduction techniques for CMOS latched comparators. IEEE Trans. Circuits Syst. II **53**(7), 541–545 (2006)
24. J. Lohstroh, Worst-case static noise margin criteria for logic circuits and their mathematical equivalence. IEEE J. Solid-State Circuits **18**(6), 803–807 (1983)
25. E. Seevinck, F.J. List, J. Lohstroh, Static-noise margin analysis of MOS SRAM cells. IEEE J. Solid-State Circuits **22**(5), 748–754 (1987)

26. A.J. Bhavnagarwala, X. Tang, J.D. Meindl, The impact of intrinsic device fluctuations on CMOS SRAM cell stability. IEEE J. Solid-State Circuits **36**(4), 658–665 (2001)
27. A. Nikoozadeh, B. Murmann, An analysis of latch comparator offset due to load capacitor mismatch. IEEE Trans. Circuits Syst. II **53**(12), 1398–1402 (2006)
28. B. Razavi, *Design of Analog CMOS Integrated Circuits* (McGraw-Hill, New York, 2001)
29. M.J.M. Pelgrom, A.C.J. Duinmaijer, A.P.G. Welbers, Matching properties of MOS transistors. IEEE J. Solid-State Circuits **24**(5), 1433–1440 (1989)
30. B. Razavi, B.A. Wooley, Design techniques for high-speed, high-resolution comparators. IEEE J. Solid-State Circuits **27**(12), 1916–1926 (1992)
31. K.-L.J. Wong, C.-K.K. Yang, Offset compensation in comparators with minimum input-referred supply noise. IEEE J. Solid-State Circuits **39**(5), 837–840 (2004)
32. E.L. Wong, P.A. Abshire, M.H. Cohen, Floating gate comparator with automatic offset manipulation functionality. IEEE International Symposium on Circuits and Systems, vol. 1, pp. 529–532 (2004)
33. E. Rodriguez-Villegas, A 0.9V offset compensated FGMOS comparator, IEEE International Symposium on Circuits and Systems, vol. 3, pp. 2160–2163 (2005)
34. D.G. Chen, A. Bermak, A low-power dynamic comparator with digital calibration for reduced offset mismatch. IEEE International Symposium on Circuits and Systems, pp. 1283–1286 (2012)
35. J. Lu, J. Holleman, A low-power high-precision comparator with time-domain bulk-tuned offset cancellation. IEEE Trans. Circuits Syst. I **60**(5), 1158–1167 (2013)
36. Y.L. Wong, M.H. Cohen, P.A. Abshire, A 1.2-GHz comparator with adaptable offset in 0.35-μm CMOS. IEEE Trans. Circuits Syst. I **55**(9), 2584–2593 (2008)
37. J. He, S. Zhan, D. Chen, R.L. Geiger, Analyses of static and dynamic random offset voltages in dynamic comparators. IEEE Trans. Circuits Syst. I **56**(5), 911–919 (2009)
38. H.J.M. Veendrick, The behaviour of flip-flops used as synchronizers and prediction of their failure rate. IEEE J. Solid-State Circuits **15**(2), 169–176 (1980)
39. H. Weinrichter, F. Hlawatsch, *Stochastische Grundlagen nachrichtentechnischer Signale* (Springer, New York, 1991)
40. P.M. Figueiredo, Comparator metasability in the presence of noise. IEEE Trans. Circuits Syst. I **60**(5), 1286–1299 (2013)
41. P. Nuzzo, F. De Bernardinis, P. Terreni, G. Van der Plas, Noise analysis of regenerative comparators for reconfigurable ADC architectures. IEEE Trans. Circuits Syst. I **55**(6), 1441–1454 (2008)
42. S. Park, M.P. Flynn, A regenerative comparator structure with integrated inductors. IEEE Trans. Circuits Syst. I **53**(8), 1704–1711 (2006)
43. A. Boni, C. Morandi, S. Padoan, A 2.5-V BiCMOS comparator with current-mode interpolation. IEEE J. Solid-State Circuits **34**(6), 892–897 (1999)
44. A. Boni, G. Chiorboli, C. Morandi, Dynamic characterisation of high-speed latching comparators. IET Electron. Lett. **36**(5), 402–404 (2000)
45. B. Goll, H. Zimmermann, Simple creation of half and full frequency, inverted and non-inverted clock signals with maximum 10ps delay time differences in 120nm CMOS, Austrochip 2006, pp. 143–148 (2006)

Chapter 3
State of the Art

Although there are many papers on comparators in micron and submicron CMOS technology, in the following only deep-submicron and nanometer comparators are summarized. Often only flash ADCs are present in the literature without characterizing the comparators mentioned accurately and completely. Therefore not all interesting properties are described in the papers and could not be extracted in the following description of the state of the art.

Figure 3.1 shows the core of a 4-stage comparator in CML, which is described in [1], with two additional latches and a final digital flip-flop. *CP* (non-inverted) and *CN* (inverted) are the clock signals and *OP* (non-inverted) and *ON* (inverted) are output nodes. To reduce the regenerative time constant of the comparator, an additional differential inductor load has been added. Each 11.88 nH inductor occupied an area of $32 \times 32\,\mu\text{m}^2$ with a quality factor of $Q_L = 0.68$ at 4 GHz and a coupling coefficient of less than 0.005. To reduce kickback to the input transistors $N0$ and $N1$ have been added. The comparator was implemented in a 4-bit flash ADC in a 0.18 μm CMOS technology, where the supply voltage was 1.8 V in the analog and 2.1–2.5 V in the digital part. The ADC worked at a maximum clock frequency of 4 GHz.

In [2, 3] a 3-stage CML comparator (see Fig. 3.2) with a clock frequency of 10 GHz with additional two amplifiers in a 0.11 μm CMOS process with a supply voltage of 1.2 V was presented, which is used in a demultiplexer (DEMUX) to extract every 4th bit of a 40 Gb/s data stream. With an input differential voltage of $\Delta V_{in} = 1\text{V}_{pp}$ a $BER < 10^{-12}$ was measured with a PRBS $2^{23} - 1$. To get a higher eye pattern opening a bandwidth modulation technique was proposed, where in the clocked amplifier an additional clocked feedback amplifier in negative feedback is added to modulate the input impedance. So in the reset phase of the regenerative stage a low input impedance of the clocked amplifier (feedback amplifier turned on) provides assistance. In the regeneration phase of the regenerative stage a high input impedance of the clocked amplifier provides high gain where sufficient bandwidth also remains. The measured power consumption of the comparator with the output drivers was 37 mW. The simulated sum of all CML tail currents of the comparator itself was 4.2 mA.

B. Goll and H. Zimmermann, *Comparators in Nanometer CMOS Technology*,
Springer Series in Advanced Microelectronics 50, DOI 10.1007/978-3-662-44482-5_3

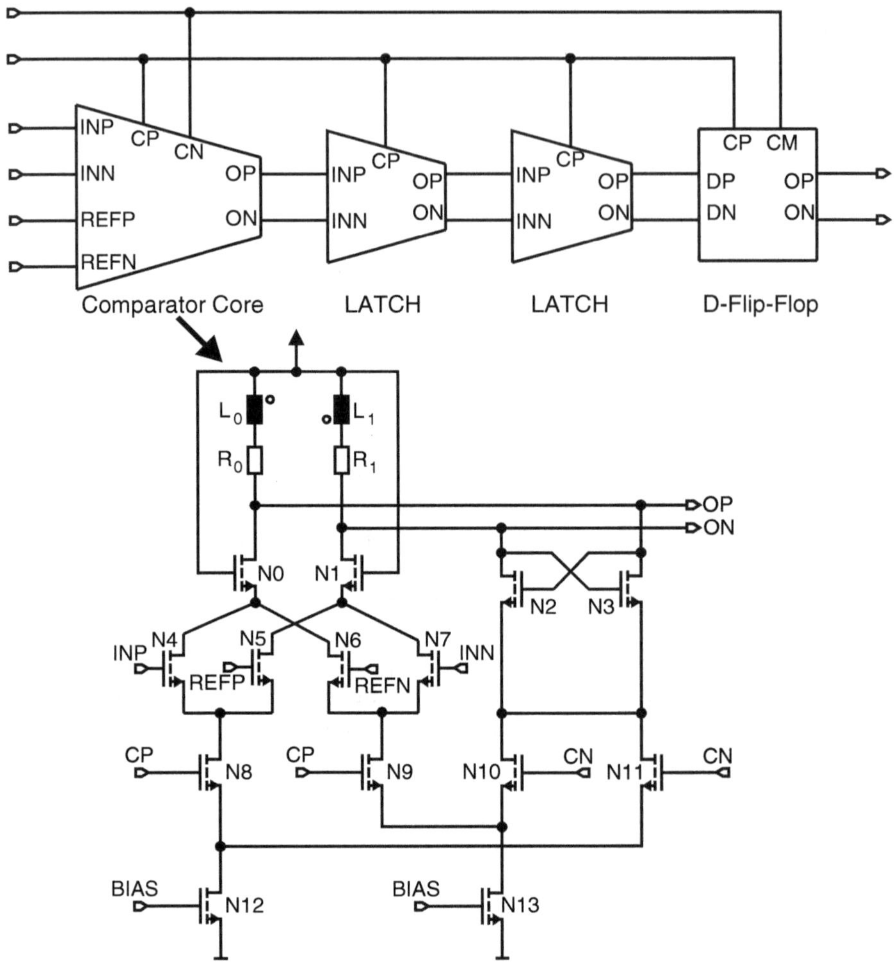

Fig. 3.1 CML-comparator with inductor load [1]

In [4] offset compensation at a widely used comparator, which is depicted in Fig. 3.3, was treated.In opposite to the structure in Fig. 2.7a additional reset switches $P4$, $P5$, $P6$ and $P7$ were added to achieve a faster reset and to avoid a hysteresis. A static offset compensation, where also offset changes due to supply noise were treated, was done by compensating the mismatch by applying the appropriate error currents, which could be digitally adjusted (not shown in Fig. 3.3), at nodes A and B. The comparator itself was built in a 0.18 μm CMOS process with a supply voltage of $V_{DD} = 1.8\,\text{V}$. The operating clock frequency was 1.4 GHz and the power consumption was 350 μW. The standard deviation of the offset was $\sigma_{OS} = 95\,\text{mV}$ without compensation and $\sigma_{OS} = 16\,\text{mV}$ with compensation.

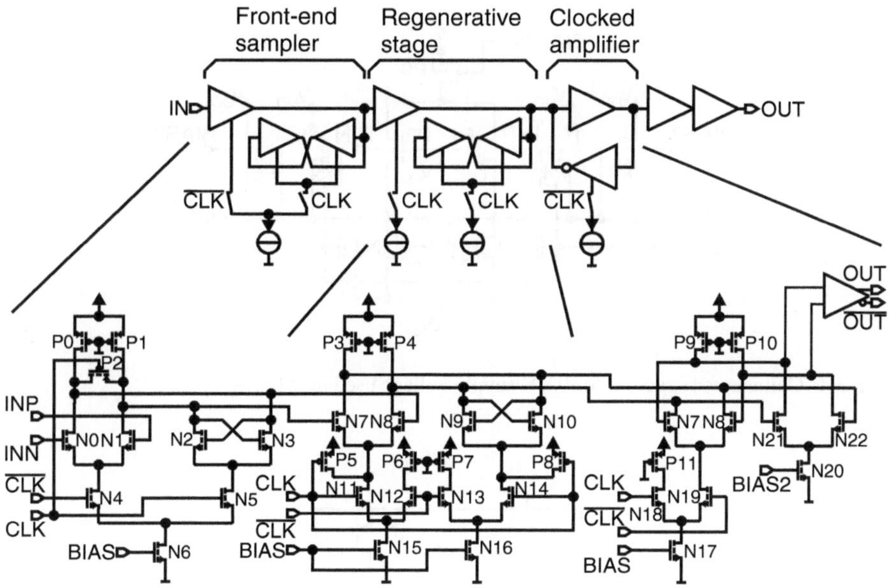

Fig. 3.2 CML comparator for a 40 Gb/s DEMUX [2, 3]

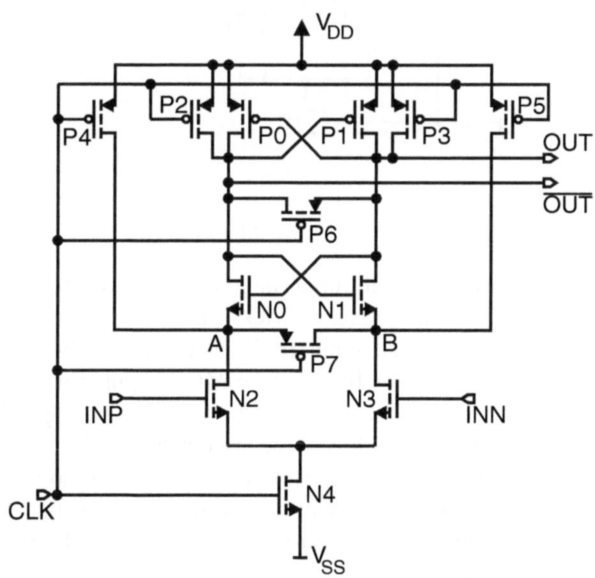

Fig. 3.3 Comparator structure used in [4]

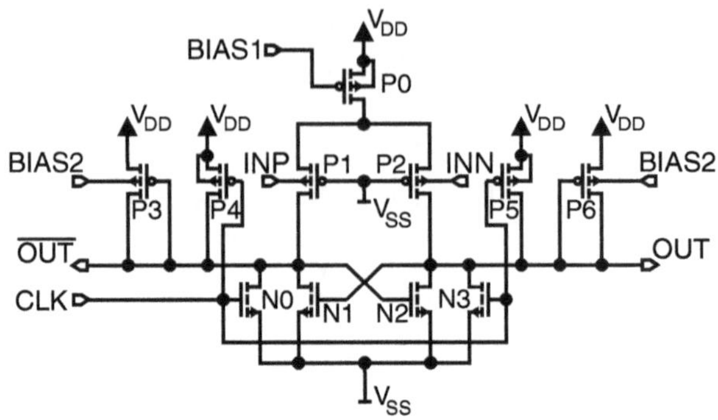

Fig. 3.4 Comparator structure for 0.8 V supply voltage used in [5]

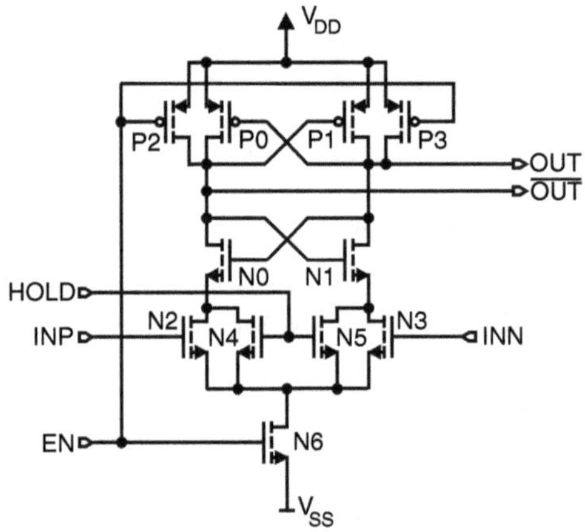

Fig. 3.5 Latch-type voltage sense amplifier for SRAM used in [6–8]

A 1-bit quantizer with rail-to-rail input range for a $\Sigma\Delta$-modulator with a supply voltage of $V_{DD} = 0.8\,\text{V}$ in a $0.18\,\mu\text{m}$ CMOS process was presented in [5] (see Fig. 3.4). To achieve a rail-to-rail input range at 0.8 V supply voltage, the input voltages were applied at the n-wells of the p-MOS input transistors $P1$ and $P2$ to cause an initial current difference for the latch $N1$ and $N2$. To speed up the circuit, transistors $P3, P4, P5$ and $P6$ are added. In the reset phase ($CLK = V_{DD}$), both output nodes are pulled towards V_{SS} by transistors $N0$ and $N3$. Due to simulation results the quantizer had a sensitivity of $10\,\mu\text{V}$ at a clock frequency of 10 MHz and $100\,\mu\text{V}$ at 12 MHz, where only metastability and no noise were considered. The simulated power consumption was $134\,\mu\text{W}$.

In [6–8] yield and speed optimization of the latch-type voltage sense amplifier, depicted in Fig. 3.5, was performed for an implementation in a SRAM with a $0.13\,\mu\text{m}$ CMOS technology with a nominal supply voltage of $V_{DD} = 1.5\,\text{V}$. Measurement results showed, that the sense amplifier operated down to a supply voltage of 0.7 V with a delay time in the comparison phase of higher than 11 ns. At a supply voltage of

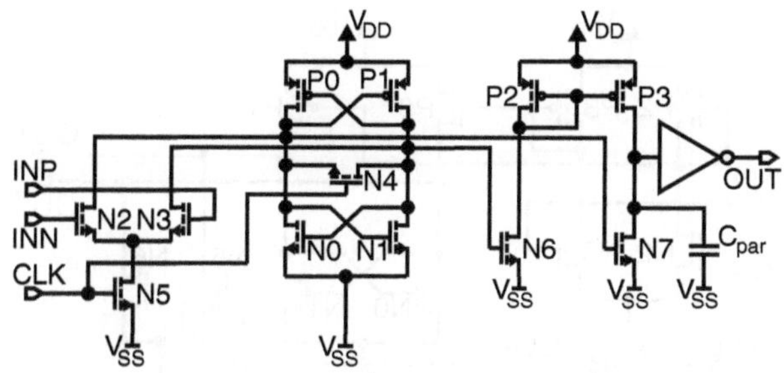

Fig. 3.6 Comparator for a 6-bit flash ADC used in [9]

$V_{DD} = 1.5\,\text{V}$ the delay time was higher than 1.5 ns. Due to theoretical investigations, the yield and the offset depends on the input common-mode voltage, where the best results occurred at about $0.6V_{DD}$ while the delay time was nearly constant compared to higher input common-mode voltages. So at a supply voltage of $V_{DD} = 1.5\,\text{V}$ a standard deviation of the offset of $\sigma_{OS} = 19\,\text{mV}$ was measured at an input common-mode voltage of 1.5 V and $\sigma_{OS} = 8.5\,\text{mV}$ at 1.05 V. No power consumption was stated for the sense amplifier.

In Fig. 3.6 the circuit of the comparator of a 4 GHz, 6-bit flash ADC is depicted. Not shown is the pre-amplifier with a gain of 2 in front to enhance the resolution. Transistors $N6, N7, P2$ and $P3$ work as a differential amplifier and an analog memory cell. During the tracking phase ($CLK = high$, transistors $N4$ and $N5$ are on) both outputs of the latch (transistors $P0$, $P1$, $N0$ and $N1$) are equal and the decision of the previous regeneration phase remains stored in the parasitic capacitance C_{par}. No power consumption was stated for the comparator. The technology used was a standard 0.13 μm CMOS process with a supply voltage of $V_{DD} = 1.5\,\text{V}$.

Figure 3.7 shows the comparator, which is implemented in a 1.3 GHz, 6-bit flash ADC in a 0.25 μm CMOS technology with a supply voltage of $V_{DD} = 1.8\,\text{V}$ [10]. In principle the function of the comparator is similar to that described in Fig. 2.4. The

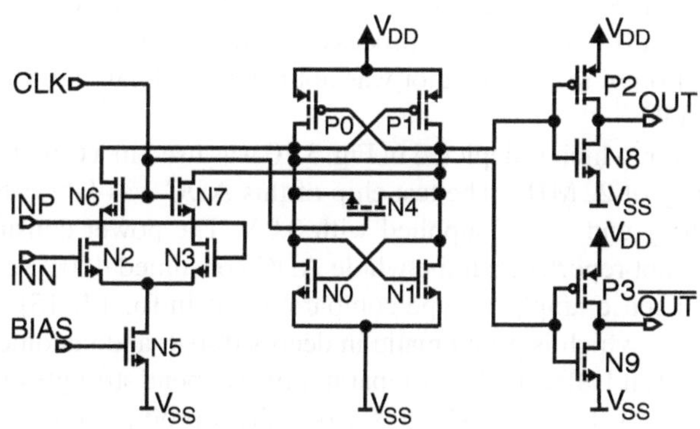

Fig. 3.7 Comparator for a 6-bit flash ADC used in [10]

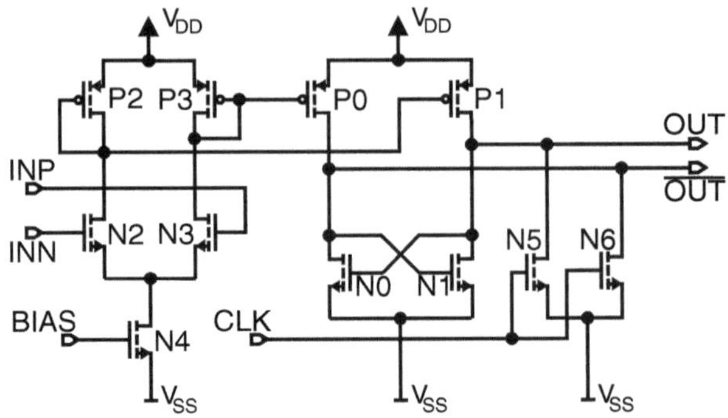

Fig. 3.8 Comparator for a 6-bit flash ADC used in [11]

additional switches $N6$ and $N7$ were added to reduce kickback noise and inverters $P2$-$N8$ and $P3$-$N9$ are used as buffers.

A comparator [11] for a 1.6 GHz, 6-bit flash ADC in a 0.13 μm CMOS process with a supply voltage of $V_{DD} = 1.5\,\text{V}$ is depicted in Fig. 3.8. The circuit structure is similar to that in Fig. 2.5. The difference is, that the output nodes are pulled to V_{SS} in reset phase by transistors $N5$ and $N6$ so that a logical voltage level is defined. The standard deviation of the offset of the comparator alone was $\sigma_{OS} = 80\,\text{mV}$. With the help of cascaded, offset compensated pre-amplifiers, which are similar to that in Fig. 2.15a, with an overall gain of about 40, the overall offset was reduced to $\sigma_{OS} = 2\,\text{mV}$. In the paper no power consumption of one comparator was stated.

In [12] a 4-bit flash ADC in a 0.18 μm CMOS technology with a nominal supply voltage of 1.8 V is described. The comparator had four stages, where the first stage was a simple differential amplifier with two pairs of n-MOS input-transistors instead of only one pair to compare two differential input voltages. The following stages were a cascade of three CML latches, where one of them is depicted in Figs. 2.6 and 3.9 for example. The simulated voltage swing at the outputs of the comparator was ±0.4 V. The simulated clock frequency was 5 GHz and the standard deviation of the offset was $\sigma_{OS} = 67\,\text{mV}$. The simulated sum of the tail currents of the pre-amplifier and the following three CML latches in series was 1 mA. The test chip of the ADC was fabricated in a 0.13 μm CMOS process and the supply voltage was 0.8 V. The power consumption of the comparator was not mentioned, but the whole flash ADC consumed 480 μW.

The comparator, which is depicted in Fig. 3.10 was used in a 6-bit flash ADC with a clock frequency of 22 MHz. The test chip of this ADC was fabricated in 0.13 μm CMOS technology and it was supplied with 0.8 V. The power consumption of the comparator was not reported, but the whole ADC consumed 480 μW.

Conventional sense-amplifier type comparators as in [6, 14, 15] require a large voltage headroom, which is problematic in deep-sub-μm and nanometer CMOS. In addition the speed and offset of this comparator type depend strongly on the common-mode voltage V_{cm} of the input, which is especially a problem for applications with

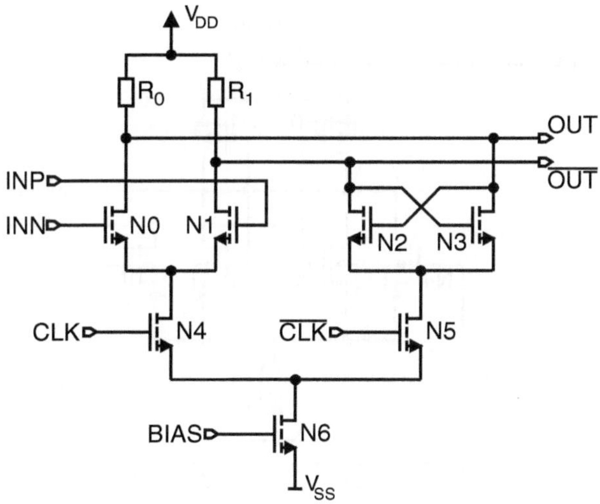

Fig. 3.9 CML latch used in [12]

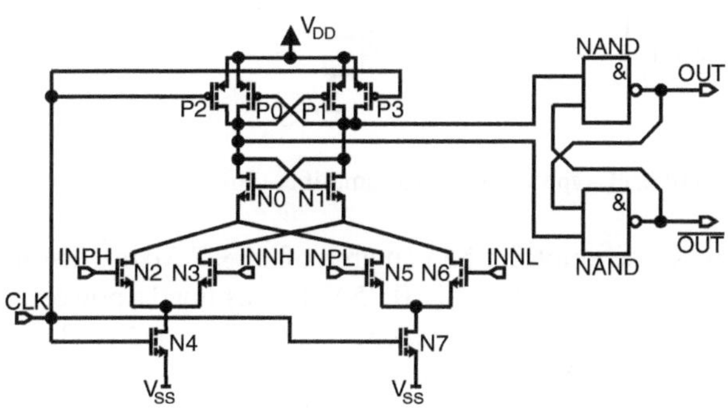

Fig. 3.10 Comparator for a 6-bit flash ADC used in [13]

wide common-mode range as for instance analog-digital converters. As an improvement, a so-called double tail sense amplifier was introduced in [16]. This sense amplifier uses one tail for the input stage and a second one for the latch. Figure 3.11 shows the corresponding circuit diagram. In this topology less drain-source voltages have to fit into the supply voltage and double tail sense amplifiers can work with lower supply voltage. A large current can be used in the latch (*P*4 has a large width) for fast latching independent of common-mode voltage V_{cm} at the input nodes *INP* and *INN* and a small current in the input stage (*N*6 has a small width) can be used for a low offset. At $V_{DD} = 1.2$ V and $f_{clk} = 3$ GHz and for an drop of V_{cm} from 1.0–0.7 V, the delay increased only by 7 ps for the double-tail sense amplifier compared to an increase by 25–60 ps for the conventional topology. At $V_{DD} = 1.0$ V, the delay of the double-tail sense amplifier was 15 ps larger versus 29 ps for the conventional topology. The double-tail sense amplifier was fabricated in a 1.2 V 90 nm CMOS

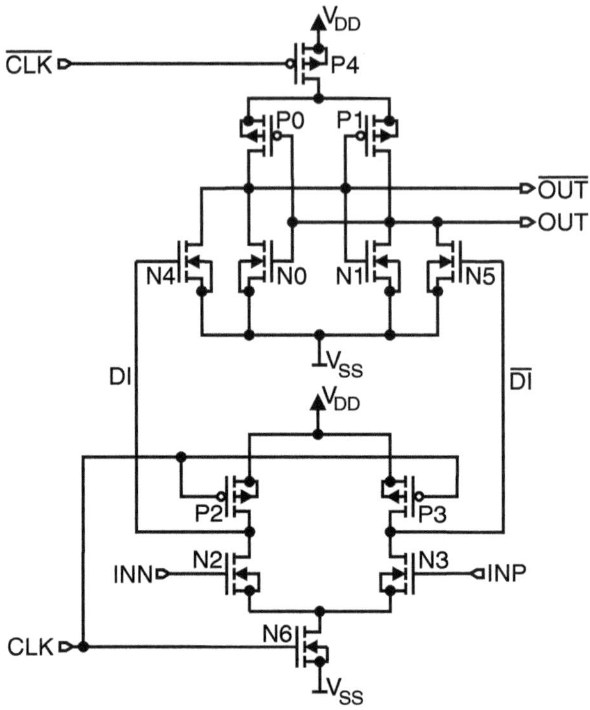

Fig. 3.11 Double-tail latch-type voltage sense amplifier [16]

technology. This sense amplifier had an area of $11 \times 7.5\,\mu\text{m}^2$. Its offset voltage σ_{os} was 8 mV for $V_{cm} = 1.1$ V and $V_{cm} = 0.75$ V. It consumed a power of 113fJ/decision for $\delta V_{in} = 50$ mV at 1 GHz clock frequency and $V_{DD} = 1.2$ V. The power consumption at 1 GHz was 113 µW and 225 µW at 2 GHz clock frequency. The equivalent input noise of the sense amplifier was $1.5\,\text{mV}_{rms}$. A mathematical analysis of the double-tail sense amplifier can be found in [17].

A charge compensation technique for comparator offset control was suggested in [18]. A dynamic offset compensation is obtained by a timing control principle. Figure 3.12 shows the comparator consisting of a preamplifier and cross-coupled inverters that act as regenerative latch. The preamplifier amplifies the differential input signal. The PMOS transistors $P1$ and $P2$ convert the output voltage signal of the preamplifier into current signals. The resulting current signals are inserted into the regenerative stage. During the reset signal *RES* is at a high voltage level so that the reset transistor *RES* is on, the two outputs of the latch are connected. The two MOS capacitors *RESa* and *RESb* are essential for the offset compensation. The time delay between the negative edge of the *RES* clock and the positive edge of the *RESb* clock is fix. The position of the edge of the *RESa* clock, however, is made variable by an analog-to-timing converter, where the delay time is set by a simple transconductance stage and an inverter [18]. In such a way a smaller or larger charge can be transferred from the channel of the *RES* transistor during switching it off to the node between the

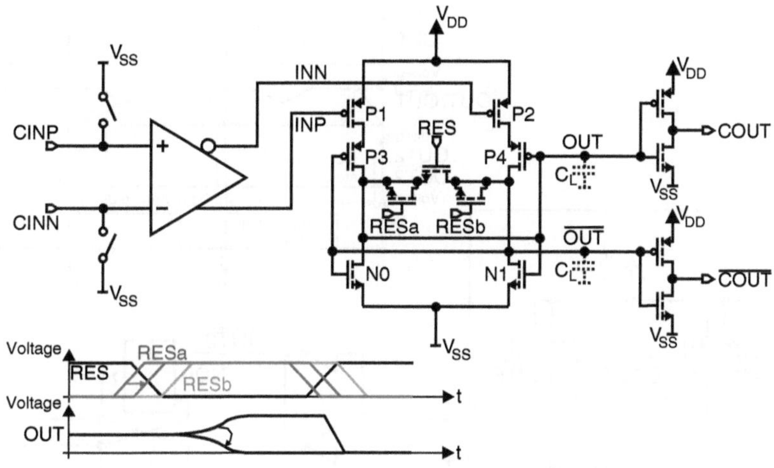

Fig. 3.12 Comparator with dynamic offset control [18]

drain of $P3$ and the drain of $N0$. With setting the input voltage of the transconductance stage in an appropriate manner therefore the offset voltage of the comparator can be compensated. This 1 GHz comparator was fabricated in 1.2 V 65 nm CMOS. It occupied a chip area of $65 \times 25\,\mu m^2$, of which the offset compensation circuit took 21 and 12 % of the 380 μW power consumption of the complete comparator circuit. With a change of the input voltage of the transconductance stage of 500 mV, the delay of the RESa clock could be changed by 50 ps, which enabled the compensation of 60 mV comparator offset voltage.

A tri-level comparator was suggested in [19] to relax the necessary speed of the comparator and to increase the DAC's resolution by one bit. With the help of this tri-level comparator, operation of a successive-approximation-register (SAR) ADC at an extremely low supply voltage of 0.5 V was possible leading to an excellent power efficiency. Usually applied regenerative comparators [20] exhibit the phenomenon of metastability, which can lead to false operation of the ADC. For small input signals, the comparator enters the metastable state and the comparison needs a very long time [21] especially in nanometer CMOS due to the low supply voltages. In 40 nm CMOS with 1 V supply voltage, for 4 mV LSB, the decision time was 100 ps. For a supply voltage of 0.5 V, however, the transistors operate in the subthreshold regime at very low currents and the decision time increased to about 100 ns. By forward biasing the source-well junction, i. e. by reducing the threshold voltage via the backgate effect, the decision time could be reduced to 40 ns, which is however still huge. As long as the decision is not finished, the comparator output is in a metastable state and the operation of the ADC is slowed down. To improve this situation, a tri-level decision technique employing two comparators was suggested in [22]. Compared to this approach, in [19] only one comparator was necessary for tri-level decision avoiding chip area for offset compensation of the two-comparator solution in [22]. The proposed three-level comparator used three output levels: high, middle, and

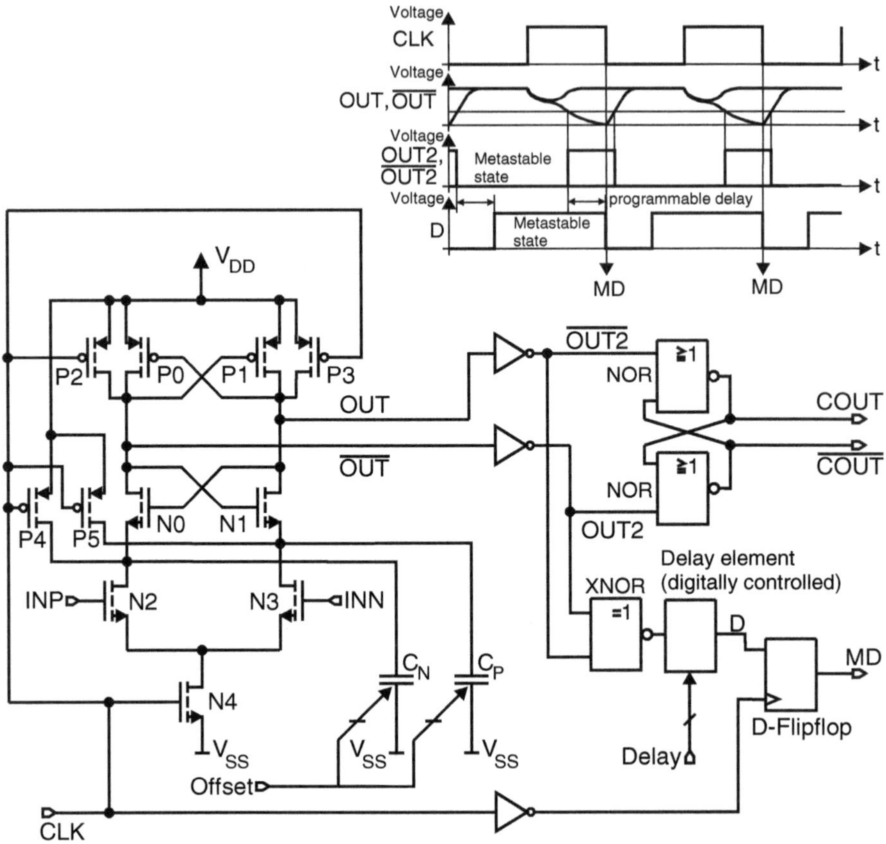

Fig. 3.13 Tri-level comparator [19]

low. The range of the middle state was chosen as half of the LSB voltage of the DAC within the SAR ADC. Figure 3.13 shows the circuit and the behavior of the tri-level comparator. The preamplifier consists of an NMOS differential amplifier and two cross-coupled inverters in the drain branches of the two NMOS transistors. Offset control is implemented via two digitally tunable capacitors (C_N and C_P). The metastability detector (MD) consists of an XNOR gate, a digitally adjustable delay element, and a D-flipflop.Within the reset phase, both outputs of the regenerative latch are pre-charged to V_{DD}. As a consequence both inputs of the XNOR gate are *low* and its output is *high*. When the clock is set *high*, the comparison starts and the voltage difference betweenn the two outputs of the regenerative latch increases until one output reaches *high*, the other one reaches *low*, and the decision is finished. Then the XNOR output changes from *high* to *low*. The interval between the clock's rising edge and the XNOR's transition corresponds to the decision time. The tunable delay element adds a further delay and the state of the comparator is determined via the negative edge of the clock by the D-flipflop. When the output of the D-flipflop is low,

Table 3.1 State of the art of comparators

Technology	Topology/ Application	Supply voltage	Clock	Sensitivity	Offset	Power consumption	References
0.18 µm CMOS	4-stage comparator: core comparator (CML, inductor loads) +2 latches + digital flip flop in 4-bit flash ADC	1.8 V analog 2.1–2.5 V digital	4 GHz	n.m.	n.m.	n.m.	[1]
0.11 µm CMOS	3-stage comparator (3rd stage consists of 2 additional amplifiers) in CML with bandwidth modulation technique in 40 GB/s DEMUX	1.2 V	10 GHz	BER $< 10^{-12}$ for $\Delta U_{in} = 1 V_{pp}$ measured	n.m.	37 mW measured with output drivers Simulated sum of CML tail currents of comparator itself: 4.2 mA	[2, 3]
0.25 µm CMOS	Comparator with pre-amplifier in a 6-bit flash ADC	1.8 V (analog part)	1.3 GHz	n.m.	n.m.	n.m.	[10]

(continued)

Table 3.1 (continued)

Technology	Topology/ Application	Supply voltage	Clock	Sensitivity	Offset	Power consumption	References
0.18 μm CMOS	Investigation and comparison of a comparator with and without offset compensation, where also offset changes due to supply noise are treated	1.8 V	1.4 GHz	n.m.	$3\sigma_{OS} = 95\,\text{mV}$ without offset compensation $3\sigma_{OS} = 16$ mV with offset compensation	350 μW	[4]
0.18 μm CMOS	1-bit quantizer with rail-to-rail input range for $\Sigma\Delta$-modulators	0.8 V	10–12 MHz	10 μV at 10 MHz (simulated) 100 μV at 12 MHz (simulated)	n.m.	134 μW	[5]
0.13 μm CMOS	Yield, speed optimization of a latch-type voltage sense amplifier for SRAM	1.5 V, works down to 0.7 V	Delay time in comparison phase higher than 11 ns at 0.7 V supply Delay time in comparison phase higher than 1.5 ns at 1.5 V	n.m.	1.5 V supply, $\sigma_{OS} = 19\,\text{mV}$ at an input common-mode voltage of 1.5 V 1.5 V supply, $\sigma_{OS} = 8.5\,\text{mV}$ at an input common-mode voltage of 1.05 V	n.m.	[6–8]
0.13 μm CMOS	Comparator with pre-amplifier in a 6-bit flash ADC	1.5 V	4 GHz	n.m.	n.m.	n.m.	[9]

(continued)

Table 3.1 (continued)

Technology	Topology/ Application	Supply voltage	Clock	Sensitivity	Offset	Power consumption	References
0.13 µm CMOS	Comparator with pre-amplifier in a 6-bit flash ADC (offset compensated pre-amplifiers plus comparator latch)	1.5 V	1.6 GHz	n.m.	$\sigma_{OS} = 80\,mV$ of comparator latch, $\sigma_{OS} = 2\,mV$ with offset compensated pre-amplifiers (gain about 40)	n.m.	[11]
0.18 µm CMOS	4-bit flash ADC, pre-amplifier + 3 CML latches, ±0.4 V swing at outputs (simulated)	1.8 V	5 GHz	n.m.	$\sigma_{OS} = 67\,mV$ (comparator simulation)	Sum of tail currents: 1 mA × 1.8 V (simulated)	[12]
0.13 µm CMOS	Pre-amplifier and comparator in 6-bit/22 MHz flash ADC	0.8 V	22 MHz	n.m.	n.m.	n.m.	[13]
90 nm CMOS	Sense amplifier	1.2 V	1 GHz, 2 GHz	measured RMS noise voltage $V_{rms} = 1.5\,mV$	$\sigma_{OS} = 13\,mV$ (simulated)	113 µW@1 GHz, 225 µW@2 GHz, 92fJ/decision	[16]
65 nm CMOS	Comparator with pre-amplifier	1.2 V	1 GHz	n. m.	down to 20 mV (measured)	380 µW@1 GHz	[18]
40 nm CMOS	SAR-ADC with tri-level comparator	0.5 V	15–50 MHz[a]	n.m.	n.m.	1.2 µW@ 1.1MS/s[b], 6.3fJ/conversion-step[b]	[19]

n.m. = not mentioned, sim. = simulated

[a] for tri-level comparator

[b] for SAR-ADC with tri-level comparator

the decision is finished. When the output of the D-flipflop is high, the input level is in the middle range and the comparator is in the metastable state. The middle level range can be controlled by the tunable delay element. The SAR ADC was fabricated in 40 nm CMOS technology. The comparison time could be reduced to about 20 ns for an SNR improvement of zero. For a decision time of about 40 ns the SNR of the comparator could be improved by 6 dB. The tri-level comparator demonstrated experimentally the improvement of the SAR ADC's resolution by 3 dB.

Table 3.1 shows a summary of clocked, regenerative comparators from the literature, which are described in this chapter.

References

1. S. Park, Y. Palaskas, M.P. Flynn, A 4 GS/s 4b Flash ADC in 0.18 μm CMOS, in IEEE International Solid-State Circuits Conference, February 2006, pp. 570–571
2. Y. Okaniwa, H. Tamura, M. Kibune, D. Yamazaki, T.-S. Cheung, J. Ogawa, N. Tzartzanis, W.W. Walker, T. Kuroda, A 40 Gb/s CMOS clocked comparator with bandwidth modulation technique. IEEE J. Solid-State Circuits **40**(8), 1680–1685 (2005)
3. Y. Okaniwa, H. Tamura, M. Kibune, D. Yamazaki, T.-S. Cheung, J. Ogawa, N. Tzartzanis, W.W. Walker, T. Kuroda, A 0.11 μm CMOS clocked comparator for high-speed serial communications, in IEEE Symposium on VLSI Circuits, June 2004, pp. 198–201
4. K.-L.J. Wong, C.-K.K. Yang, Offset compensation in comparators with minimum input-referred supply noise. IEEE J. Solid-State Circuits **39**(5), 837–840 (2004)
5. M. Maymandi-Nejad, M. Sachdev, 1-bit quantiser with rail to rail input range for sub-1 V $\Delta\Sigma$ modulators. IET Electron. Lett. **39**(12), 894–895 (2003)
6. B. Wicht, T. Nirschl, D. Schmitt-Landsiedel, Yield and speed optimization of a latch-type voltage sense amplifier. IEEE J. Solid-State Circuits **39**(7), 1148–1158 (2004)
7. B. Wicht, T. Nirschl, D. Schmitt-Landsiedel, A Yield-Optimized Latch-Type SRAM Sense Amplifier, in IEEE European Solid-State Circuits Conference, September 2003, pp. 409–412
8. B. Wicht, J.-Y. Larguier, D. Schmitt-Landsiedel, A 1.5 V 1.7 ns 4 k ×" 32 SRAM with a Fully-Differential Auto-Power-Down Current Sense Amplifier, in IEEE International Solid-State Circuits Conference, Februrary 2003, pp. 462–463
9. C. Paulus, H.-M. Blüthgen, M. Löw, E. Sicheneder, N. Brüls, A. Courtois, M. Tiebout, R. Thewes, A 4 GS/s 6b Flash ADC in 0.13 μm CMOS, in IEEE Symposium on VLSI Circuits, June 2004, pp. 420–423
10. K. Uyttenhove, M. Steyaert, A 1.8 V 6-Bit 1.3 GHz flash ADC in 0.25 μm CMOS. IEEE J. Solid-State Circuits **38**(7), 1115–1122 (2003)
11. C. Sandner, M. Clara, A. Santner, T. Hartnig, F. Kuttner, A 6bit, 1.2 GSps low-power flash-ADC in 0.13 μm digital CMOS. IEEE J. Solid-State Circuits **40**(7), 1499–1505 (2005)
12. S. Sheikhaei, S. Mirabbasi, A. Ivanov, A 4-Bit 5 GS/s Flash A/D Converter in 0.18 μm CMOS, in IEEE International Symposium on Circuits and Systems, vol. 6 (2005) pp. 6138–6141
13. J.H.-C. Lin, B. Haroun, An Embedded 0.8 V/480 μW 6b/22 MHz Flash ADC in 0.13 μm Digital CMOS Process using Nonlinear Double-Interpolation Technique, in IEEE International Solid-State Circuits Conference, February 2002, pp. 308–309
14. B. Nikolic et al., Improved sense-amplifier-based flip-flop: design and measurements. IEEE J. Solid-State Circuits **35**(6), 876–884 (2000)
15. K.-L.J. Wong, C.-K.K. Yang, Offset compensation in comparators with minimum input-referred supply noise. IEEE J. Solid-State Circuits **39**(5), 837–840 (2004)
16. D. Schinkel, E. Mensink, E. Klumperink, E. van Tuijl, B. Nauta, A Double-Tail Latch-Type Voltage Sense Amplifier with 18 ps Setup-Hold Time, in IEEE International Solid-State Circuits Conference, February 2007, pp. 314–605

17. S. Babayan-Mashhadi, R. Lotfi, Analysis and design of a low-voltage low-power double-tail comparator. IEEE Trans. Very Large Scale Integr. **22**(2), 343–352 (2014)
18. X. Zhu, Y. Chen, M. Kibune, Y. Tomita, T. Hamada, A Dynamic Offset Control Technique for Comparator Design in Scaled CMOS Technology, in IEEE Custom Integrated Circuits Conference, 2008, pp. 495–498
19. A. Shikata, R. Sekimoto, T. Kuroda, H. Ishikuro, A 0.5 V 1.1M S/sec 6.3fJ/conversion-step SAR-ADC with tri-level comparator in 40 nm CMOS. IEEE J. Solid-State Circuits **47**(4), 1022–1030 (2012)
20. T. Kobayashi, K. Nogami, T. Shirotori, Y. Fujimoto, A current-controlled latch sense amplifier and a static power-saving input buffer for low-power architecture. IEEE J. Solid-State Circuits **28**(4), 523–527 (1993)
21. A. Rodriguez-Vazquez, F. Medeiro, E. Janssens, *CMOS Telecom Data Converters* (Kluwer Academic, Boston, MA, 2003)
22. J. J. Kang, M. P. Flynn, A 12b 11M S/s Successive Approximation ADC with Two Comparators in 0.13 μm CMOS, in IEEE Symposium VLSI Circuits Digest, June 2009, pp. 240–241

Chapter 4
Nanometer CMOS Technology

In this chapter the CMOS technologies used for design and fabrication of comparator test chips described in this book are introduced. The MOS transistor characteristics in nanometer CMOS are shown. Furthermore the consequences of the nanometer hell of physics [1] on transistor speed and gain, supply voltage and signal headroom, low-frequency noise, matching, gate leakage current and linearity of analog circuits are addressed.

4.1 120 nm CMOS Technology

The technology, which was used for the designed test chips belonged to a 0.13 μm CMOS generation foundry [2–5]. It is a platform technology for embedded DRAM and designed for SRAM, logic, mixed-signal and mixed-voltage I/O applications. Due to the fact, that the minimum lithographic image is 120 nm for gates, here this technology is called 120 nm CMOS technology or 120 nm CMOS process. The cross section of this technology is shown in Fig. 4.1. In principle it is a twin well CMOS process on a non-epi p-substrate, where for isolation a shallow trench isolation (STI) is used. The p-well of a n-MOS transistor is electrically connected to the p-substrate and leads therefore to the back-gate effect, when the source is not connected to V_{SS}, because the whole p substrate must be connected with V_{SS}. In opposite to that for p-MOS transistors the separated n-well may be connected with the source thus avoiding the back-gate effect. To lower the resistance, a Co salicided (self-aligned silicided) layer for n^+ and p^+ polysilicon and diffusion areas is added. The planarized passivation and interlevel dielectrics have the same relative dielectric constant as silicon oxide. A so-called stud contact (tungsten plug) to connect the poly-silicon gate or diffusion region of transistors to metal level M1 is marked with CA and the wiring level contacts are named V1, V2, V3, VL, VQ and VV. For the metal layers, copper is used. Layers M1–M4 are thin and tighter-pitch copper levels and ML and MQ are thick copper levels with a relaxed pitch. The last metal level LB is also used for pads and made of AlCu [2].

B. Goll and H. Zimmermann, *Comparators in Nanometer CMOS Technology*,
Springer Series in Advanced Microelectronics 50, DOI 10.1007/978-3-662-44482-5_4

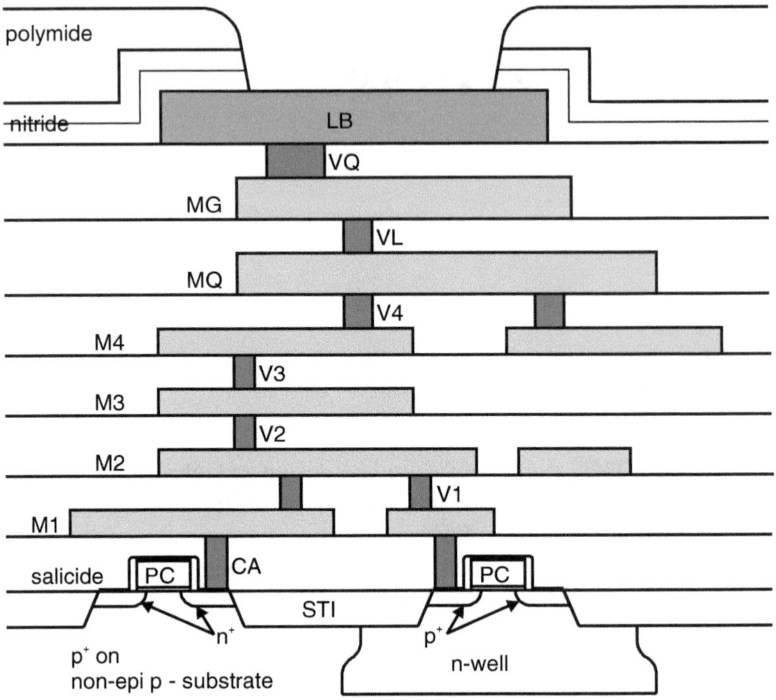

Fig. 4.1 Cross section of the used 0.12 μm CMOS process

Optional transistors of the process are for a nominal supply voltage of 1.5 V surface channel transistors (standard oxide thickness) with a low threshold voltage $V_t \approx 0.29\,\text{V}$ (low-V_t transistors), a regular threshold voltage $V_t \approx 0.35\,\text{V}$ (regular-V_t transistors) and a high threshold voltage $V_t \approx 0.46\,\text{V}$ (high-V_t-transistors). Beside that the process also provides analog surface channel transistors with thicker gate oxide for supply voltages of 2.5 V or 3.3 V. To save power and additional process steps it is a challenge to implement analog circuit blocks with 1.5 V supply voltage, where only low-, regular- and high-V_t transistors are used. A further step may be also an implementation with e.g. only regular-V_t transistors. The characteristics of n-MOS and p-MOS transistors with low-, regular- and high-V_t are depicted in Figs. 4.2 and 4.3. The characteristics of a long-channel transistor with gate length $L = 600\,\text{nm}$ is compared with a short-channel transistor with $L = 120\,\text{nm}$. The width W was chosen, that approximately the same drain current I_D flows in saturation through the regular-V_t transistor. In Fig. 4.4 the input characteristics of the transistors are shown, where instead of Figs. 4.2a and 4.3a the square root of the drain current is used to distinguish different areas more easily. The area, where the gate-source voltage V_{GS} is below the threshold voltage, but already high enough, that a depletion region at the surface of the silicon is built up, but no well defined conducting channel exists, is called weak inversion [6]. In weak inversion the drain current (subthreshold current)

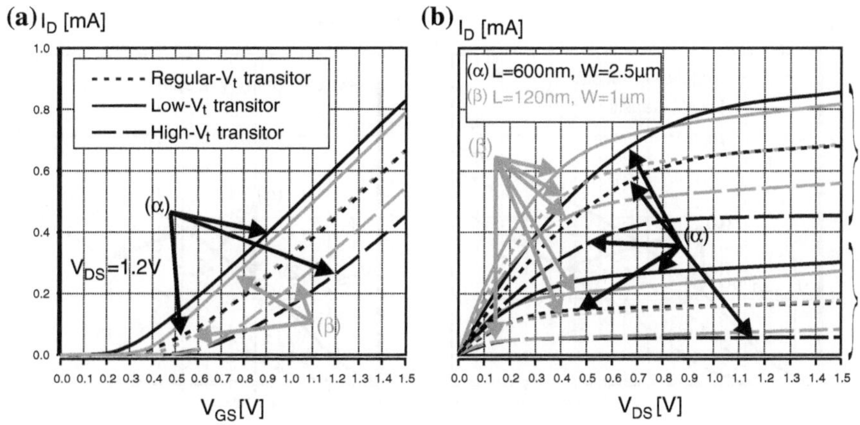

Fig. 4.2 Characteristics of n-MOS transistors **a** Input characteristics. **b** Output characteristics

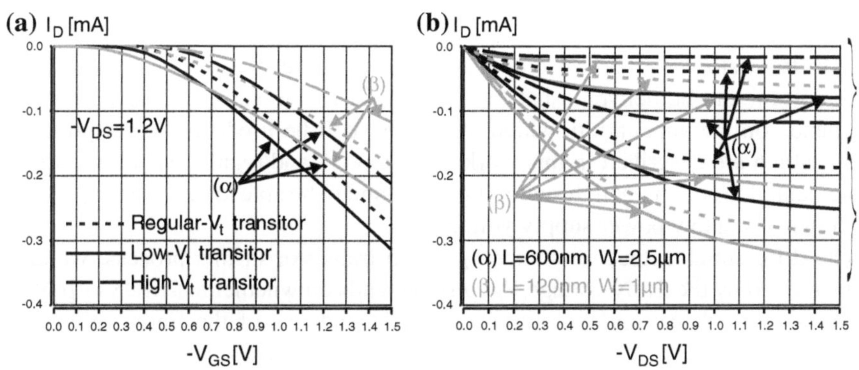

Fig. 4.3 Characteristics of p-MOS transistors **a** Input characteristics. **b** Output characteristics

is exponentially dependent on V_{GS}. After a transition region (moderate inversion) around the threshold voltage the transistors have a quadratic dependence on V_{GS} (like an ideal MOS transistor). This can be seen in the linear region in Fig. 4.4 for gate-source voltages higher than V_t. At higher V_{GS} the linear dependence of the square root of the drain current degenerates due to velocity saturation of the carriers in the channel, where the drift velocity of carriers saturates towards a constant saturation speed at higher horizontal electrical field along the channel (higher V_{DS}) instead of the linear dependence at lower fields. In the output characteristics (see Figs. 4.2b and 4.3b) the effect of velocity saturation can be seen, that current saturation begins at a lower drain-source voltage V_{DS} than the border $V_{DS} = V_{GS} - V_t$ of an ideal MOS transistor. Another effect is, that in principle long-channel transistors have a higher differential output resistance $r_{DS} = (V_A + V_{DS})/I_D$ (V_A = Early voltage) in saturation than short-channel transistors due to the higher influence of channel-length modulation [6]. In the used process low-V_t transistors have the lowest and

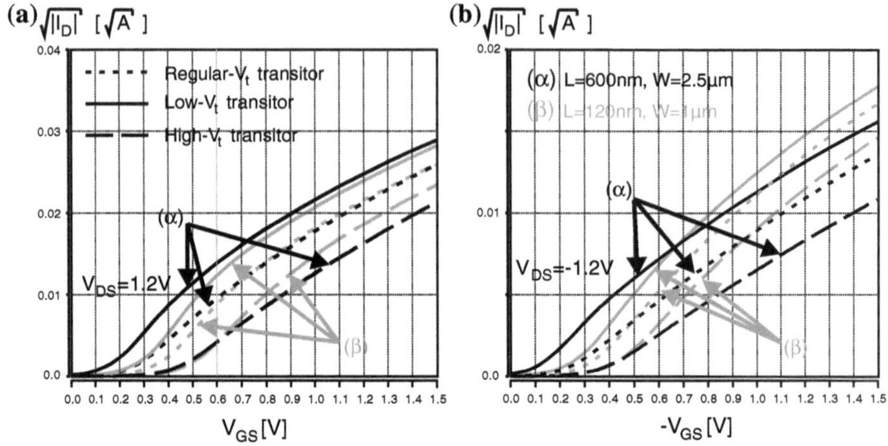

Fig. 4.4 Input characteristics of transistors. **a** N-MOS transistors. **b** P-MOS transistors

high-V_t transistors the highest output resistances. Instead with low-V_t transistors more current can be delivered but they also have a high subthreshold current. So it depends on the application, which transistors will be used. In the designed test chips, which are described in this work, only low-V_t and regular-V_t transistors are implemented to achieve a high switching speed and to be able to design comparators, which work at a low power supply voltage.

The differential output resistance r_{DS} of a transistor in this process, which is characterized by the Early voltage V_A, also depends on whether a n-MOS or p-MOS transistor is used. An approximation of the dependence of the Early voltage V_A on the channel length L of a n-MOS and a p-MOS transistor is shown in Fig. 4.5. In this CMOS process, p-MOS transistors in principle have a higher Early voltage

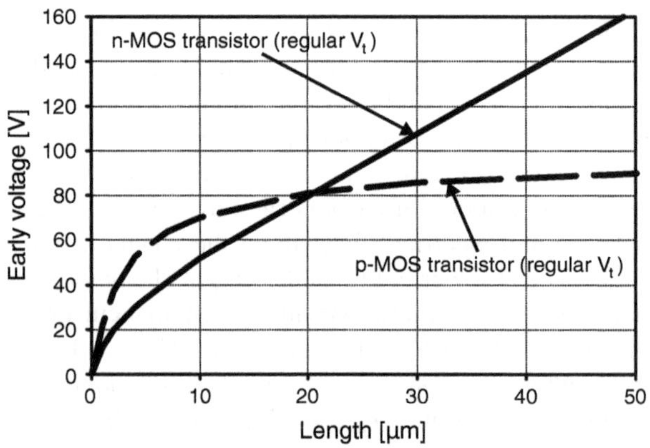

Fig. 4.5 Approximation of early voltage versus channel length L for $W = 10\,\mathrm{m}$, $V_{GS} = 0.5\,\mathrm{V}$ and $V_{DS} \approx 0.7\,\mathrm{V}$

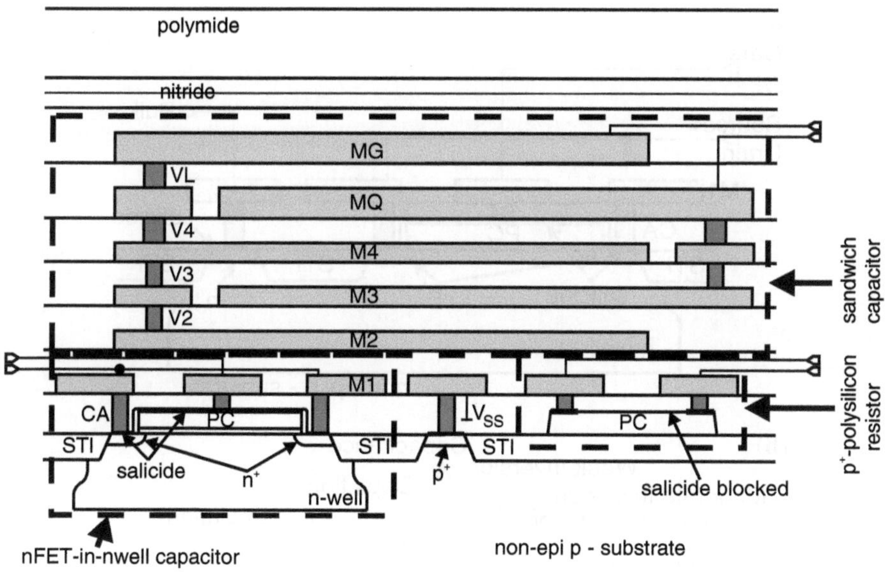

Fig. 4.6 Sandwich capacitor, nFET-in-n-well capacitor and p^+-polysilicon resistor in 0.12 μm CMOS

and therefore a higher output resistance than n-MOS transistors at lower channel lengths. For a minimal channel length of $L = 0.12\,\mu m$ Early voltages of below 10 V have to be considered for low-V_t, regular-V_t and high-V_t transistors. The type of resistor, which was implemented in the designed chips, is the p^+-polysilicon resistor (see Fig. 4.6), where the salicidation is blocked to enhance the Ohm/square value. The implemented capacitors were sandwich capacitors and nFET-in-nwell capacitors (see Fig. 4.6). Sandwich capacitors are built, when two or more metal layers are used as capacitor plates and overlap with a distinct area, which defines the capacitance. The dielectric is the gate oxide between the metal layers. The advantage of sandwich capacitors are good matching properties and a high linearity. Disadvantageous is, that the capacitance/square value is low. So for block capacitors for the power lines mostly nFET-in-nwell capacitors with a thicker gate oxide to have a higher robustness against overvoltage were chosen, which also have a higher capacitance/square-value. This type of capacitors is built by a poly-gate area over a n-well area, which is connected to V_{SS} to when the capacitance of electron accumulation below the gate is used.

A further important electronic part is a voltage-controlled capacitor, which is used in this work as a variable capacitive load for inverters to form a voltage-controlled delay line, because the delay time of an inverter cell depends beside other parameters on the load capacitance at the output node (see also Sect. 2.6). Figure 4.7 shows a p-MOS transistor, which is used as a voltage-controlled capacitor (p-MOS varactor) [7]. For voltage-controlled delay lines the inversion-mode p-MOS varactor is used in this work, where the accumulation region is removed by connecting the bulk to V_{DD}, because the tuneable capacitance range is higher than for the standard

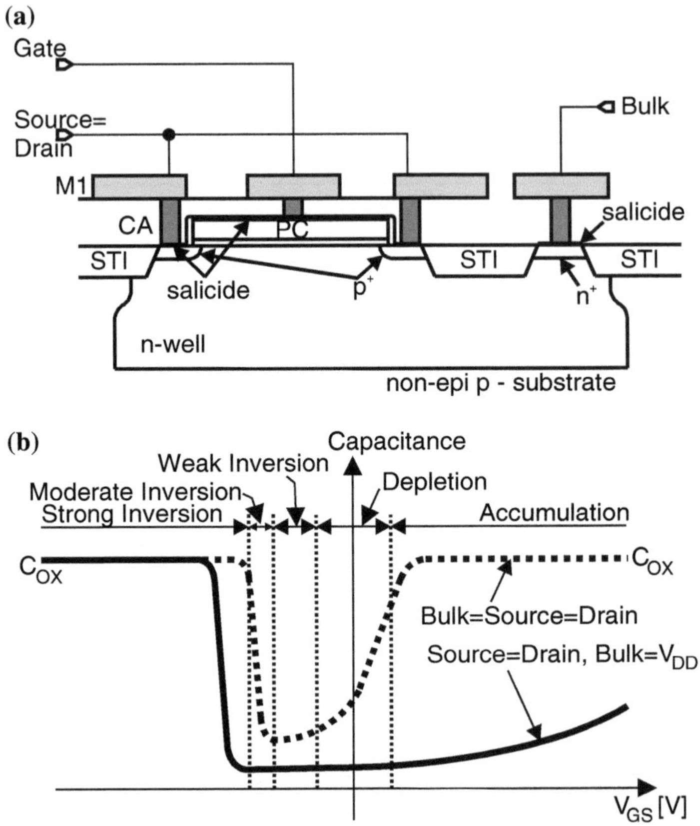

Fig. 4.7 A p-MOS transistor (*S* source, *D* drain, *G* gate, *B* bulk) used as a voltage-controlled capacitor (p-MOS varactor) [7] **a** A p-MOS varactor. **b** Characteristics of a standard p-MOS capacitor (S=D=B) and an inversion-mode p-MOS capacitor (S=D, B=VDD)

p-MOS varactor, which may operate in inversion or in accumulation depending on the gate-source voltage V_{GS}.

4.2 65 nm Low-Power CMOS Technology

The 65 nm bulk CMOS technology used was developed for logic, SRAM, mixed-signal, and mixed-voltage I/O applications as well as for embedded DRAM applications. Besides the 65 nm CMOS base process, there exists a 65 nm low-power (LP) CMOS process. The 65 nm low-power process offered MOSFETs with an approximately 100 mV higher threshold voltage than the base process and hence a lower off-state leakage current. Both options have three different oxide thicknesses to support several supply voltages from 1.2 to 3.3 V [8].

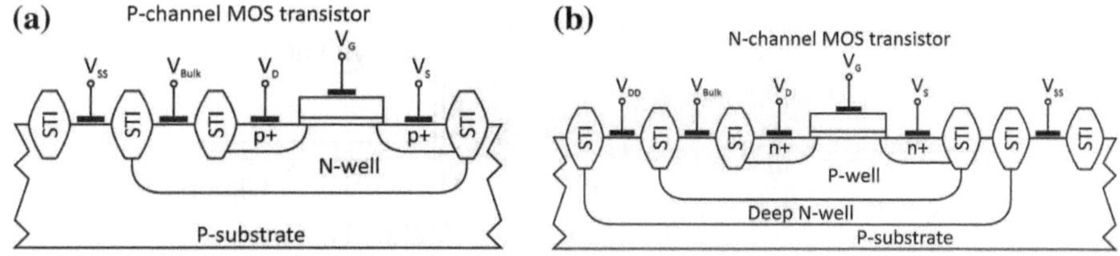

Fig. 4.8 Cross sections of MOSFET transistors in 65 nm triple-well CMOS. **a** P-channel MOSFET transistor. **b** N-channel MOSFET transistor

In the following circuits only the 65 nm CMOS low-power technology is used, thus only important characteristics of the low-power CMOS are highlighted. Of course it is more challenging to develop analog circuits in the low-power version than in the base process. The 65 nm CMOS technology used is a triple well process on a p-substrate. The PMOS transistor is located in an isolated N-well and the NMOS transistor is placed in an insulated P-well, which is itself embedded in a deep N-well. Cross-sections of the transistors are depicted in Fig. 4.8a, b. Active devices are separated by a shallow trench isolation (STI). Stress engineered devices are offered and the stress is caused by the location and the dimension of the STI. Mechanical stress increases the carrier mobility in p-channel MOSFETs and deteriorates the carrier mobility in n-channel MOSFETs [9].

The nominal supply voltage in the 65 nm low-power CMOS process is 1.2 V and the maximum supply voltage is 1.32 V. The technology cross section of the 65 nm CMOS process is shown in Fig 4.9. It should be noted that the aspect ratios are not displayed correctly. The process uses a P-substrate, wherein N-well, P-well, and a deep N-well are embedded. Inside the wells are the appropriate N+ and P+ diffusion regions [10]. The gate oxide isolates the polysilicon gates from the transistor channels. The spacers ensure the isolation between gate and source/drain areas. Self-aligned silicide, also called salicide, forms the interconnect between the semiconductor and metal layers. The contacts connect the salicide to the metal layer 1 (M1). Metal M1–M4 are 1x thin metal layers, which are connected by the vias V1–V3. Via V0_2B connects 1x thin metal M4 to thick metal (2x) M1_2B. And finally Via V0_4B contacts the last copper metal layer M1_4B, which is a 4x thick metal layer. In total there exist 6 copper metal layers. The last metal layer LB is used for wiring and wirebonding. Metal LB is an aluminum wire level and is connected to M1_4B by the via VV. For bondpads the oxide and nitride film is opened as well as the polyimide passivation layer [11].

4.2.1 Transistor Characteristics

The progress of scaling is noticeable in several transistor parameters and properties. Figure 4.10 shows the output characteristics of an NMOS transistor with varied gate length L in the 65 nm CMOS technology. The transistor width W is 5 μm and the

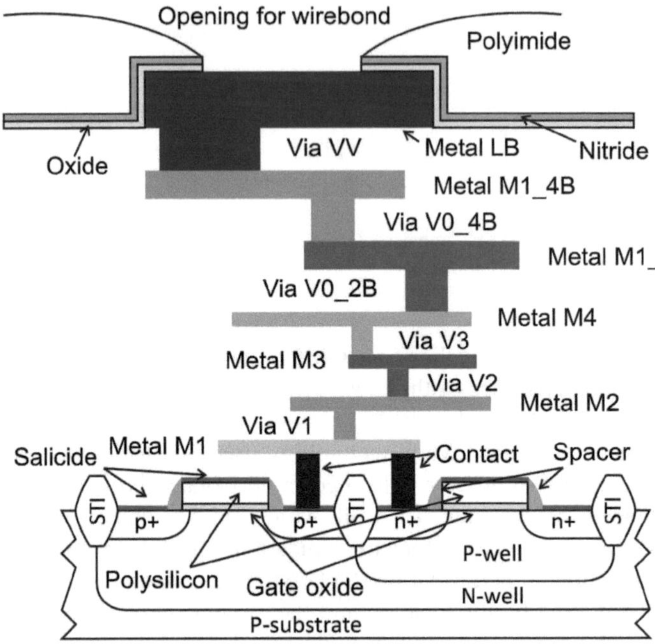

Fig. 4.9 Cross section of the 65 nm low-power CMOS process

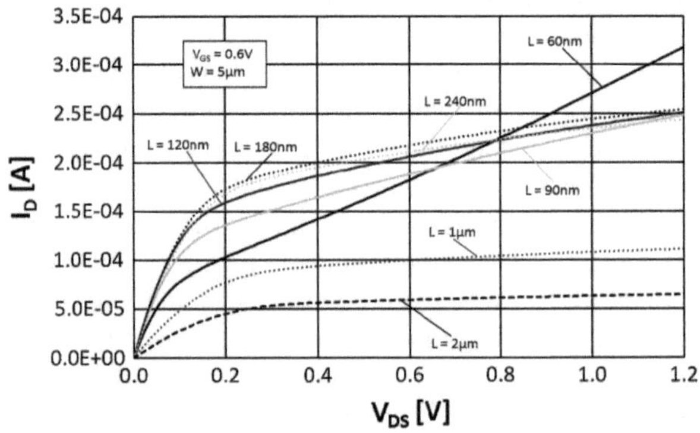

Fig. 4.10 NMOS output characteristics at varied gate length L

gate source voltage V_{GS} of 600 mV is applied. The Early voltage declines with decreasing gate length L and is in a range of a few volts. At the gate length of 60 nm the Early voltage is only about 0.31 V. At a constant transistor width the drain current should increase at smaller gate lengths. In Fig. 4.10 it is different because of the threshold-voltage roll-off [12]. The threshold-voltage roll-off is caused by the short-channel effect and the fringing-field effect and results in a rising threshold voltage at decreasing gate lengths. Figure 4.11 depicts the output conductance g_{DS}

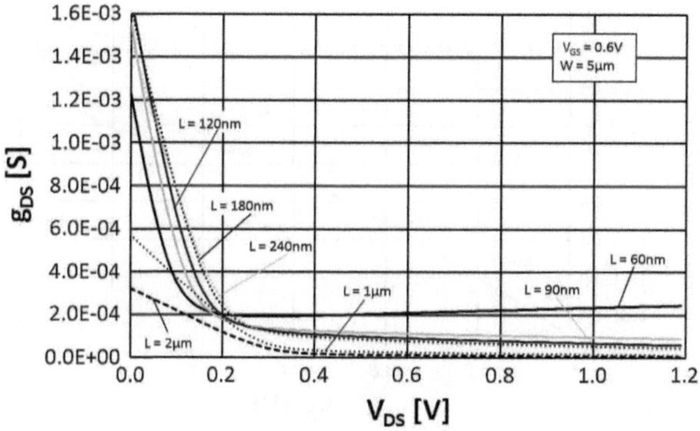

Fig. 4.11 NMOS output conductance g_{DS} at varied gate length L

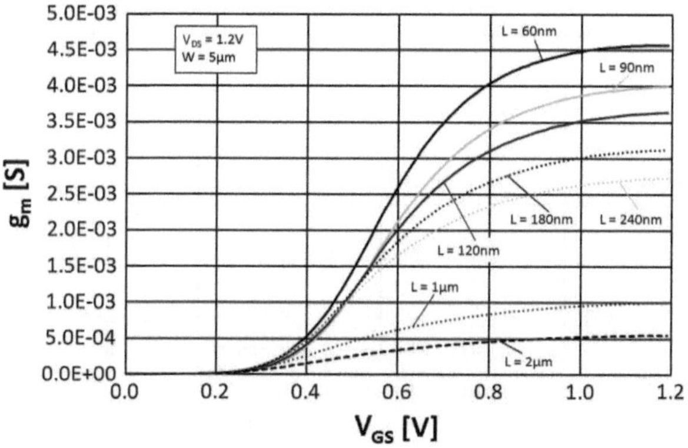

Fig. 4.12 NMOS transconductance g_m at varied gate length L

against the drain-source voltage V_{DS} which is decreasing with increasing gate lengths (for V_{DS} >0.4 V).

The transconductance g_m of an NMOS transistor with the gate width of 5 µm is shown in Fig. 4.12. V_{DS} is 1.2 V. g_m is decreasing with larger lengths L as well as g_{DS}.

Figure 4.13 depicts the output characteristics of a PMOS transistor with various gate lengths and 5 µm gate width. The gate-source voltage is fixed at −600 mV. The threshold-voltage roll-off takes effect as well, but is not so obvious. The Early voltage of a PMOS minimum gate-length transistor is in the same range as of an NMOS transistor. The 60 nm PMOS transistor has an Early voltage of only about 0.26 V. Figure 4.14 shows the output conductance g_{DS}.

Figure 4.15 shows the transconductance of PMOS transistors while gate lengths L are varied. The PMOS transistor has a gate width of 5 µm and is biased with a V_{DS} of 1.2 V. Again a smaller L improves the transconductance g_m.

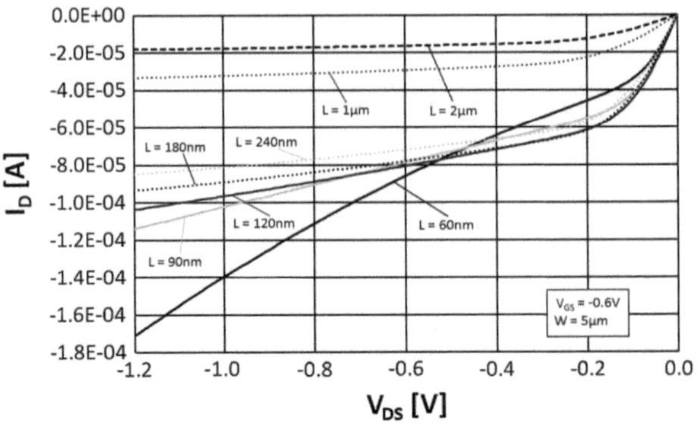

Fig. 4.13 PMOS output characteristics at varied gate length L

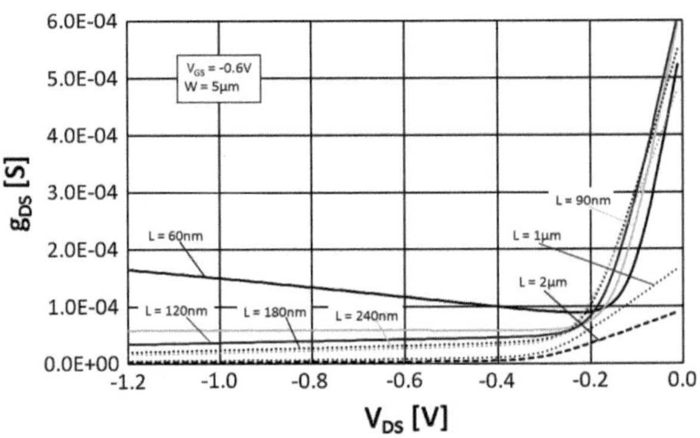

Fig. 4.14 PMOS output conductance g_{DS} at varied gate length L

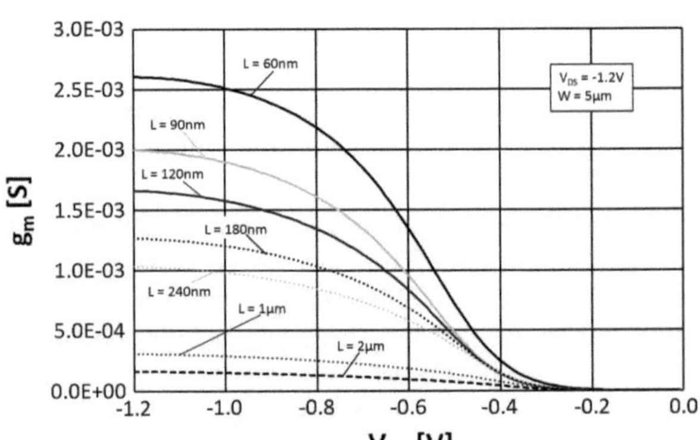

Fig. 4.15 PMOS transconductance g_m at varied gate length L

4.3 Consequences of Device Properties on Circuit Design

Comparators are part of many mixed-signal integrated circuits or systems on chip (SoC). The effects of device properties on analog nanometer CMOS circuits are severe. An outline of challenges and benefits is given in the following.

4.3.1 *Transistor Speed and Gain*

Major effect of scaling in digital circuits is the improvement of speed and packing density. Analog circuits benefit from the increasing speed as well. Parasitic capacitances are reduced and the gate resistance is increased due to the gate length scaling. The transit frequency of the transistors (f_t) and the maximum oscillation frequency increase with ongoing scaling. The transistor speed and the transistor intrinsic gain are in direct competition. High-speed transistors lead to a low output resistance and consequently to a low intrinsic gain. Both can be improved by increasing V_t or decreasing V_{GS} to operate in moderate inversion. However, moderate inversion is accompanied with a drop of f_t and oscillation frequency [13].

4.3.2 *Supply Voltage and Signal Headroom*

Digital circuits benefit from a small supply voltage and have a reduced power consumption and increased logic speed. In analog circuits a reduced supply voltage leads to a small signal range and a limited signal headroom and may cause a larger layout area. Newer technologies may induce an increased power consumption [13]. At a constant power consumption the performance deteriorates when newer technologies are used because of their lower supply voltage [14]. Due the low supply voltage of deep-sub-μm and nanometer analog circuits additional circuit blocks (e.g. due to more amplifier stages to achieve the same gain as with conventional circuits in sub-μm technology and/or more replica circuits) the power dissipation increases to keep a constant circuit performance. The increase of power becomes more relevant as the supply voltage is in the range of the threshold voltage V_t. Low V_t devices are sometimes available for analog circuits but cannot be used in the digital domain due to the subthreshold currents. The limited signal range and signal headroom demand accurate analog designs and a good noise performance to maintain the signal-to-noise ratio (SNR) and dynamic range.

4.3.3 *Low-Frequency Noise*

Flicker noise or 1/f noise is generated by the fluctuation of the total number of carriers and the fluctuation of the mobility of the carriers in the transistor channel

[15]. It attains increasing observance in analog design due to the shrinking feature sizes in CMOS technology. 1/f noise increases inversely proportional to the gate area and has a considerable amount in deep submicron and nanometer CMOS technology. A common expression of the 1/f noise density is given in [16]:

$$\overline{v_{fn}^2} = \frac{KF_F}{WLC_{ox}^2} \frac{\Delta f}{f} \tag{4.1}$$

In 4.1 KF_F is a parameter, nearly independent of the technology. W is the gate width and L represents the gate length. C_{ox} is the specific gate oxide capacitance. It is notable, that almost all technology effects are included in C_{ox}^2.

4.3.4 Matching

Matching is an important criterion in analog circuits, especially in differential structures. The matching of V_t, g_m, saturation current (I_{dsat}), and the device matching is mainly dependent on the precision of the manufacturing process. In deep submicron technologies additional factors become significant. Voltage matching, which is usually described by the difference of V_t of two identical transistors, becomes less sensitive to the device dimensions, when technology scales [17]. Matching can be improved by increasing the device dimensions.

4.3.5 Gate Leakage Current

The gate leakage current is mainly limited by digital considerations, such as static power consumption. In nanometer CMOS technology gate leakage currents have to be considered in analog circuit design. The gate leakage current depends on the oxide thickness, the gate-source voltage V_{GS}, the gate-drain voltage (V_{GD}), and the gate area [17]. The gate leakage current affects the input bias currents, the gate leakage mismatch and the shot noise due to the gate current. In order to minimize the gate leakage high-k dielectrics for the gate insulator are developed [13].

4.3.6 Linearity

Distortions in analog circuits in nanometer CMOS technology are mainly caused by the increased influence of the series resistance and velocity saturation. By scaling the CMOS technology, i.e. due to reduced supply voltage, the voltage headroom decreases and in analog bias conditions scaling worsens the linearity [18].

References

1. H. De Man, *From the Heaven of Software to the Hell of Nanoscale Physics: An Industry in Transition* (Keynote Slides, ACACES, 2007)
2. C11N Design Manual, Infineon/ALTIS
3. F. Schlögl, Fully Differential Operational Amplifiers in Deep-Sub-μm CMOS Technology, Ph.D. thesis, Vienna University of Technology (2004), pp. 11–15
4. K. Schneider, Burst-Mode Receivers for Passive Optical Networks, Ph.D. thesis, Vienna University of Technology (2004), pp. 18–21
5. K. Schneider, H. Zimmermann, *Highly Sensitive Optical Receiver* (Springer, Berlin, 2006)
6. P.R. Gray, P.J. Hurst, S.H. Lews, R.G. Meyer, *Analysis and Design of Analog Integrated Circuits* (Wiley, New York, 2001)
7. P. Andreani, S. Mattissson, On the use of MOS varactors in RF VCOs. IEEE J. Solid-State Circuits **35**(6), 905–910 (2000)
8. Z. Luo, A. Steegen, M. Eller, R. Mann, C. Baiocco, P. Nguyen, L. Kim, M. Hoinkis, V. Ku, V. Klee, F. Jamin, P. Wrschka, P. Shafer, W. Lin, S. Fang, A. Ajmera, W. Tan, D. Park, R. Mo, J. Lian, D. Vietzke, C. Coppock, A. Vayshenker, T. Hook, V. Chan, K. Kim, A. Cowley, S. Kim, E. Kaltalioglu, B. Zhang, S. Marokkey, Y. Lin, K. Lee, H. Zhu, M. Weybright, R. Rengarajan, J. Ku, T. Schiml, J. Sudijono, I. Yang, C. Wann, High performance and low power transistors integrated in 65 nm bulk CMOS technology. IEEE International Electron Devices Meeting, 2004, pp. 661–664
9. S. Maeda, Y-S. Jin, J-A. Choi, S-Y. Oh, H-W. Lee, J-Y. Yoo, M-C. Sun, J-H. Ku, K. Lee, S-G. Bae, S-G. Kang, J-H. Yang, Y-W. Kim, K-P. Suh, Impact of mechanical stress engineering on Flicker noise characteristics. Symposium on VLSI Technology, June 2004, pp. 102–103
10. B. Tavel, M. Bidaud, N. Emonet, D. Barge, N. Planes, H. Brut, D. Roy, J.C. Vildeuil, R. Difrenza, K. Rochereau, M. Denais, V. Huard, P. Llinares, S. Bruyere, C. Parthasarthy, N. Revil, R. Pantel, F. Guyader, L. Vishnubotla, K. Barla, F. Arnaud, P. Stolk, M. Woo, Thin oxynitride solution for digital and mixed-signal 65 nm CMOS platform. IEEE International Electron Devices Meeting, 2003, pp. 643–646
11. P. Bai, C. Auth, S. Balakrishnan, M. Bost, R. Brain, V. Chikarmane, R. Heussner, M. Hussein, J. Hwang, D. Ingerly, R. James, J. Jeong, C. Kenyon, E. Lee, S.-H. Lee, N. Lindert, M. Liu, Z. Ma, T. Marieb, A. Murthy, R. Nagisetty, S. Natarajan, J. Neirynck, A. Ott, C. Parker, J. Sebastian, R. Shaheed, S. Sivakumar, J. Steigerwald, S. Tyagi, C. Weber, B. Woolery, A. Yeoh, K. Zhang, M. Bohr, A 65 nm logic technology featuring 35 nm gate lengths, enhanced channel strain, 8 Cu interconnect layers, Low-k ILD and 0.57 μm^2 SRAM cell. IEEE International Electron Devices Meeting, 2004, pp. 657–660
12. F. Ji, J.P. Xu, J.J. Chen, H.X. Xu, C.X. Li, P.T. Lai, A compact threshold-voltage model of MOSFETs with stack high-k gate dielectric. IEEE International Conference of Electron Devices and Solid-State Circuits, 2009, pp, 236–239
13. A. Mercha, W. Jeamsaksiri, J. Ramos, S. Jenei, S. Decoutere, D. Linten, P. Wambacq, Impact of scaling on analog/RF CMOS performance. IEEE Solid-State and Integrated Circuits Technology, Oct 2004, pp. 147–152
14. A.-J. Annema, B. Nauta, R. van Langevelde, H. Tuinhout, Analog circuits in ultra-deep-submicron CMOS. IEEE J. Solid-State Circuits **40**(1), 132–143 (2005)
15. J.-H. Lee, S.-Y. Kim, I. Cho, S. Hwang, J.-H. Lee, 1/f Noise characteristics of sub-100 nm MOS transistors. J. Semicond. Technol. Sci. **6**(1), 38–42 (2006)
16. W.M.C. Sansen, *Analog Design Essentials* (Springer, New York, 2006)
17. L. Lanny, L. Trond Ytterdal, C. Wulff, K. Martin, Analog circuit design in nanoscale CMOS technologies. Proc. of the IEEE **97**(10), 1687–1714 (2009)
18. R. van Langevelde, L.F. Tiemeijer, R.J. Havens, M.J. Knitel, R.F.M. Roes, P.H. Woerlee, D.B.M. Klaassen, RF-distortion in deep-submicron CMOS technologies. IEEE International Electron Devices Meeting, 2000, pp. 807–810

Chapter 5
Measurement Circuits and Setup

In this chapter circuits being necessary on chip to be able to measure the performance of the investigated comparators are described. First a buffer with a very low input capacitance and a 50 Ω output resistance is introduced, which bridges the gap between on-chip comparator and measurement equipment. An on-chip temperature sensor follows, which allows to monitor the temperature of the comparator chip. Furthermore, a transfer stage, delay time measurements of the comparator, clock driver, and voltage-controlled delay line are described. The complete on- and off-chip measurement setup inclusive microcontroller board follow.

5.1 A 10 GHz Voltage Buffer in 0.12 μm CMOS Technology

For the rise times of clocked, regenerative comparators in 0.12 μm CMOS technology, time durations of less than 100 ps have to be considered, because of reduced gate lengths of MOS transistors and smaller parasitic capacitances. Disadvantageous of the used deep-sub-micron 0.12 μm CMOS technology is, that e.g. hot carrier injection, oxide breakdown and junction breakdown lead to a lower power supply voltage (1.5 V ± 10 %) to meet reliability and lifetime issues. Furthermore the transconductances of MOS transistors are limited due to e.g. velocity saturation and limited overdrive voltages. To measure an analog waveform of an on-chip clocked, regenerative comparator with an off-chip high-frequency oscilloscope with 50 Ω input resistance, a voltage buffer has to be added on-chip to the design. Such a buffer shall be designed by using available transistors of the CMOS process and requires a high bandwidth and a low input capacitance to minimize its influence on the comparator.

To meet these demands several stages are needed. A chain of simple source followers [1] cannot be used, because the gain of one stage is significantly smaller than one. This is due to the back-gate effect and the in general relative small transconductances of MOS transistors in 0.12 μm CMOS compared to a bipolar-junction transistor of special BiCMOS (Bipolar & CMOS) processes. Moreover level shifters, e.g. complementary source followers, have to be added, which increases the number of stages

B. Goll and H. Zimmermann, *Comparators in Nanometer CMOS Technology*,
Springer Series in Advanced Microelectronics 50, DOI 10.1007/978-3-662-44482-5_5

and reduces therefore the overall gain. Additionally a source follower with p-MOS transistors introduces more parasitic capacitances, even to the previous stage (higher gate-source capacitance), compared to its n-MOS counterpart with similar transconductance, because of the higher *W/L*-ratio needed. In [2] a voltage follower, which is insensitive to the fabrication process and which is able to drive a resistive load is described. This design has got a gain near unity when using it with a capacitive load, but it also cannot be cascaded because of the reasons already stated. A further voltage follower with high performance and rail-to-rail input is introduced in [3]. When several stages of it are used, this design consists of many transistors and would not save chip area. Furthermore the two input transistors increase the capacitive load of the previous node. Unity-gain buffers based on an inverter structure [4] are quite fast, but they also use two complementary input transistors (higher capacitive load of the previous stage) and are sensitive to process tolerances. In [5] a unity-gain amplifier with reduced gain error is proposed. With varying of transistor sizes and choosing an appropriate compensation capacitance the resonance and the damping factor of the transfer characteristic can be adjusted. When this amplifier is cascaded, the first stage should have a low input capacitance and with the load capacitance at the input of the next stage the bandwidth is predetermined in principle. Therefore justification of the bandwidth would be constricted.

This section describes a multi-stage voltage buffer in 0.12 μm CMOS, where regular-V_t and low-V_t transistors were used [6]. It is advantageous to use mostly low-V_t transistors, because they are able to deliver more current for higher bandwidth, even with their worse output resistances, which reduce in principle gain. For current sources and active loads in opposite mostly regular-V_t transistors were used due to their higher output resistance. The designed voltage buffer has got a small input capacitance (about 3 fF) and a bandwidth of 10 GHz.

5.1.1 Circuit Description

The block diagram of the realized five-stage voltage buffer is depicted in Fig. 5.1. The buffer consists of four stages followed by a source follower, which defines the fifth stage. The overall schematic of the buffer is shown in Fig. 5.2. In stage 2, stage 3,

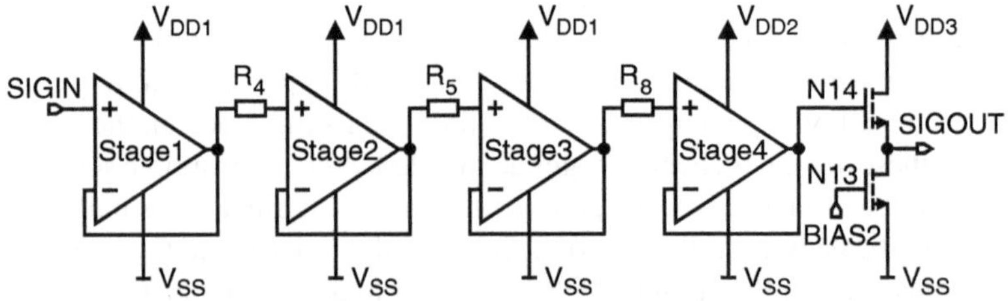

Fig. 5.1 Block diagram of the voltage buffer [6]

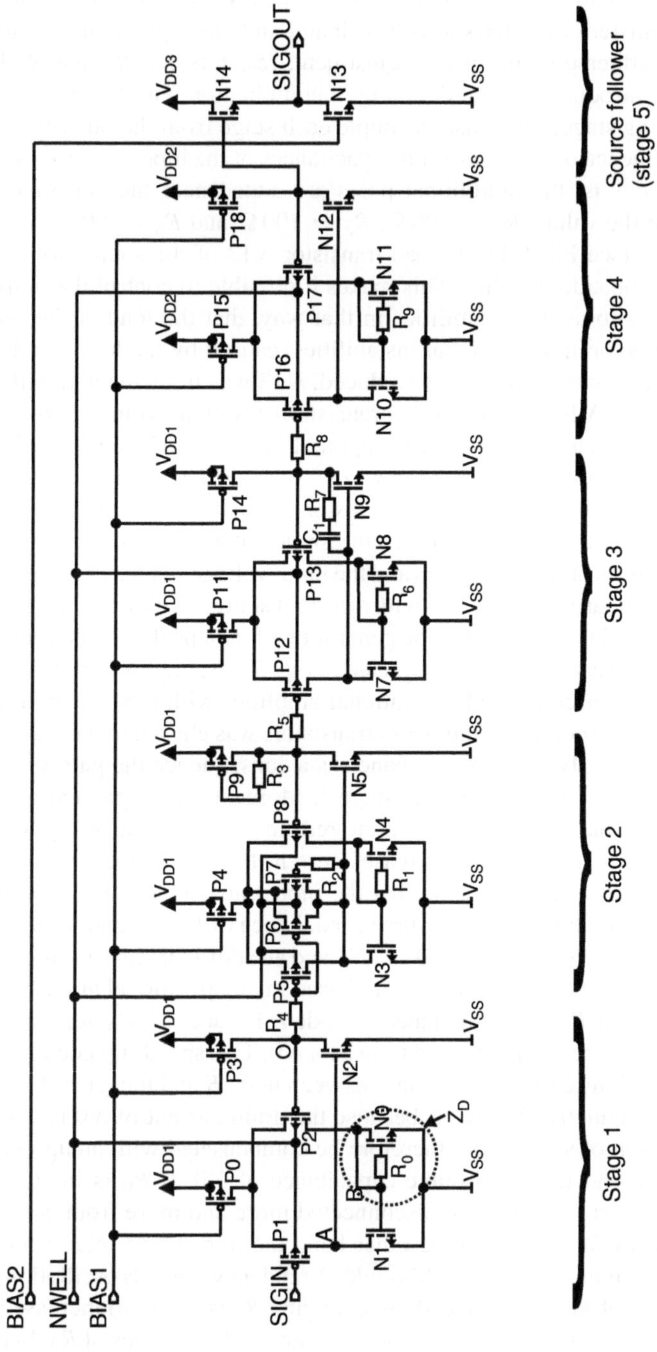

Fig. 5.2 Schematic of the voltage buffer

and stage 4, gain peaks at high frequencies at the individual transfer characteristics are forced to enhance the bandwidth. These peaks are adjusted in the way, that the overall transfer characteristic is broadband and has a good enough flatness. To introduce a further possibility for adjustment, resistors R_4, R_5 and R_8 have been added which introduce a slight damping in combination with the input capacitance of the following stage. They also decouple each stage from the other to some extent so that the influence of the gate-drain capacitances of the input transistors is reduced. Disadvantageous is, that additional poles are added near the cut-off region. The resistors have the values $R_4 = 400\,\Omega$, $R_5 = 100\,\Omega$ and $R_8 = 90\,\Omega$. With the bias voltage *BIAS2* (see Fig. 5.1) the load transistor $N13$ of the source follower can be adjusted from outside the chip. This makes it possible to control the output current to some degree to avoid instabilities in that way, that the load at the output node *SIGOUT* can be optimized so that instabilities, caused by the loop via the parasitic gate-source capacitance of $N14$ are reduced. $N13$ was implemented with a regular-V_t transistor and $N14$ was a low-V_t transistor. Also bias voltage *BIAS*1 could be adjusted from off-chip. The buffer is supplied via three different pads V_{DD1}, V_{DD2} and V_{DD3} to reduce influence of bond wires. At *NWELL* (see Fig. 5.2) a bias voltage for the n-wells of the input p-MOS transistors, which is lower than the appropriate source potential is applied to lower their threshold voltages. This increases slightly the transconductances of the transistors and the input voltage range. The bias network for *NWELL* is a standard bias network with a current source, where the current is defined by resistor R_{10} [1] (see Fig. 5.3). The generation of gain peaks can be illustrated with the schematic of stage 1, which is depicted in Fig. 5.2. Stage 1 is in principle a simple, as voltage follower connected, operational amplifier with p-MOS input transistors $P1$ and $P2$. This structure with p-MOS transistors was chosen to have the possibility to increase the supply voltage to enhance as a consequence the gate-source voltage of $P1$ and $P2$ to be able to speed up stage 1. Moreover the separated n-wells of $P1$ and $P2$ could be biased via pin *NWELL* to reduce the threshold voltages of the input p-MOS transistors, which also increases current flow and speed. These transistors are designed with a small size, so that the input capacitance of the voltage buffer, which is approximately the gate-source capacitance of $P1$, is small (about 3 fF). The active load consists of transistor $N0$ and $N1$ and an additional resistor R_0. This type of active load was first introduced in [7] and used in an operational amplifier in [8]. By increasing resistor R_0 the capacitance of node B is successively separated from the effective gate-source capacitance of transistor $N0$. This speeds up node B at moderate values of R_0, because of a slight delay between node B and the gate of transistor $N0$ is introduced (inductive behavior, because the drain current of $N0$ is delayed). The transfer function of stage 1 can therefore be compensated with an appropriate time constant of R_0 and the gate-source capacitance of $N0$. If R_0 is increased further, the gate of transistor $N0$ will be disconnected more and more from node B. Then a too high resistor R_0 causes on the other hand the impedance seen from node B to increase, which then slows down this node. A tendency towards oscillation especially for high values of R_0 is introduced. So changing R_0 is a way to increase bandwidth and control gain peaks at high frequency for intermediate values of R_0. In Fig. 5.4 the small-signal equivalent circuit of stage 1 without feedback and disconnected from

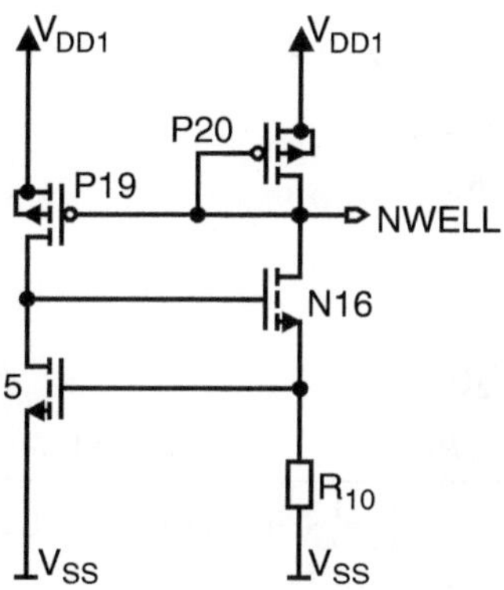

Fig. 5.3 N-well bias voltage generation

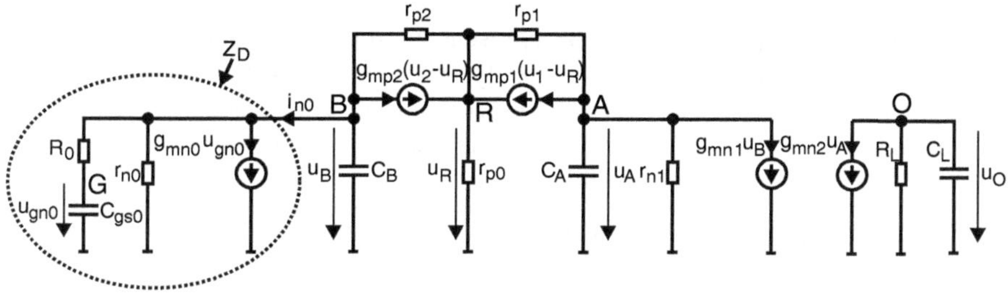

Fig. 5.4 Small-signal equivalent circuit of stage 1 in open loop (no connection of node *O* to the gate of *P*2) and cut off connection from node *O* to R_4

R_4 is shown. The transistor output resistances are represented by r_{p0}, r_{p1}, r_{p2}, r_{n1} and $R_L = r_{n2}||r_{p3}$ and the corresponding transconductances are represented by g_{mp1}, g_{mp2}, g_{mn1} and g_{mn2}. Z_D is the impedance of the system $N0$, R_0 (seen from node *B*). C_A, C_B and C_L are the capacitive loads of nodes *A*, *B* and output *O* respectively. With the assumption, that r_{p0}, r_{p1}, r_{p2} and r_{n1} (diode connection of $N0$ has in principle a low impedance) are infinite, and $g_{mp1} = g_{mp2}$, the open-loop gain G_1 can be calculated with 5.1–5.5 with Laplace transformation and can be expressed as shown in 5.6 with $Z_D(s)$ of 5.7.

$$R: g_{mp2}(u_2 - u_R) + g_{mp1}(u_1 - u_R) + \frac{u_B - u_R}{r_{p2}} + \frac{u_A - u_R}{r_{p1}} = \frac{u_R}{r_{p0}} \tag{5.1}$$

$$A: \frac{u_A - u_R}{r_{p1}} + g_{mp1}(u_1 - u_R) + g_{mn1}u_B + \frac{u_A}{r_{n1}} + sC_A u_A = 0 \tag{5.2}$$

$$B: \frac{u_B - u_R}{r_{p2}} + g_{mp2}(u_2 - u_R) + \frac{u_B}{Z_D} + sC_B u_B = 0 \tag{5.3}$$

$$O : g_{mn2}u_A + \frac{u_O}{R_L} + sC_L u_O = 0 \tag{5.4}$$

$$G : u_{gn0}C_{gs0} = \frac{u_B - u_{gn0}}{R_0} \tag{5.5}$$

$$G_1(s) = \frac{U_O(s)}{U_1(s) - U_2(s)}$$

$$G_1(s) \approx \frac{g_{mp1}g_{mn2}r_{n1}R_L}{2} \frac{\frac{1}{Z_D(s)} + g_{mn1} + sC_B}{\left(\frac{1}{Z_D(s)} + sC_B\right)(1 + sC_A r_{n1})(1 + sR_L C_L)} \tag{5.6}$$

$$Z_D(s) = \frac{U_B(s)}{I_{n0}(s)} \approx \frac{1 + sR_0 C_{gs0}}{g_{mn0}\left(1 + s\frac{C_{gs0}}{g_{mn0}}\right)} \tag{5.7}$$

In 5.1–5.6, u_1 and u_2 (see also Fig. 5.4) represent the small-signal gate-source voltages of input transistors $P1$ and $P2$ respectively ($U_1(s)$ and $U_2(s)$ are the appropriate Laplace transformation). The overall gain $A_1(s)$ of stage 1 with feedback, but without connection to resistor R_4 (see Fig. 5.2) can be calculated with 5.8.

$$A_1(s) = \frac{G_1(s)}{1 + G_1(s)} \tag{5.8}$$

As it can be seen in 5.6–5.8, $A_1(s)$ can be controlled by choosing an appropriate R_0, where g_{mn0} is the transconductance and C_{gs0} the effective gate-source capacitance of transistor $N0$. How $|A_1(f)| = |A_1(s = j\omega)|$ is altered, when different R_0 are implemented, is shown in Fig. 5.5. By finding an appropriate value for R_0, gain peaks at different higher frequencies of A_1 can be introduced to be able to adjust A_1 to some extent. A further advantage of adjusting A_1 with R_0 is that process tolerances have only a slight influence. Within the process tolerances of R_0 $A_1(s)$ does not change strongly. The value of R_0 was chosen to be 7 kΩ, so that only a very small gain peak occurs. A strong resonance would be coupled back to the input and influence the measured signal somewhat. The schematic of stage 2 is also depicted in Fig. 5.2. In principle stage 2 acts in the same way as stage 1. Transistor $P6$ causes an imbalance in the differential amplifier, which slightly increases the gain above unity. This was made to compensate the gain errors of the other stages and especially of the source follower. With $R2$, $P7$ and $R3$, $P9$ the gain peak, created by $R1$ and $N4$ can be enhanced and shifted a little bit to higher frequencies. Stage 3 and stage 4 were constructed in a similar way as stage 1. The only difference is an additional compensation path R_7, C_1 in stage 3. The simulated individual transfer characteristics of stage 1 to stage 4 are depicted in Fig. 5.6. The bandwidth of stage 1 is smaller than the bandwidths of stage 2 and stage 3, because due to the small transconductances of input transistors $P1$ and $P2$ (see Fig. 5.2), a trade-off in the size of $N2$ was made to have on the one hand a small capacitance at nodes A, B and on the other hand a sufficient g_{mn2} to reduce gain error. Furthermore together with $P3$ an appropriate output resistance had

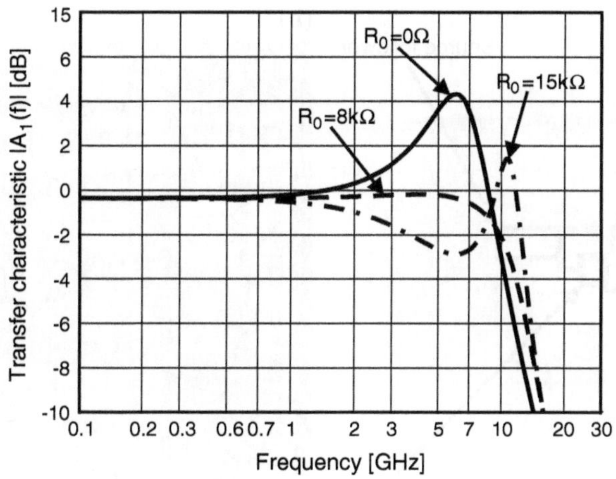

Fig. 5.5 Simulation results of gain $|A_1(f)| = |A_1(s = j\omega)|$ of separated stage 1 with feedback, where resistor R_0 is varied [6]

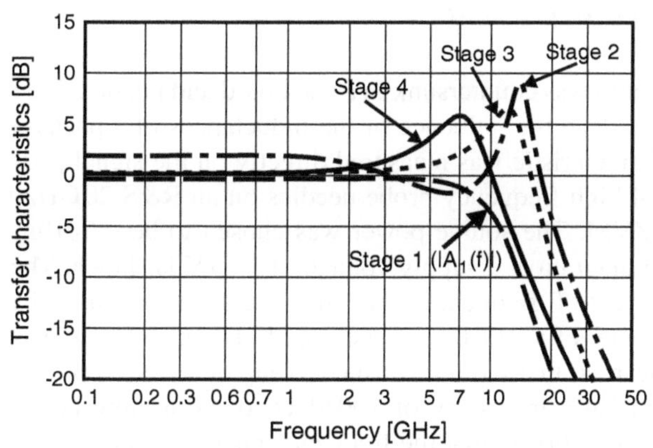

Fig. 5.6 Simulation of individual transfer characteristics of stages 1–4 [6]

to be adjusted, because of the higher input capacitance of stage 2. Also stage 4 has got a small bandwidth, because this stage has to drive the input capacitance of the final source follower. These losses of bandwidths are compensated by gain peaking of stage 2 and stage 3. Figure 5.7 shows a layout plot and a microphotograph of the test chip. Due to the passivation layer and the planarisation of the metal layers in this 0.12 μm CMOS technology a chip photo does not show any details. The area of the whole test chip amounts to $877 \times 907\,\mu\text{m}^2$. Thereof $16 \times 17\,\mu\text{m}^2$ is dedicated to stage 1, $32 \times 38\,\mu\text{m}^2$ to stage 2, $61 \times 52\,\mu\text{m}^2$ to stage 3, and $7{,}970\,\mu\text{m}^2$ to stage 4. The size of the source follower (stage 5) amounts to $130 \times 40\,\mu\text{m}^2$.

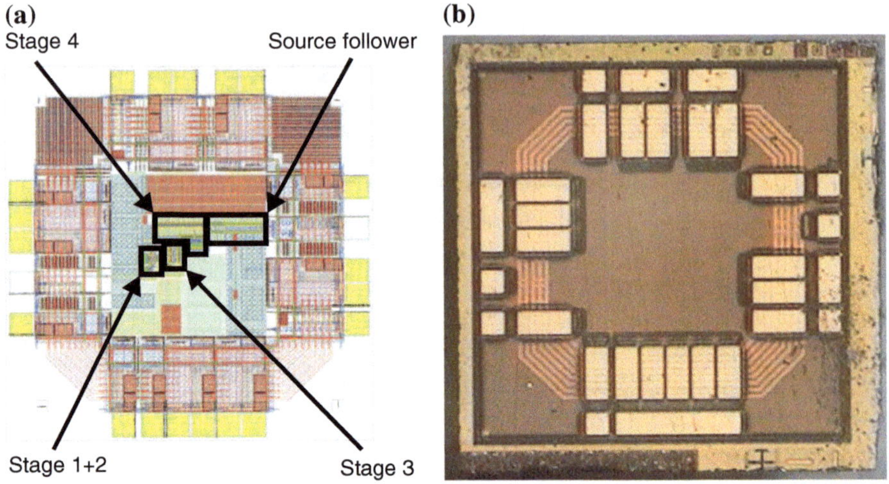

Fig. 5.7 The test chip with the voltage buffer. **a** Layout plot [95], **b** Microphotograph

5.1.2 Measurement Results

The fabricated chip was countersunk in a test board and bonded directly to the board connections to reduce the influence of the inductances of a package (see Fig. 5.8). The transfer characteristic was measured directly on the signal pads (see Fig. 5.8a) with the help of high frequency probe needles on an R&S 20 GHz vector network analyzer type ZVM. The source power was chosen to be −6 dBm. In Fig. 5.9 the measured bode diagram at a supply voltage of 1.65 V is shown. The voltage buffer has got a −3 dB cut-off frequency of 10 GHz at a supply voltage of 1.65 V. At 1.5 V a cut-off frequency of 9.2 GHz was measured. The phase of the voltage buffer is sufficiently linear, so that a quite constant group delay is given. The measurement of Fig. 5.9 began at a frequency of 100 MHz to see a more detailed view. Other measurements have shown, that the frequency characteristic below 100 MHz is flat and no resonances occur.

To verify linearity and to measure large-signal behavior, each signal pad (input and output) of the test chip was bonded with four parallel wires to 50 Ω micro-strip lines on the test board. On this board a 50 Ω termination was introduced at the input line as near as possible to the pad. Series resistors were introduced on the board between the 50 Ω termination and the bond wire to the chip pad to damp oscillations caused by the inductance of the bond wire and the capacitance of the pad. This type of termination was quite sufficient for a single-tone measurement and a step response measurement due to simulations and reflexion measurements. Unfortunately no on-chip 50 Ω termination pad was available at this moment for the test chip, which would be a better termination. In later test chips with comparators self-designed 50 Ω input pads were used. The single-tone measurement was done with a 3 GHz

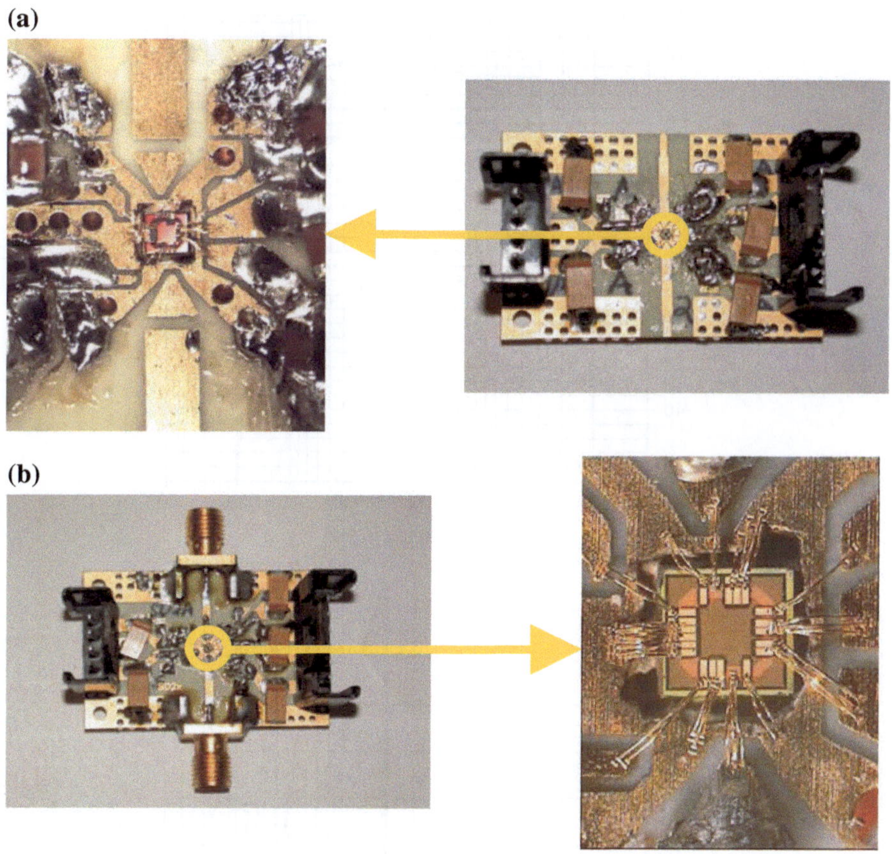

Fig. 5.8 Test boards with voltage buffer chips. **a** Test board for measurements with high-frequency probe needles, **b** Test board for linearity and step response measurements

spectrum analyzer HP E4402B. The measurement frequency was at 900 MHz, which was generated by a network analyzer of the type HP 8753E. A typical single-tone measurement is shown in Fig. 5.10, where the input power P_I to the test chip was −5 dBm. The slight damping of the output power P_f (900 MHz) is due to SMA cables, SMA connectors and the microstrip lines on the test board. In Fig. 5.10 the second harmonic (P_{2f}) at 1.8 GHz and the third harmonic (P_{3f}) at 2.7 GHz can be seen. In Table 5.1 the suppression P_{2f}/P_f of the second harmonic (1.8 GHz) and P_{3f}/P_f of the third harmonic (2.7 GHz) in reference to the power P_f at 900 MHz is shown in dB, where a supply voltage of 1.65 V was applied. Damping variation of cables, connectors and slight differences of the overall transfer characteristic of the measured voltage buffer at different frequencies were neglected.

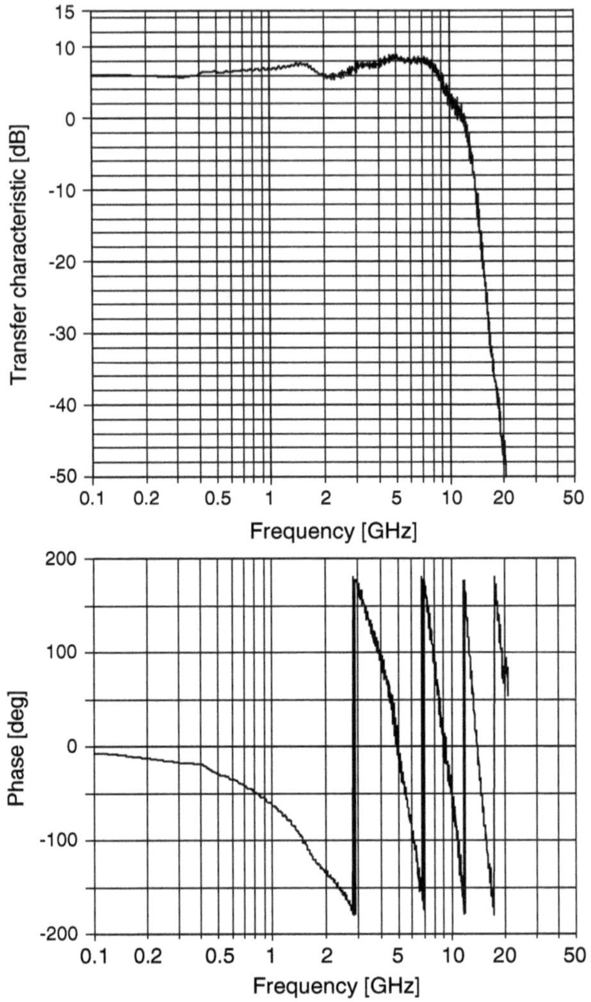

Fig. 5.9 Bode diagram of the voltage buffer at a supply voltage of 1.65 V. Because no 50 Ω termination was implemented at the input pad, an offset of 6 dB occurs in the frequency characteristics [6]

The measured response to a rectangular input signal is shown in Fig. 5.11. It is not an exact step response measurement due to a certain slope of the input signal at *SIGIN*. The power dissipation of the voltage buffer was measured to be 170 mW at a supply voltage of 1.65 V and an output load of 50 Ω.

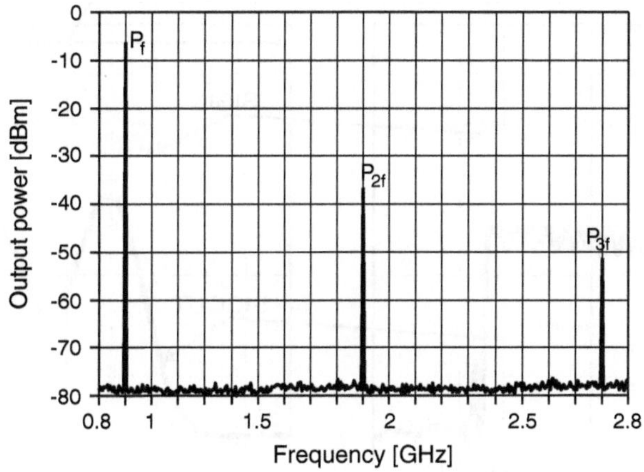

Fig. 5.10 Single tone measurement with an input frequency of 900 MHz [6] (input power $P_I = -5$ dBm)

Table 5.1 Suppression P_{2f}/P_f of the second harmonic (1.8 GHz) and P_{3f}/P_f of the third harmonic (2.7 GHz) [6]

Input power at 900 MHz P_I [dBm]	Suppression at 1.8 GHz P_{2f}/P_f [dB]	Suppression at 2.7 GHz P_{3f}/P_f [dB]
0	−21.8	−31.6
−5	−30.3	−45
−10	−35.4	−53.1
−15	−40.7	−63.2
−20	−45	−64.6

5.1.3 Redesign of the Voltage Buffer for Usage in a Test Chip with a Comparator

The problem of the 10 GHz analog voltage buffer, which was described in the sections before is that it needs a quite large chip area only to measure one distinct signal with a high-frequency oscilloscope, where 50 Ω loads have to be driven. So a redesign was done, where two signals could be selected depending on an additional digital input. In Fig. 5.12 the principle block diagram of the comparator, connected with the voltage buffer, where in stage 1, signals *OUT* or $\overline{OUT}$ can be selected (via inputs *SIGIN*1 and *SIGIN*2) with pin *DIGBUF* (via inputs *DIGP* and *DIGN*) to be delivered to *SIGOUT*. To select *SIGIN*1, digital pins *DIGN* = 1.65 V and *DIGP* = V_{SS} = 0 V (*DIGBUF* = V_{SS} = 0 V). To switch *SIGIN*2 to *SIGOUT*, *DIGN* = V_{SS} = 0 V and *DIGP* = 1.65 V are set (*DIGBUF* = 1.65 V). To compensate load capacitance changes at inputs *SIGIN*1 and *SIGIN*2, which would affect the offset of the comparator depending on *OUT* or $\overline{OUT}$ is selected, a replica of stage 1 of the voltage buffer is

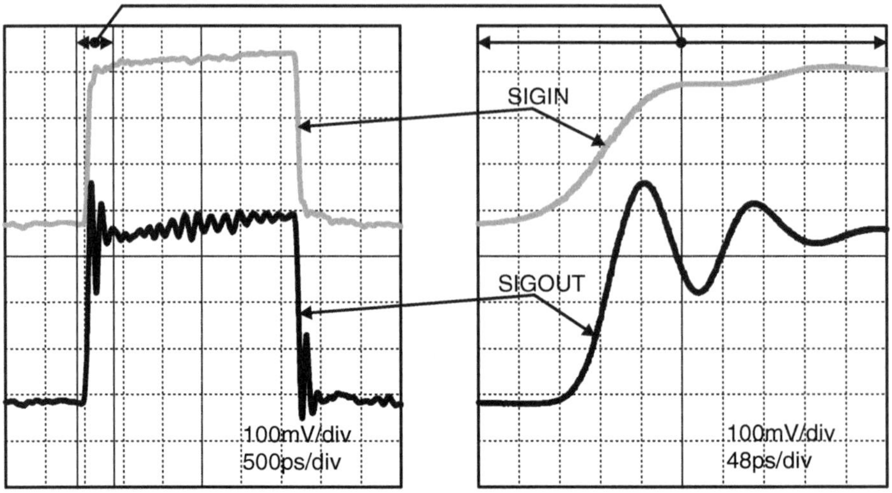

Fig. 5.11 Oscilloscope pictures to show the large-signal characteristics of the buffer

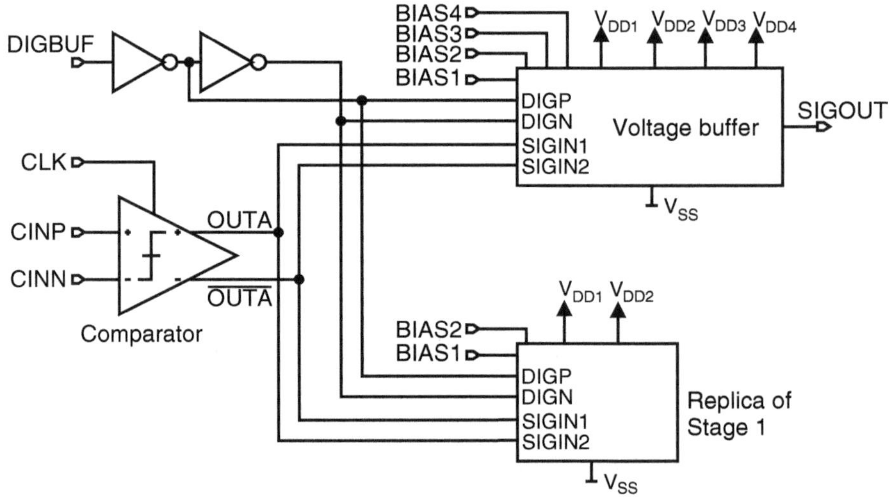

Fig. 5.12 Principle block diagram to connect both output nodes *OUT* and $\overline{OUT}$ of the comparator to one analog voltage buffer, where the first stage was designed to select the appropriate signals with pin *DIGBUF* (at pins *DIGP* and *DIGN*). To always have the same capacitive load at both outputs of the comparator, a replica of only stage 1 of the voltage buffer is added

added. To illustrate functionality of the voltage buffer, the block diagram is depicted in Fig. 5.13. For simplification, the transistors, which act as switches are represented with switch symbols, where the appropriate transistor names are added. The position of the switches, which are shown in Fig. 5.13, are for selection of signal *SIGIN*1. The complementary position of all switches would select *SIGIN*2. In principle the

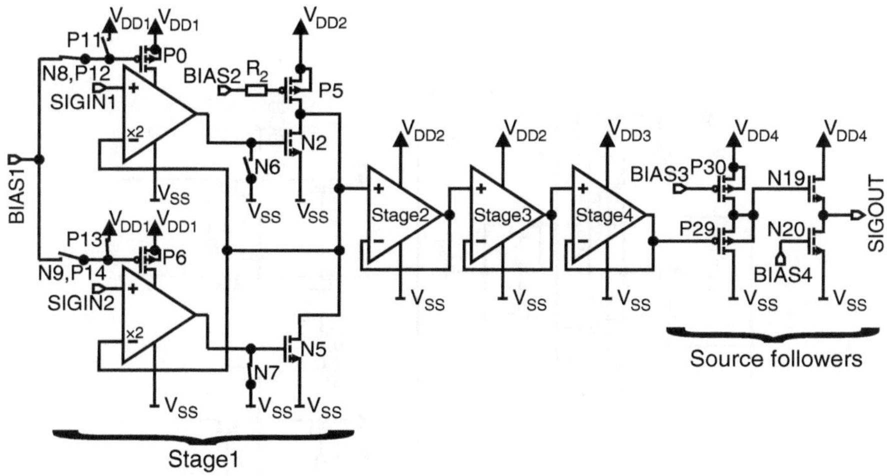

Fig. 5.13 Block diagram of the redesigned voltage buffer. The position of the switches, which are shown here, are for selection of signal *SIGIN*1. The complementary position of switches would select *SIGIN*2. Because of the higher load capacitance after stage 1, the input signal is attenuated by about a factor of 2 to meet bandwidth requirements of about 10 GHz

gate of tail transistor *P*0 of the buffer for *SIGIN*1 in stage 1, which is switched on, is connected to bias voltage *BIAS*1 so that a current can flow (transmission gate *N*8, *P*12 on, switch *P*0 off). The buffer in stage 1, to where *SIGIN*2 is connected to, is switched off. Here the gate of tail transistor *P*6 is connected to V_{DD1} so that *P*6 is switched off and no current can flow through. *BIAS*1 is separated by the switched off transmission gate *N*9, *P*14. Furthermore transistor *N*5 is switched off, because the gate is pulled down to V_{SS} by switch *N*7. Transistor *P*5 works as an active load and builds with transistor *N*2 an amplifier, when *SIGIN*1 is selected. In the other case, if *SIGIN*2 is selected, transistor *P*5 builds with transistor *N*5 an amplifier. Because of the now higher load capacitance after stage 1 in comparison to the voltage buffer, described in the sections before, the input signal (voltage level) is attenuated by a factor of about two (6 dB) to meet overall bandwidth requirements of about 10 GHz. The bias voltages *BIAS*1, *BIAS*2, *BIAS*3 and *BIAS*4 are applied to a pad from outside the chip. The source followers are added to drive the pad capacitance and 50 Ω load. To compensate the gain error of these source followers the attenuation in stage 1 is somewhat smaller than two.

The whole schematic of the voltage buffer is depicted in Fig. 5.14. The bias voltages at pins *NWELL*1 and *NWELL*2 are generated with a standard circuit [1], where p-MOS diode connected transistors *P*31 and *P*33 have been added (see Fig. 5.15) to reduce further the n-well bias voltage to an appropriate level below the source potential of the input transistors, but not too low so that the p-source-n-well diodes do not conduct. The schematic in Fig. 5.15 represents two bias circuits, one for *NWELL*1 and one for *NWELL*2, where the circuit for *NWELL*1 is supplied with V_{DD1} and *NWELL*2 with V_{DD2}. The reference current through resistor R_{11} is defined by the

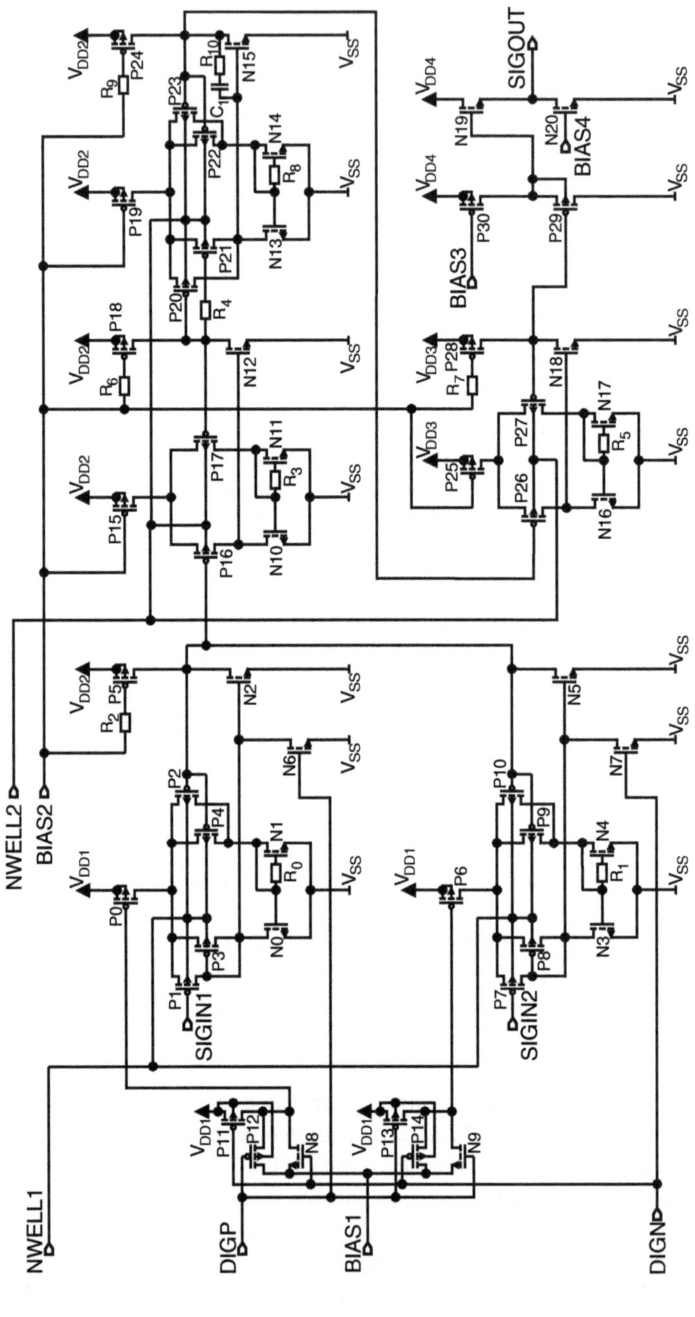

Fig. 5.14 Complete schematic of the redesigned voltage buffer

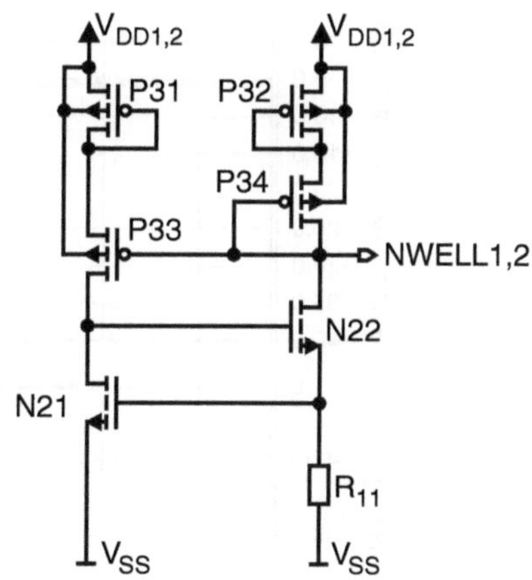

Fig. 5.15 Bias voltage generation for n-well connections *NWELL*1 and *NWELL*2 (one circuit for each connection)

threshold voltage of transistor $N21$ divided by R_{11}, if *W/L* of $N21$ is assumed to be large enough. The functionality of switch transistors $N6, N7, P11, P13$ (see Fig. 5.14) and transmission gates $N8$, $P12$ and $N9$, $P14$ is already explained with the help of block diagram in Fig. 5.13. The difference to the circuit of stage 1, which is already described in Sect. 5.1.1, are additional input transistors $P3$ and $P4$, where $P3$ is diode connected to decrease the output resistance and therefore enhance speed. But the trade-off is, that the small signal amplification is reduced too. The attenuation of a factor of about two is done via feedback to transistors $P2$ and $P4$, which together have a higher *W/L* than $P1$. Furthermore supply voltage V_{DD1} of the input differential amplifiers is separated to be able to raise this voltage from e.g. 1.65–1.75 V to enhance the input voltage range. Supply voltages V_{DD2} to V_{DD4} are designed for 1.65 V. Resistors R_2, R_6, R_7 and R_9 reduces the overall capacitive loads at the output node of a stage, because then the influence of the gate-drain capacitance of active load transistors $P5$, $P18$, $P28$ and $P24$ respectively are reduced. Resistor R_4 is added, so that firstly the influence of the additional gate capacitance of transistor $P21$ is reduced and secondly that the amplification is decreased for higher frequencies (R_4 in connection with the gate capacitance of $P21$) to avoid high resonances. Resistors R_0, R_1, R_3, R_5 and R_8 of the active n-MOS loads are added to generate resonances as described in Sect. 5.1.1.

The simulated transfer characteristics of the complete voltage buffer for a supply voltage of $V_{DD2} = V_{DD3} = V_{DD4} = 1.65$ V is depicted in Fig. 5.16 for the cases of $V_{DD1} = 1.65$ V and $V_{DD1} = 1.75$ V. The resonance peak at frequencies higher than 10 GHz is not as high as it appears in the simulation due to the experience of the measurements on the 10 GHz voltage buffer in Sect. 5.1.2. So a −3 dB boarder frequency of 10 GHz is expected. The other characteristic values, e.g. linearity are similar to the voltage buffer, measured in Sect. 5.1.2. The block diagram and the schematic of the replica of stage 1 is depicted in Fig. 5.17.

A layout plot of the voltage buffer is shown in Fig. 5.18, where the size is approximately $500 \times 400\,\mu m^2$ without the the long connection wires towards the bottom of

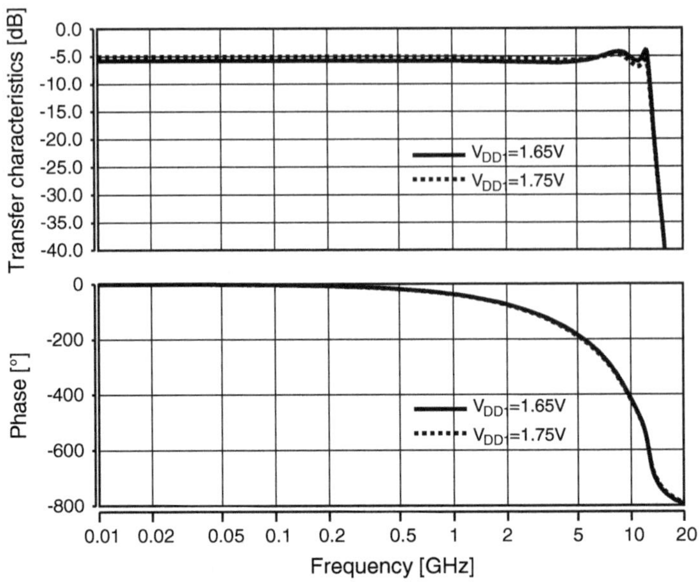

Fig. 5.16 Overall simulated transfer characteristics of the redesigned voltage buffer for $V_{DD1} = 1.65\,\text{V}$ and $V_{DD1} = 1.75\,\text{V}$

the picture. Because of time run short no test chip of the switchable voltage buffer was produced and measured, because the new, switchable buffer showed a similar high-frequency behavior in simulation as the measured voltage buffer of Sects. 5.1.1 and 5.1.2. So the switchable buffer was implemented in some test chips for comparators as shown in Fig. 5.12.

5.2 Temperature Measurement

In common on-chip temperature measurements are done as explained in [9] and in an application note from Maxim [10].

The widely used approach for measuring the temperature is to use a diode as a temperature sensor. There, two currents are forced through the diode in conducting direction with a typical ratio of $I_H : I_L = 10 : 1$ and each voltage drop is measured. This voltage drop can be described with 5.9, where V_H is the diode voltage while I_H is flowing and V_L is the voltage drop of the diode when I_L is applied to the diode.

$$V_H - V_L = \eta \frac{k\vartheta}{e}\left(\ln\left(\frac{I_H}{I_L}\right)\right), \quad V_H - V_L = 1.986 \times 10^{-4} \cdot \eta\vartheta \ \text{ for } \ \frac{I_H}{I_L} = 10 \tag{5.9}$$

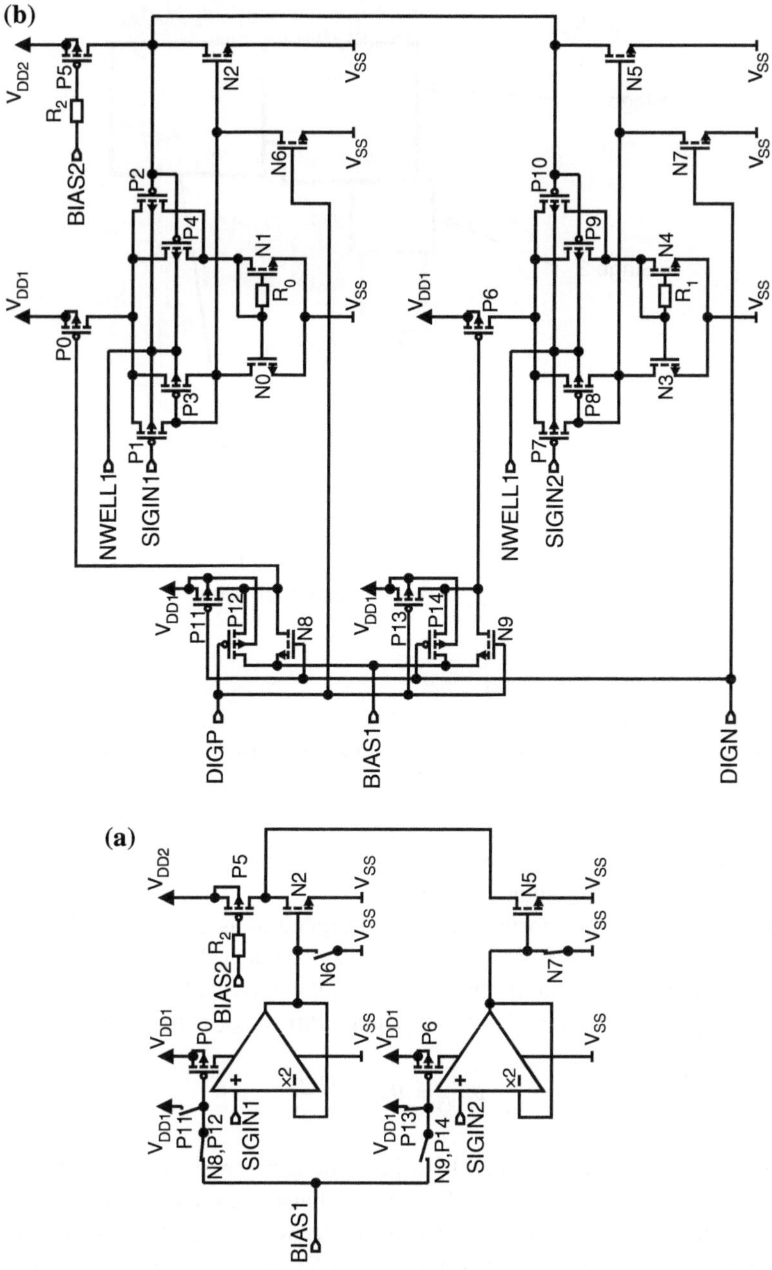

Fig. 5.17 Replica of stage 1 of the voltage buffer (see also Fig. 5.12). **a** Block diagram, **b** Schematic

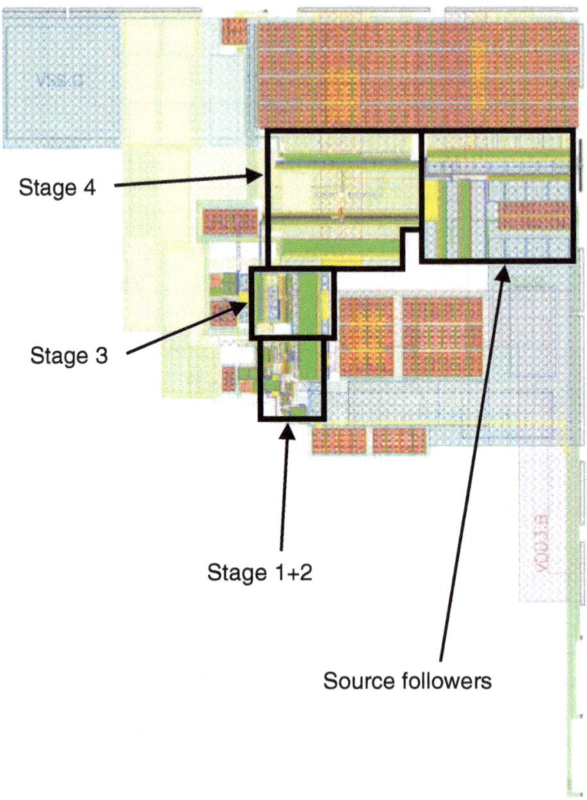

Fig. 5.18 Layout plot of the redesigned voltage buffer in 0.12 μm CMOS technology

ϑ is the temperature in Kelvin (K), $k = 1.38 \times 10^{-23}$ J/K is the Boltzmann's constant, η is the ideality factor of the diode (about 1, but varies with processing) and $e = 1.602 \times 10^{-19}$ C is the elementary charge.

Most temperature monitoring chips are designed to assume an ideality factor of the diode of $\eta_{nom} = 1.008$ for an accurate measurement. If the real ideality factor (η_{real}) of a diode is different to this value, then the corrected, true temperature ϑ_{cor} can be distinguished with 5.10, when a temperature ϑ_{meas} is measured.

$$\vartheta_{cor} = \vartheta_{meas} \frac{\eta_{nom}}{\eta_{real}} \tag{5.10}$$

An additional error is contributed by a parasitic series resistance R_S of the diode. Then $V_H - V_L$ has an offset of $R_S(I_H - I_L)$ and also the measured temperature has an offset. The on-chip temperature sensor diode, which was implemented in later, more advanced test chips for a measurement of a comparator, is shown in Fig. 5.19.

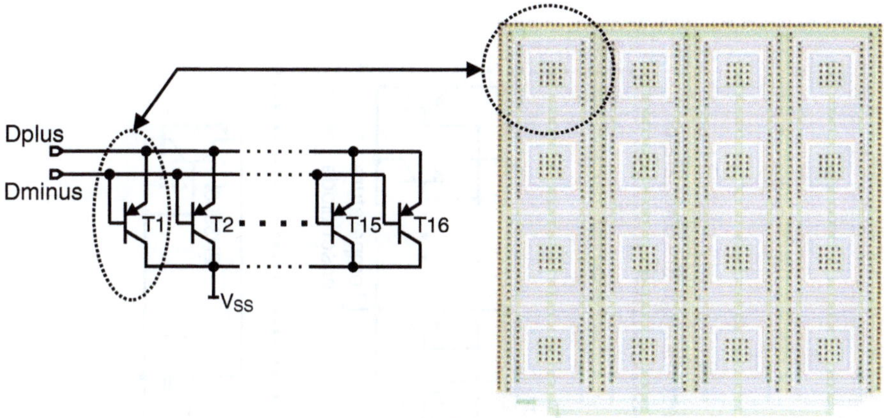

Fig. 5.19 Schematic and layout of the implemented temperature sensor diode

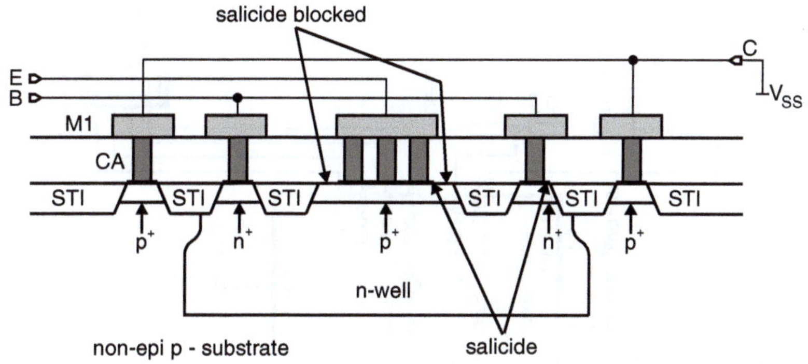

Fig. 5.20 Cross section of a pnp-transistor in the used 0.12 μm CMOS process

The cross section of the pnp transistor, which is possible to be designed in this 0.12 μm CMOS process and which was used as temperature sensor diode is depicted in Fig. 5.20, where the substrate is the collector and connected to V_{SS}.

In the used 0.12 μm CMOS technology, a pnp-transistor cell has a very low differential current amplification factor β, but such a transistor cell was sufficient to build a diode block of 16 base-emitter diodes (collector connected to V_{SS}) for the measurement of the on-chip temperature to get a value for control reasons. To reduce the series resistance, an array of 4×4 of such transistor cells were implemented, where the whole side length was 31 μm.

The schematic of the whole temperature measurement part is depicted in Fig. 5.21 (see also Sect. 5.5). The temperature sensors are handled with the monitoring chip IC5 (MAX6655MEE [11]). It is controlled by a microcontroller PIC18F452-I/P via a System Management Bus (SMB), which is an evoluted Inter-IC bus (I2C-bus). The address for SMB communication for IC5 is fixed with jumpers JP3 and JP4, where

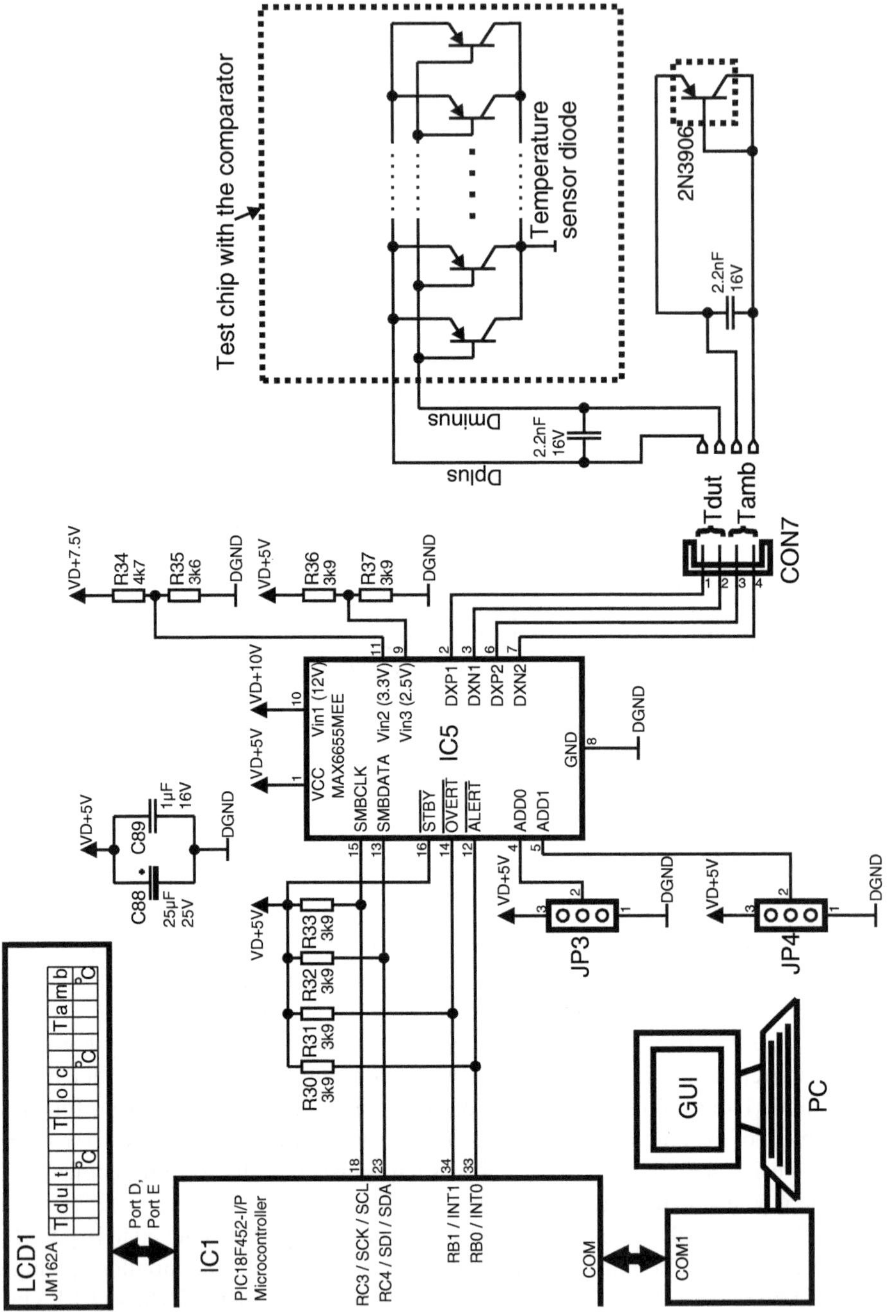

Fig. 5.21 Schematic of temperature measurement part

pins *ADD*0 and *ADD*1 are connected to *DGND* to define the 7 bit device address 0b0011000(0) of IC5. Pin $\overline{STBY}$ is connected to *high* so that IC5 is not in standby mode. At IC5, output pin $\overline{OVERT}$ (open drain) become *low* when a measured temperature is higher than a programmed value and output pin $\overline{ALERT}$ (open drain) is switched to *low* by IC5, if e.g. a measured temperature or voltage is higher or lower than a programmed value or if e.g. a remote diode for temperature measurements is fault. Pin $\overline{OVERT}$ is connected with $RB1/INT1$ and pin $\overline{ALERT}$ is connected with $RB0/INT0$ of IC1. Communication with IC5 is done by the internal I2C bus muster of IC1 between pins $RC3/SCK/SCL$ of IC1 and pin *SMBCLK* (clock line) of IC5 and between $RC4/SDI/SDA$ of IC1 and *SMBDATA* of IC5 (data line). Between pins *DXP*1 and *DXN*1 the on-chip diode cell array of the test chip with the comparator is plugged on (connector CON7) to measure Tdut. The ambient temperature sensor (pnp-transistor 2N3906, which is diode connected) for *Tamb* is also plugged to connector CON7 between pins *DXP*2 and *DXN*2 of IC5. Inputs *Vin*1 to *Vin*3 are for monitoring of different supply voltages, which is also provided by IC5. Furthermore also the internal temperature *Tloc* of the monitoring chip IC5 can be measured and read out by the microcontroller IC1. At LCD1 all the measured temperatures *Tdut*, *Tloc* and *Tamb* are displayed. Controlling LCD1 is done via Port D and Port E of the microcontroller IC1. With the graphical user interface (GUI) the measured temperatures can be read out with the serial port (COM1↔COM, RS232) by a computer (PC), because in IC1 also a Universal Asynchronous Receiver Transmitter (UART) is implemented.

5.3 Transfer Stage and Delay Time Measurement of the Comparator

In newer test chips for investigation of a comparator transfer stages are used, which are designed for high data rates and for using the voltage levels of CMOS logic. During the reset phase of the comparator a transfer stage holds the decision of the previous comparison phase, because a clock period is divided into a reset- and a comparison phase of the comparator. Holding the logical level during reset is important, because the time duration of a data bit is extended in this way to a whole clock period (The reset phase is removed from the output data stream). So firstly the bandwidth of the data stream is reduced for the output drivers and secondly the occurrence of the same voltage level on the inverted and non-inverted outputs of the test chip are avoided, which would appear during reset phase for a considerable long time. Thirdly measurement of the bit error rate (BER) with the used BER analyzer station needs a non-return-to-zero (NRZ) data stream, where a voltage level of a bit lasts for the whole clock period and changes directly to the voltage level of the next bit in the following clock period. A principle block diagram of a comparator with the transfer stages and the output drivers is shown in Fig. 5.22, where also the different delay times of the appropriate circuit blocks are depicted. Here the voltage level to which

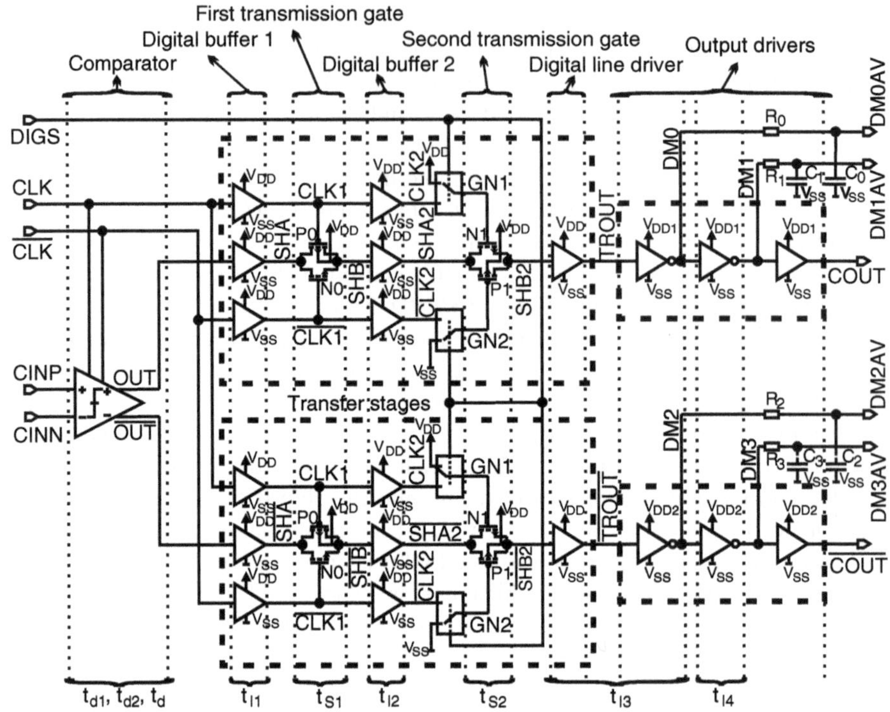

Fig. 5.22 Block diagram of a comparator with transfer stages and output drivers for the test chip

a delay time between two signals is referred is $V_{DD}/2$ (time point at 50 % of V_{DD} at the edge of the signal referred to the time point at 50 % of V_{DD} at the appropriate edge of the reference signal), where rise and fall times of the edges are assumed to be the same. Especially the delay time t_d of the comparator is here only defined as the time till one output of the comparator reaches $V_{DD}/2$, where the reset time t_{Re} is not considered in t_d. The delay time t_d of the comparator is here an indication for the time duration of a decision of the comparator, which is most interesting for applications. For the two different input voltage differences (*CINP-CINN*) (offset neglected) in Fig. 5.22 the delay times of the comparator are denoted by $t_d = t_{d1}$ and $t_d = t_{d2}$, where for t_{d1} a larger *CINP-CINN* is assumed. So in common $t_{Re} \leq t_{d1} \leq t_{d2}$ is valid. The other delay times t_{S1} and t_{S2} are referred to the appropriate transmission gates $N0$, $P0$ and $N1$, $P1$ respectively and t_{I1}, t_{I2}, t_{I3} and t_{I4} are the delay times of the appropriate buffers as shown in Fig. 5.22. There are two transfer stages shown in Fig. 5.22, one connected to the output node OUT of the comparator and the second connected to $\overline{OUT}$. For simplicity in Fig. 5.22 the node and transistor names in the transfer stages (inside the dotted areas, which mark transfer stages), which leads in principle the same wave form (only shifted by a clock period) or have the same function are named also in the same way. But in Fig. 5.22 e.g. node $CLK1$ in the transfer stage connected to OUT is not the same physical node as node $CLK1$ in the second transfer stage to where $\overline{OUT}$ is applied. Transmission gate $N0$, $P0$ is placed after digital buffer 1 (two inverters) and is switched on after an appropriate delay time (t_{l1}) with $CLK1 = V_{SS} = low$ and $\overline{CLK1} = V_{DD} = high$, when the comparator is in comparison phase ($CLK = V_{SS} = low$, $\overline{CLK} = V_{DD} = high$). In another case, when the comparator would compare at $CLK = V_{DD}$, then signals CLK and $\overline{CLK}$ at the input of a transfer stage should be interchanged in the design. So during transistors $N0$ and $P0$ are switched on, node *SHB* is loaded with the logical value at node *SHA*, which is generated via digital buffer 1 by the comparator when *CINP* is compared with *CINN*. After comparison, when *CLK* switches to V_{DD} (reset phase of the comparator), transmission gate $N0$, $P0$ is switched off with $CLK1 = V_{DD}$ and $\overline{CLK1} = V_{SS}$ and the voltage level at node *SHB* is hold, because it is stored dynamically in the parasitic capacitances at this node. After an additional driver (digital buffer 2), which consists of a chain of six inverters, a second transmission gate $N1$, $P1$ is placed. With pin *DIGS* it can be chosen, whether transistors $N1$, $P1$ are turned always completely on ($DIGS = V_{DD}$) or only turned on (after an appropriate delay time t_{l2} of the digital buffer between nodes *SHB* and *SHA2*), when the first transmission gate $N0$, $P0$ is turned off ($DIGS\ V_{SS}$). In the case of $DIGS\ V_{SS}$, node $CLK2$ is connected to the gate of $N1$ (node $GN1$) and $\overline{CLK2}$ to the gate of $P1$ (node $GP1$) in the opposite way as nodes $CLK1$ to the gate of $P0$ and $\overline{CLK1}$ to the gate of $N0$. This has the advantage that in this configuration the data stream is synchronized to the clock signal, because then, different delay times of the comparator, which occur e.g. at different input voltage differences *CINP*-*CINN* or at different input common-mode voltages $(CINP + CINN)/2$, plus the delay time of the first transmission gate are replaced by one fixed delay time with a duration of the half clock period ($t_d + t_{S1} \rightarrow T/2$). Then it is not necessary to adjust the optimal delay time between output data of the comparator and the sampling clock on the BER receiver for every BER measurement. The optimal delay only has to be adjusted before the first BER measurement and when a new adjustment is necessary because of different types of drifts (e.g. temperature drift). The output drivers (see appropriate dotted areas in Fig. 5.22) consist of a chain of inverters. DC voltages at the outputs *DM0AV*, *DM1AV*, *DM2AV* and *DM3AV* are used for measurement of the delay time of the comparator. Pre-investigations of this type of measurement of the delay time t_d of a comparator were done in [9, 12, 13].

For explanation of the principle function of the transfer stage and the output driver a waveform diagram is shown in Fig. 5.23 (see also block diagram in Fig. 5.22), where *high* has the voltage level $V_{DD} = 1.5\,\text{V}$ and *low* the level $V_{SS} = 0\text{V}$. Signals *CLK* and $\overline{CLK}$ are the inverted and non-inverted clock of the comparator. In the case of Figs. 5.22 and 5.23 the comparison phase of the comparator takes place during $CLK = V_{SS}$ and the reset phase during $CLK = V_{DD}$, where both output nodes *CLK* and $\overline{CLK}$ are pulled to V_{SS}. The input voltage difference *CINP*-*CINN* changes the sign after each clock period so that the comparator decides in one clock period at *OUT* to *high* and in the following clock period to *low*. This is necessary to be able to measure the delay time t_d of the comparator by measuring the averaged voltages of nodes *DM0*, *DM1*, *DM2* and *DM3* at outputs *DM0AV*, *DM1AV*, *DM2AV* and *DM3AV*. In Fig. 5.23 two different amplitudes of *CINP*-*CINN* are shown. For a smaller amplitude

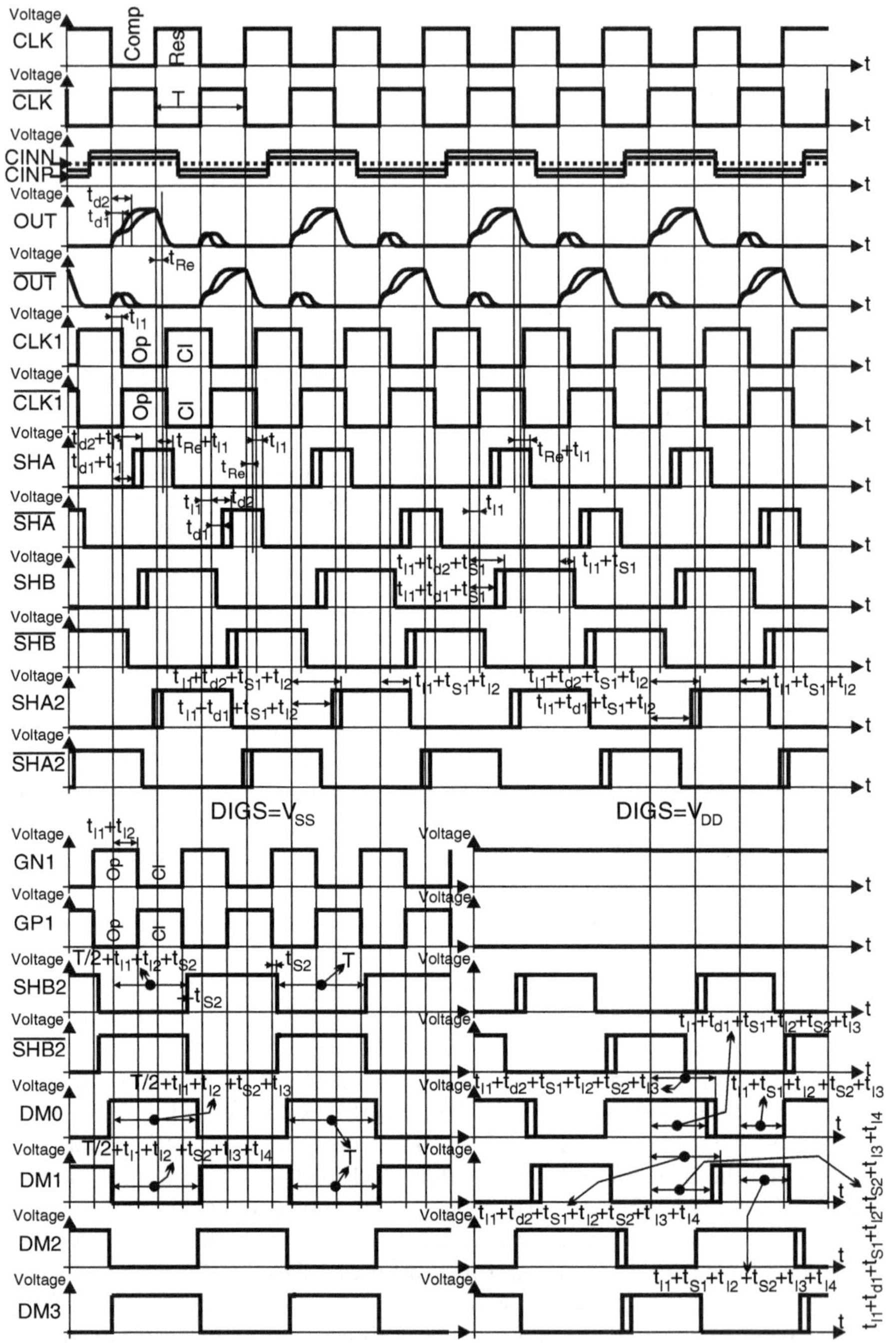

Fig. 5.23 Principle waveform diagram of the transfer stage and the output driver

the comparator needs more time for the decision ($t_d = t_{d2}$) than for a larger amplitude ($t_d = t_{d1}$). Each output of the comparator (OUT, $\overline{OUT}$) is connected to a transfer stage. In such a transfer stage the output signals OUT and $\overline{OUT}$ of the comparator are buffered with two inverters and applied to nodes SHA and $\overline{SHA}$ respectively, which are the inputs of the first transmission gate (transistors $N0$, $P0$) of the appropriate transfer stage. Signals $CLK1$ and $\overline{CLK1}$ are in principle signals CLK and $\overline{CLK}$, but delayed by the delay time t_{I1} of two inverters, which have the same size as the inverters in the buffer between nodes OUT and SHA or between $\overline{OUT}$ and $\overline{SHA}$ for delay time compensation. When $CLK1$ is *low* and $\overline{CLK1}$ is *high*, the transmission gate $N0$, $P0$ is open and the decision of the comparator is loaded (with t_{I1} delayed) to node SHB and $\overline{SHB}$ respectively, where the transmission gate itself has a delay time t_{S1}. When $CLK1$ switches to *high* and $\overline{CLK1}$ to *low*, transmission gate $N0$, $P0$ is closed and the logical state is hold at node SHB and $\overline{SHB}$ respectively. Due to the fact, that *CINP-CINN* changes polarity, node SHB changes immediately after a delay of $t_{I1} + t_{S1}$ (referred to CLK, $\overline{CLK}$) to *low*, when the transmission gate opens once again and if the comparator has decided at OUT to *high* in the previous clock period. So as a result the rising ramp of SHB is now delayed by $t_{I1} + t_d + t_{S1}$ and the falling ramp is delayed by $t_{I1} + t_{S1}$ in reference to signals CLK and $\overline{CLK}$. The average voltage of SHB is therefore proportional to the delay time t_d of the comparator. The same considerations are valid for nodes $\overline{OUT}$ and $\overline{SHB}$. Each signal $CLK1$, $\overline{CLK1}$, SHB and $\overline{SHB}$ is connected with a chain of six inverters, which all have the same size and a delay time t_{I2}. These inverters drive the second transmission gate $N1$, $P1$. Two possibilities can be chosen with pin *DIGS*.

If *DIGS* is *low* (*DIGS* V_{SS}) transmission gate $N1$, $P1$ is turned on by signals $CLK2$ and $\overline{CLK2}$, when transmission gate $N0$, $P0$ is turned off by $CLK1$ and $\overline{CLK1}$ ($CLK2$ and $\overline{CLK2}$ are delayed by t_{I2} compared to $CLK1$ and $\overline{CLK1}$). This is because $CLK1$ and $\overline{CLK1}$ are applied to the gates of $P0$ and $N0$ respectively, but $CLK2$ and $\overline{CLK2}$ are applied vice versa to the gates $GN1$ of $N1$ and $GP1$ of $P1$ respectively. Then as a consequence the logical voltage level at node $SHA2$ and $\overline{SHA2}$ is already established, when transmission gate $N1$, $P1$ is turned on. So the delay times of the data at node $DM0$ and $DM2$ results to $T/2 + t_{I1} + t_{I2} + t_{S2} + t_{I3}$ and at node $DM1$ and $DM3$ to $T/2 + t_{I1} + t_{I2} + t_{S2} + t_{I3} + t_{I4}$ referred to CLK and $\overline{CLK}$. So the overall delay time do not depend on the delay time of the comparator, which may vary. Of course so also a fix delay time is generated at the chip outputs $COUT$ and $\overline{COUT}$, which help when BER measurements are done, because so the delay time between the system clock and the data from the test chip at the bit pattern receiver (see Sect. 5.5) has not to be optimized for every bit error counting run.

If *DIGS* is *high* (*DIGS* V_{DD}), the gate $GN1$ of transistor $N1$ is set to *high* and the gate $GP1$ of $P1$ is set to *low* so that transmission gate $N1$, $P1$ is always on. Then the delay time t_d of the comparator can be determined by measuring the average DC voltages at outputs $DM0AV$, $DM1AV$, $DM2AV$ and $DM3AV$. If a delay time t_{S2} of the second transmission gate and delay times t_{I3} and t_{I4} for the following buffers (see Fig. 5.22) are considered, then the average voltage at the delay time measurement pins can be calculated with 5.11 and 5.12 (see also Fig. 5.23).

$$DM0AV = DM2AV = \frac{V_{DD}}{2T}(t_d + t_{I1} + t_{S1} + t_{I2} + t_{S2} + t_{I3} + \\ + T - (t_{I1} + t_{S1} + t_{I2} + t_{S2} + t_{I3})) = \frac{V_{DD}\,(T + t_d)}{2T} \tag{5.11}$$

$$DM1AV = DM3AV = \frac{V_{DD}}{2T}(t_{I1} + t_{S1} + t_{I2} + t_{S2} + t_{I3} + t_{I4} + \\ + T - (t_d + t_{I1} + t_{S1} + t_{I2} + t_{S2} + t_{I3} + t_{I4})) = \frac{V_{DD}\,(T - t_d)}{2T} \tag{5.12}$$

The delay time t_d can be calculated with 5.13 with the knowledge of the power-supply voltage V_{DD}, the clock period T and $\Delta V_{DM} = DM0AV - DM1AV = DM2AV - DM3AV$.

$$t_d = \frac{\Delta V_{DM}}{V_{DD}}T, \qquad \Delta V_{DM} = DM0AV - DM1AV = DM2AV - DM3AV \tag{5.13}$$

The delay time t_d, which accounts for the decision time of the comparator can be determined by only measuring DC voltages with a multimeter, which has a sufficiently high input resistance so that resistors R_0–R_3 do not influence the accuracy of the measurement. Comparing nodes *COUT* and $\overline{COUT}$ with a reference output *CREF* has the disadvantages, that in the design the reference path shall match exactly to the output paths, bond wires of different lengths influence the measurement, the cables of the measurement setup have to be deskewed and an expensive high-speed oscilloscope has to be used. Advantageous is, that the measurement of the delay time t_d is not influenced by different duty cycles of the non-inverted clock *CLK* and the inverted clock $\overline{CLK}$. This is shown in Fig. 5.24, where a duty cycle of α of *CLK* and corresponding of $(1 - \alpha)$ of $\overline{CLK}$ is assumed. Then 5.11 and 5.12 change to 5.14 and 5.15 respectively and the same measurement results appear at nodes *DM0AV*, *DM1AV*, *DM2AV* and *DM3AV*.

$$DM0AV = DM2AV = \frac{V_{DD}}{2T}(t_d + t_{I1} + t_{S1} + t_{I2} + t_{S2} + t_{I3} \\ + \alpha T + (1 - \alpha)T - (t_{I1} + t_{S1} + t_{I2} + t_{S2} + t_{I3})) = \frac{V_{DD}\,(T + t_d)}{2T} \tag{5.14}$$

$$DM1AV = DM3AV = \frac{V_{DD}}{2T}(t_{I1} + t_{S1} + t_{I2} + t_{S2} + t_{I3} + t_{I4} \\ + T - (t_d + t_{I1} + t_{S1} + t_{I2} + t_{S2} + t_{I3} + t_{I4})) = \frac{V_{DD}\,(T - t_d)}{2T} \tag{5.15}$$

If the rectangular input signal *CINP* switches not symmetrically around *CINN* (no offset of the comparator is assumed), then the measurement result from *DM0AV* and *DM1AV* differs from that of *DM2AV* and *DM3AV* respectively, because the comparator has to handle the amount of two different input voltage differences and

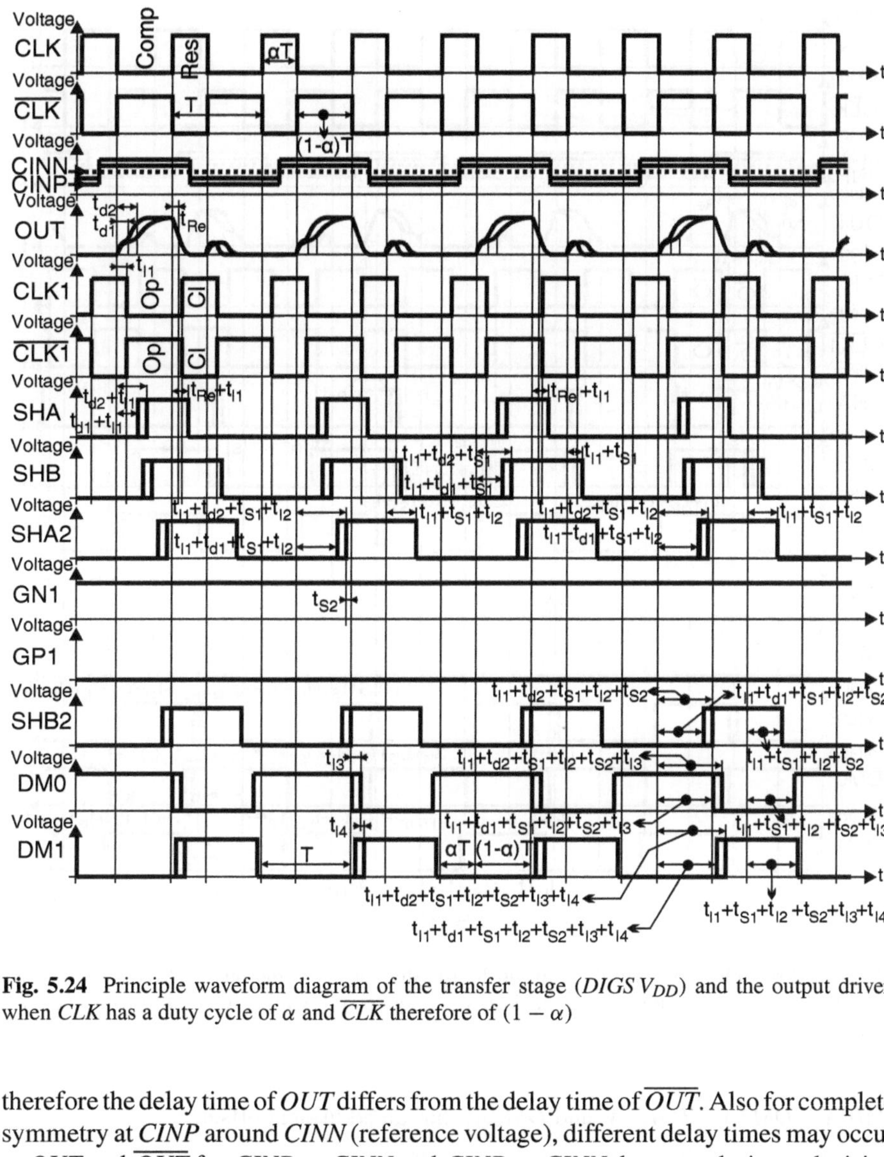

Fig. 5.24 Principle waveform diagram of the transfer stage (*DIGS* V_{DD}) and the output driver, when *CLK* has a duty cycle of α and $\overline{CLK}$ therefore of $(1-\alpha)$

therefore the delay time of *OUT* differs from the delay time of $\overline{OUT}$. Also for complete symmetry at *CINP* around *CINN* (reference voltage), different delay times may occur at *OUT* and $\overline{OUT}$ for $CINP > CINN$ and $CINP < CINN$, because during a decision of the comparator the mean input voltage $(CINP + CINN)/2$ is different for $CINP > CINN$ and $CINP < CINN$ because *CINN* is a fixed reference voltage.

The delay time measurement technique suffers in principle on the difference Δt of the delay time of the buffer between nodes *CLK* and *CLK*1 or $\overline{CLK}$ and $\overline{CLK1}$ and the buffer between nodes *OUT* and *SHA* (or $\overline{OUT}$ and $\overline{SHA}$). This can be seen schematically in Fig. 5.25, where signal *CLK*1 and $\overline{CLK1}$ is delayed with $t_{I1} + \Delta t$, while signal *SHA* is only delayed with t_{I1} in reference to signals *CLK* and $\overline{CLK}$. Then 5.11 and 5.12 change to 5.16 and 5.17 and each measurement result of nodes

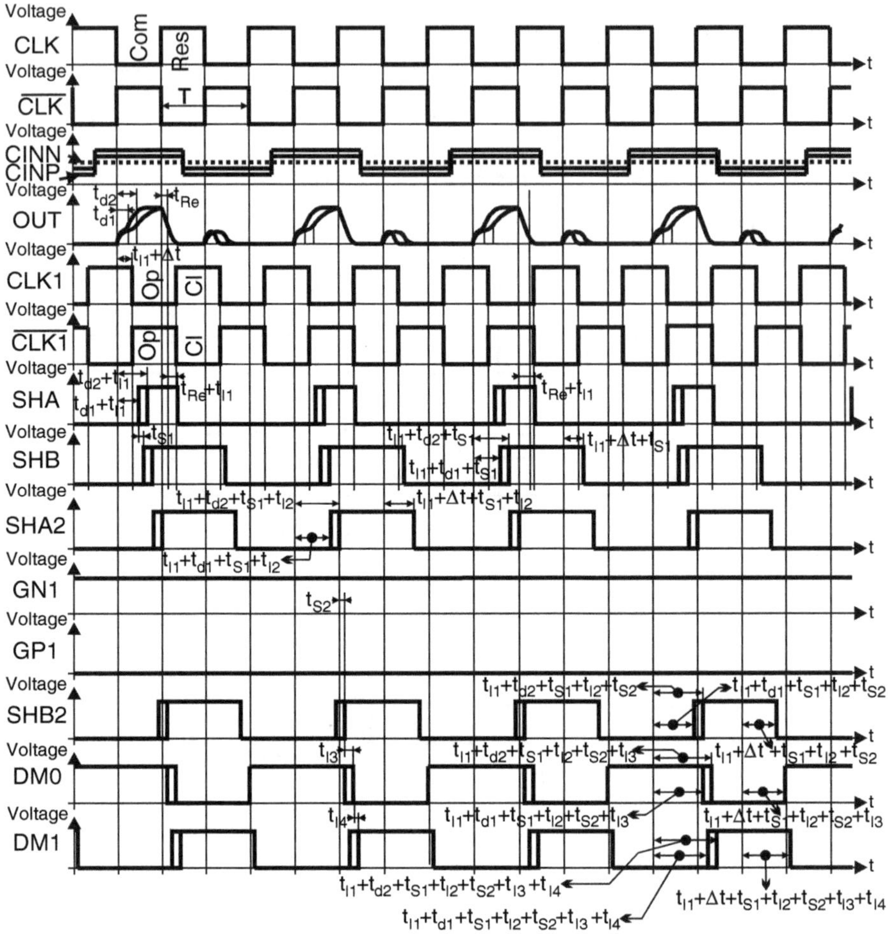

Fig. 5.25 Principle waveform diagram of the transfer stage and the output driver, when *CLK*1 and $\overline{CLK1}$ are delayed with $t_{I1} + \Delta t$ and *SHA* is only delayed with t_I referred to *CLK* and $\overline{CLK}$ ($DIGS = V_{DD}$)

*DM*0*AV*, *DM*1*AV*, *DM*2*AV* and *DM*3*AV* has an additional offset.

$$DM0AV = DM2AV = \frac{V_{DD}}{2T}(t_d + t_{I1} + t_{S1} + t_{I2} + t_{S2} + t_{I3} + T - (t_{I1} + \Delta t + t_{S1} + t_{I2} + t_{S2} + t_{I3})) = \frac{V_{DD}\,(T + t_d - \Delta t)}{2T} \tag{5.16}$$

$$DM1AV = DM3AV = \frac{V_{DD}}{2T}(t_{I1} + \Delta t + t_{S1} + t_{I2} + t_{S2} + t_{I3} + t_{I4} + T - (t_d + t_{I1} + t_{S1} + t_{I2} + t_{S2} + t_{I3} + t_{I4})) = \frac{V_{DD}\,(T - t_d + \Delta t)}{2T} \tag{5.17}$$

The occurring additional offset, which disturbs the delay time measurement, can be seen in 5.18 and is Δt.

$$t_d = \frac{\Delta V_{DM}}{V_{DD}} T + \Delta t, \quad \Delta V_{DM} = DM0AV - DM1AV = DM2AV - DM3AV \tag{5.18}$$

The whole circuit diagram of the transfer stage is shown in Fig. 5.26 and the schematic of the output buffer is depicted in Fig. 5.27. The buffer in front of the first transmission gate consists of two inverters in series to be able to deliver appropriate current for driving the parasitic capacitances at nodes *SHA* or $\overline{SHA}$ of transistors $N0$ and $P0$, which has to be designed in a sufficient size to reduce the open-resistance to be therefore able to work at even higher clock and data rates. An additional feature is that the first buffer (digital buffer 1) also blocks switching artefacts from the first transmission gate, which are coupled back to the output node *OUT* (or to $\overline{OUT}$ in the case of the other transfer stage) and influence the decision of the comparator. After a chain of six inverters the second transmission gate $N1$, $P1$ is driven. This transmission gate is controlled with the digital input pin *DIGS*.

If *DIGS* V_{DD} = *high*, then transmission gates $N2$, $P2$ and $N3$, $P3$ are switched off, node $GN1$ is pulled to V_{DD} by transistor $P32$ and node $GP1$ is pulled to V_{SS} by transistor $N33$. So transmission gate $N1$, $P1$ is switched always on and the delay time measurement of the comparator can be performed. If *DIGS* V_{DD} = *low*, then transistors $P32$ and $N33$ are switched off, node *CLK2* is connected to $GN1$ and node $\overline{CLK2}$ is connected to $GP1$ by turning on transmission gates $N2$, $P2$ and $N3$, $P3$. So the data stream from the comparator has a fixed delay time in reference to *CLK* and $\overline{CLK}$ as explained above. Transistors $N32$ and $P33$ are added to have somewhat a similar parasitic capacitance as at node *SHB*. The closed transmission gates with transistors $N4$, $P4$, $N5$, $P5$, $N6$ and $P6$ are added to have nearly the same parasitic capacitance and overall delay time as in the 4:1 multiplexer, which is explained in Sect. 5.4.

The output driver (see Fig. 5.27) consists of simply a chain of inverters, where the output is able to drive 50 Ω. It should be mentioned, that all the chains of inverters, which were designed, are not optimized to have a minimal overall delay time. Each inverter of a chain was designed to have a sufficient small rise and fall time, as it is the case when the following inverter is designed with a lower increase in size than the size, which is necessary for optimal overall delay time. This contributes less load capacitor to the output of the previous inverter so that even higher clock frequencies and data rates with CMOS logic voltage levels are possible. Each node $DM0$ and $DM1$ is connected via a 500 kΩ resistor with the output pad $DM0AV$ and $DM1AV$ respectively. This resistor value was chosen, that firstly nodes $DM0$ and $DM1$ are

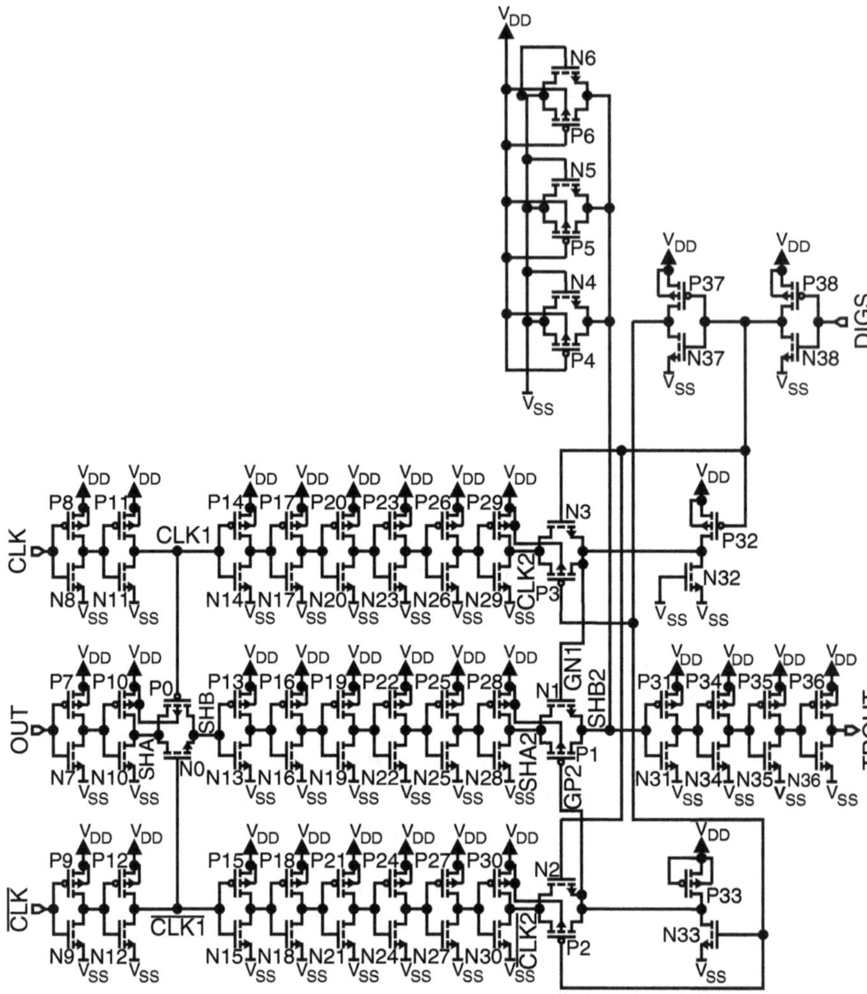

Fig. 5.26 The circuit of a transfer stage

not influenced and secondly that the average voltage at *DM0AV* or *DM1AV* can be measured with negligible error with a digital multimeter (see Sect. 5.5), which had an input resistance of 10 GΩ in voltage measurement mode. The same considerations are also valid for nodes *DM2*, *DM3*, resistors R_2, R_3 and output nodes *DM2AV* and *DM3AV*. Simulations of different waveforms are shown in Fig. 5.28, where *DIGS* V_{DD} for delay time measurement of the comparator and in Fig. 5.29, where *DIGS* V_{SS} for a fixed delay time of the data stream from the comparator. The rise and fall times of the applied test signals *CLK*, $\overline{CLK}$ and *OUT* are assumed to be 40 ps. Monte Carlo simulations of 50 samples are shown in Fig. 5.30 for clock frequencies of 250 MHz, 1 and 2 GHz, where the mean value of the voltage difference

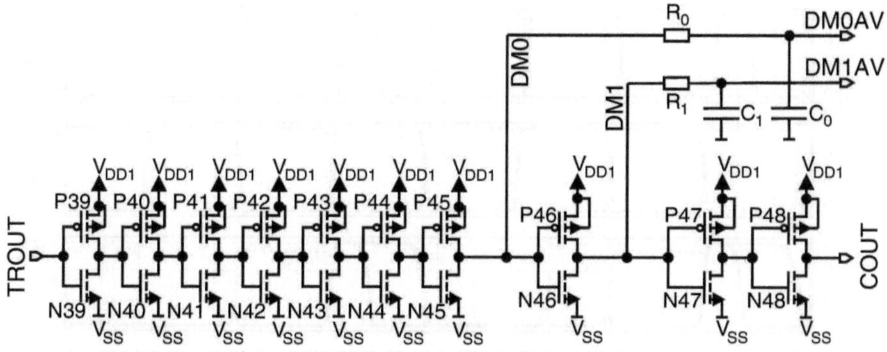

Fig. 5.27 The circuit of an output driver

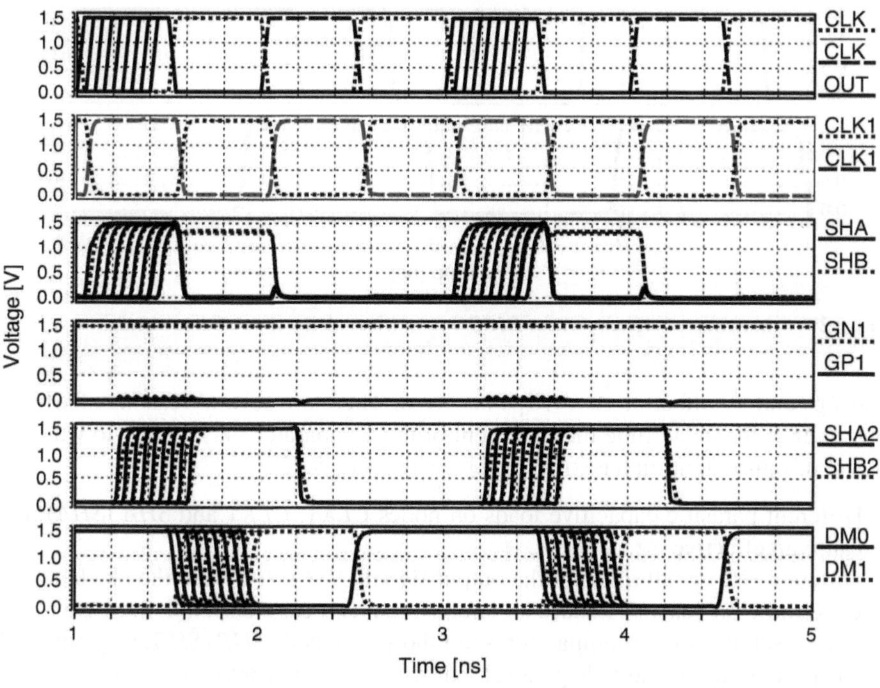

Fig. 5.28 Simulation of signals, where *DIGS* V_{DD} (delay time measurement)

$\mu(\Delta V_{DM}) = \mu(DM0AV - DM1AV)$ versus a pre-defined delay time t_d (real delay time) between the input signals *OUT* and *CLK* ($\overline{CLK}$) of the transfer stage is shown (see also waveforms in Fig. 5.28). Below the calculated mean value of the simulated measurement result of the delay time t_d, where 5.13 was used for the calculation of the simulated mean measurement result $\mu(t_d)$ out of $\mu(\Delta V_{DM})$ with $V_{DD} = 1.5\,\text{V}$. The standard deviation of the simulated measurement result $\sigma(t_d)$ varied between 6

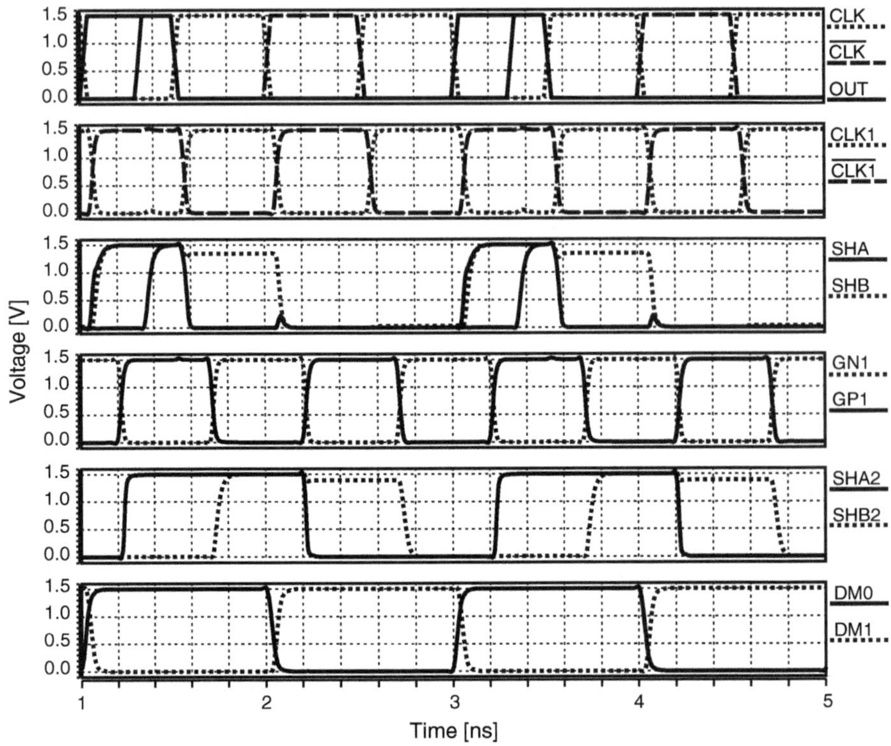

Fig. 5.29 Simulation of signals, where $DIGS = V_{SS}$ (fixed delay time of data stream)

and 7 ps and an offset of $\Delta t \approx 18$ ps occurred. There are several reasons of why an offset Δt at the delay time measurement of the comparator may appear to which in future designs attention has to be paid (see also Fig. 5.22):

- Different parasitic capacitive loads on nodes $CLK1$, $\overline{CLK1}$ and SHA ($\overline{SHA}$) may cause an effective Δt.
- Different rise or fall times of the output signal OUT ($\overline{OUT}$) and the clock signals CLK and $\overline{CLK}$ may also cause an effective Δt as it can be seen in Sect. 2.6.
- If the reset time of the comparator is too short, then node SHB ($\overline{SHB}$) may not reach an appropriate logical voltage level, but which will be corrected by the following inverter chain. This may also cause an offset Δt in the resulting data stream after the first transmission gate $N0$, $P0$.

The advantages of this delay time measurement technique are, that it shows a linear characteristic, where the slope of the measured t_d is only slightly influenced by device mismatch and that the delay time can be determined by only measuring DC voltages with a multimeter with sufficiently high input resistance. So using a highly sophisticated oscilloscope to compare the outputs of the test chip (nodes $COUT$ and $\overline{COUT}$) with a reference output $CREF$ is avoided, where additionally matching

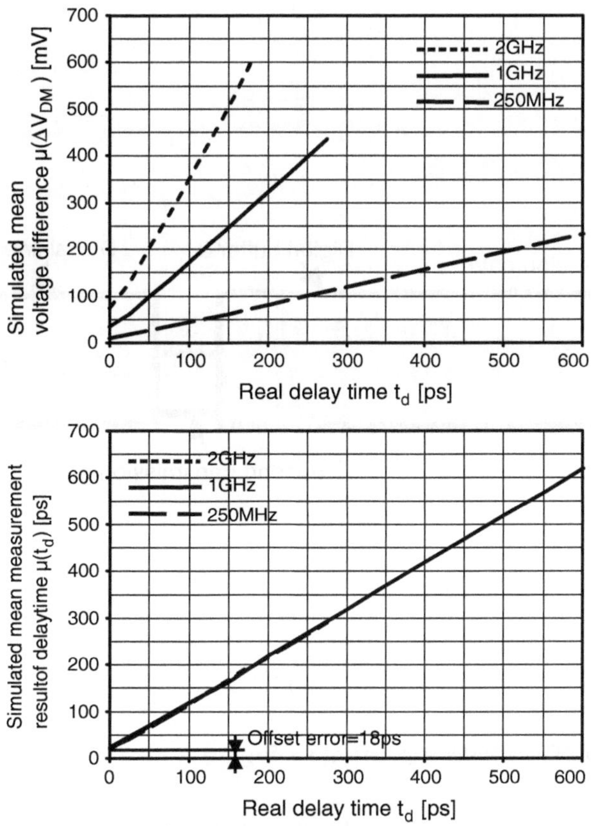

Fig. 5.30 Mean values of ΔV_{DM} and the simulated mean measurement result of delay time t_d versus a true, applied delay time t_d between *OUT* and *CLK* ($\overline{CLK}$) out of a Monte Carlo simulation of 50 samples (see also Fig. 5.28). The standard deviation of the simulated measurement result $\sigma(t_d)$ varied between 6 and 7 ps. Furthermore an offset of $\Delta t \approx 18$ ps occurred

considerations of the output paths, deskewing of different run times through the measurement cables, different values of bond wires and different load capacitances of the output drivers have to be taken into account. Different lengths of bond wires causes different series inductances and a mismatch of the output pads causes different load capacitances for the output drivers, which as a consequence influences delay time measurements, if the delay time of the comparator is extracted from the high frequency output signals of a test chip. The linear characteristics of measuring t_d with DC voltages might be also used for e.g. different types of regulation loops.

The layout of the transfer stage is shown in Fig. 5.31, where additionally some subparts are marked with a rectangular. The overall size of the transfer stage is $58 \times 56\,\mu m^2$. The layout of the output driver with additional arrays of nFET-in-nwell capacitor (see Chap. 4) to block the power supply lines are depicted in Fig. 5.32. The

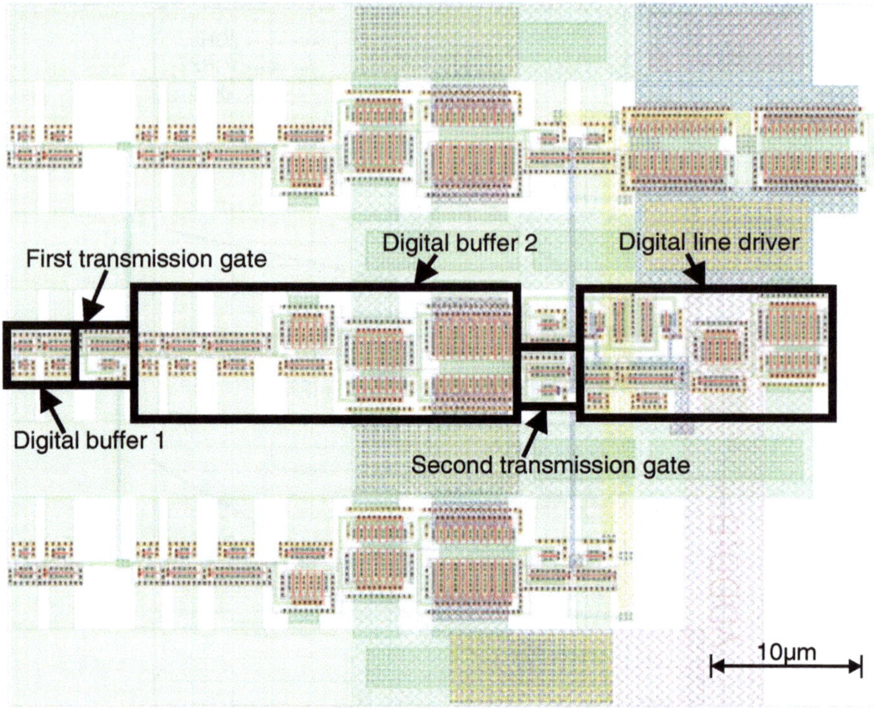

Fig. 5.31 Layout of the transfer stage (size: $58 \times 56\,\mu\text{m}^2$)

output driver with the capacitors and with including the pads for pins *DM0AV* and *DM1AV* needs an area of $212 \times 259\,\mu\text{m}^2$.

A similar solution in 65 nm CMOS for the combination of a transfer stage with an output driver and with extension for delay time measurement or fixed overall delay time is depicted in Fig. 5.33. It works in the same way like the 0.12 µm CMOS circuit described above in Fig. 5.22. The comparator to be tested is placed into the area with supply voltage level V_{Co}. So the comparator can be measured at supply voltages of below the nominal 1.2 V. Because of the transfer stage and the output driver only works properly at a supply voltage of 1.2 V, they have separate supplies, e.g. $V_{DD} = 1.2\,\text{V}$ for the transfer stage. Therfore voltage level adapters have been added. Transmission gate P6, N10 as well as P7, N11 work in a very similar way as already described in Fig. 5.22. The overall schematic of transfer stage in 65 nm CMOS can be seen in Fig. 5.34. The adapters between V_{Co} and V_{DD} are realised by two inverters connected in series (e.g. P14, N14 and P15, N15), where the second inverter (P15, N15) is supplied with V_{Co2}, a voltage level between V_{Co} and V_{DD}. The schematic of an output driver in 65 nm CMOS is depicted in Fig. 5.35.

The schematics of a transfer stage and an output driver are very similar to that shown in Figs. 5.26 and 5.27, respectively. Transient simulations have been done with a typical clocked comparator in 65 nm CMOS, where each output node (*OUT*

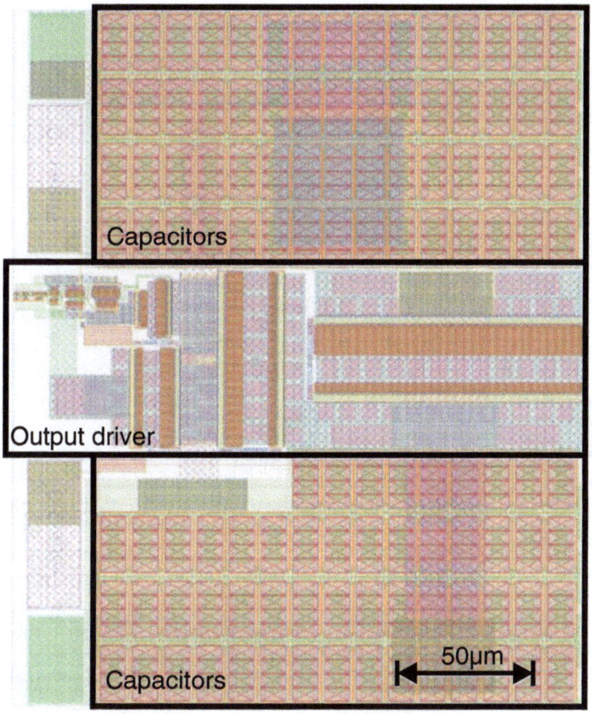

Fig. 5.32 Layout of the output driver with nFET-in-nwell capacitors to block the power supply lines (size: 212 × 259 μm^2)

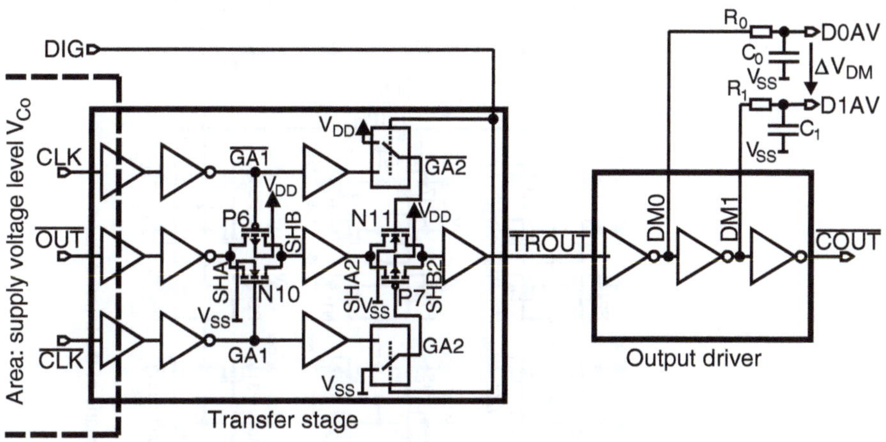

Fig. 5.33 Block diagram of a transfer stage in 65 nm CMOS technology with its output driver

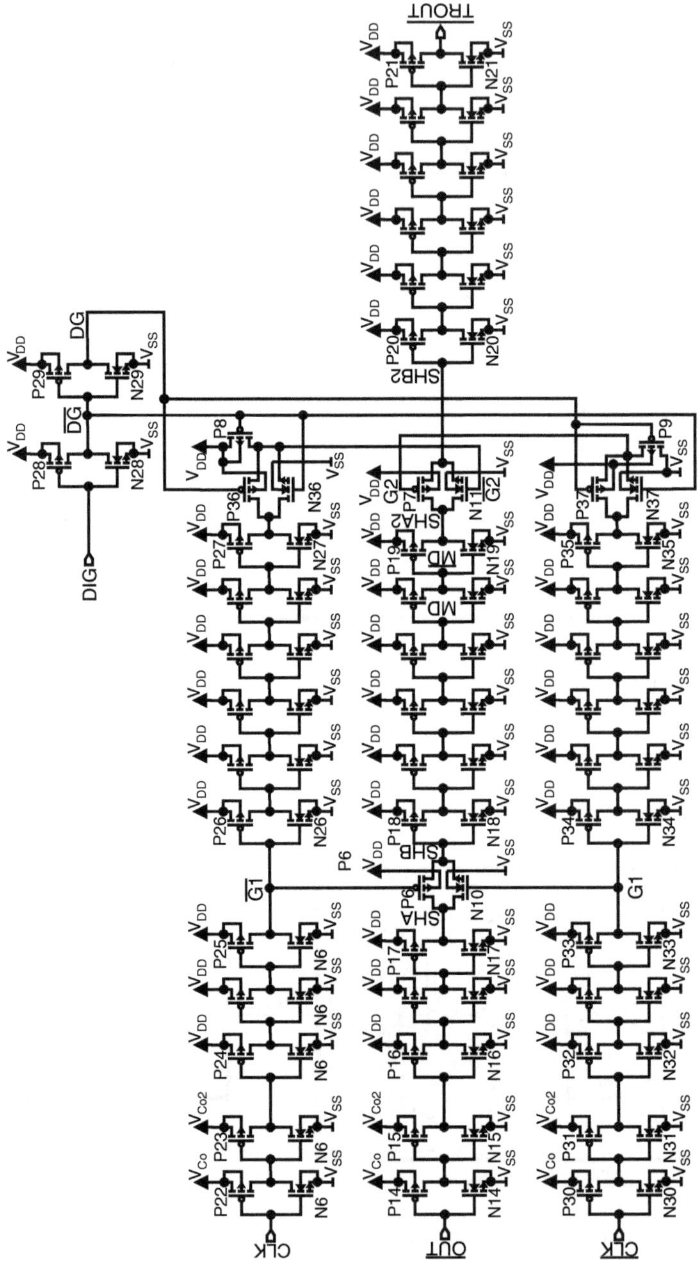

Fig. 5.34 Schematic of the transfer stage

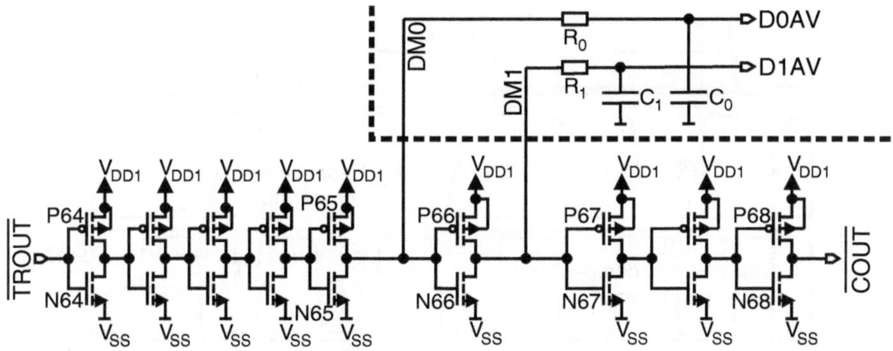

Fig. 5.35 Schematic of an output driver

and $\overline{OUT}$) are connected to a transfer stage with an following output driver. The main focus of the simulations is given to the transfer stage and the output driver, which are connected to $\overline{OUT}$ as depicted in Fig. 5.33. Figure 5.36 shows the case, where the digital pin *DIG* is set to V_{DD}. Here the delay time t_d of the comparator can be determined by measuring the mean voltage difference ΔV_{DM} of nodes $DM0$ and $DM1$ at nodes *D0AV* and *D1AV* as already discussed. For BER measurements often an overall constant delay time of the chip is advantageous. This option is turned on when setting node *DIG* to V_{SS}. The transient simulation for this case are shown in Fig. 5.37. To verify the a delay time measurement of the comparator with option $DIG = V_{DD}$, a Monte-Carlo simulations have been done, where the result is depicted in Fig. 5.38. A measured delay time has an offset error of 6 ps and a standard deviation of $\sigma \leq 4$ ps. The layout plots of the transfer stage and the output driver are shown in Figs. 5.39 and 5.40, respectively. The transfer stage consumes an area of $25 \times 76\,\mu\text{m}^2$ and the output driver needs $50 \times 165\,\mu\text{m}^2$.

5.4 Clock Driver with Voltage-Controlled Delay Lines and a Reference Output

In this section the clock driver and the reference output, which is added to the test chip to be able to monitor the different internal clock signals, are described. Additionally a voltage-controlled delay line is implemented into the clock paths mainly to be able to adjust an optimal sampling time point of the comparator for comparing the input signals *CINP* and *CINN*, because a sufficient setup time of the input signals has to be ensured. The sampling time point is determined by the edge of the clock signal (mostly the reset signal) at which comparison of the comparator starts referred to e.g. an edge of a bit in a synchronous input data signal. The block diagram of this functional block, which was implemented in newer test chips for comparators is depicted in Fig. 5.41. The input stage generates a coarse non-inverted, digital clock

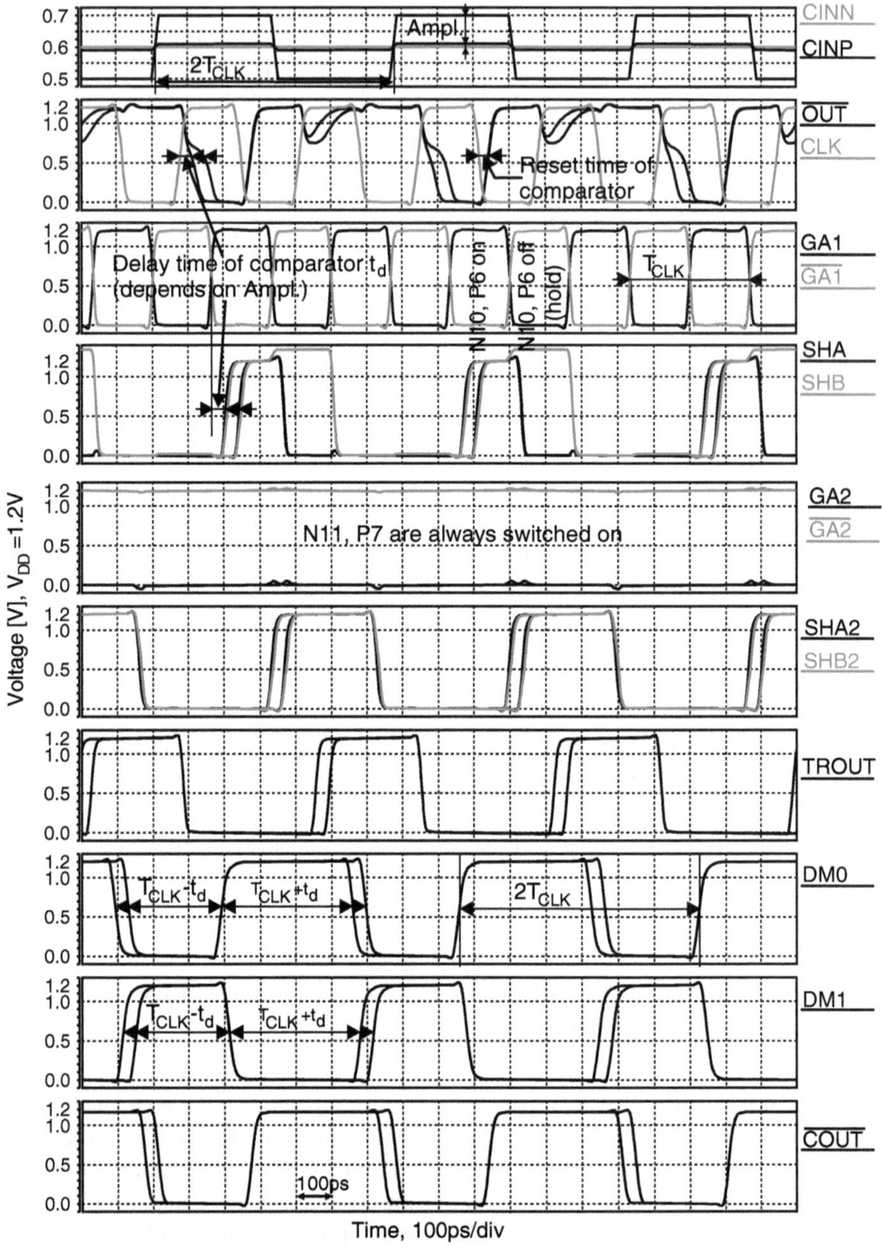

Fig. 5.36 Transient simulations of a clocked comparator (*CINP*, *CINN* are the input voltages to be compared, $\overline{OUT}$ is the inverted output of the comparator) connected to the transfer stage with an output driver. To illustrate the effect of different delay times of the comparator (t_d), a data stream with two different amplitudes is applied to *CINP*. Transmission gate N11, P7 is always switched on when $DIG = V_{DD}$ is applied. So a delay can be observed at the output of the chip $\overline{COUT}$

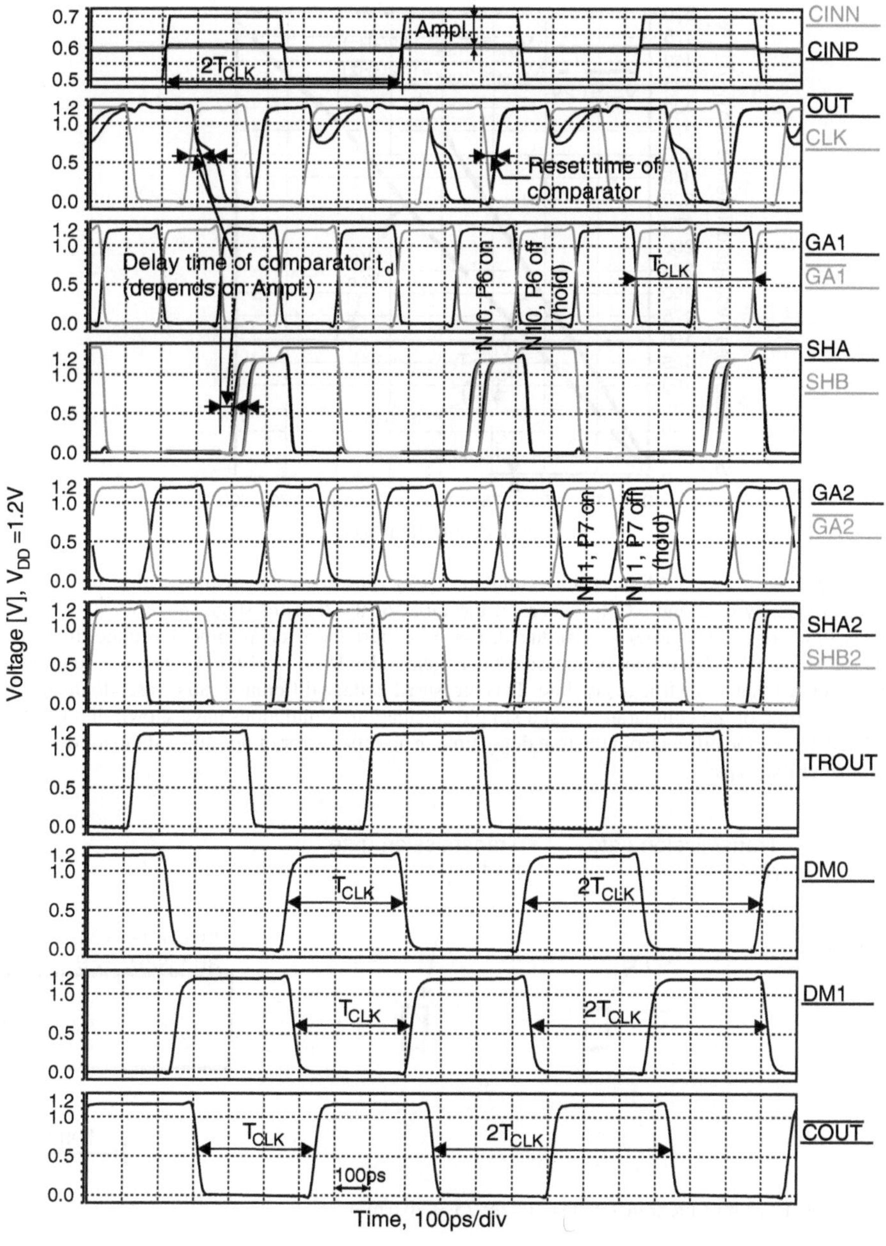

Fig. 5.37 Transient simulations of a clocked comparator (*CINP*, *CINN* are the input voltages to be compared, $\overline{OUT}$ is the inverted output of the comparator) connected to the transfer stage with an output driver. To illustrate the effect of different delay times of the comparator, a data stream with two different amplitudes is applied to *CINP*. Transmission gate N11, P7 is clocked as illustrated when $DIG = V_{SS}$ is applied. So the overall delay time of the chip is held constant at the output of the chip $\overline{COUT}$

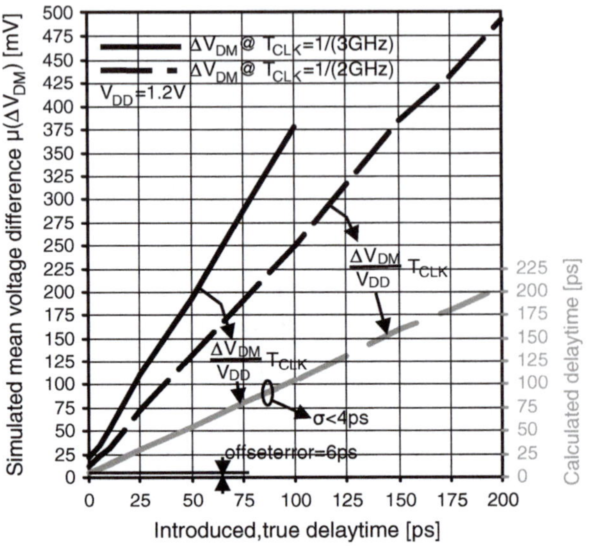

Fig. 5.38 Result of a Monte-Carlo simulation (50 runs for each distinct delay time) for a clock frequency of 2 and 3 GHz, where at the input of the transfer stage (node $\overline{OUT}$) a known delay time (true delay time) is introduced. For each introduced delay time a voltage difference $\Delta V_{DM} = D0AV - D1AV$ can be measured between nodes *D0AV* and *D1AV*, which is in an ideal case proportional to the true delay time. This measured voltage difference can be used to determine the delay time of the comparator with a simple calculation. Simulations have shown that this calculated delay time differs from the true delay time in an offset error of 6ps and a standard deviation of smaller than 4 ps

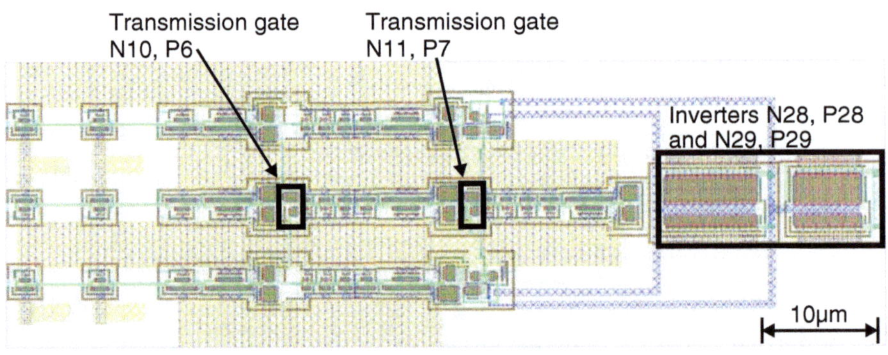

Fig. 5.39 Layout of the transfer stage (size: 25 × 76 μm^2)

DCLK and the corresponding inverted, digital clock $\overline{DCLK}$ out of a biased sine wave, which is applied to the input pad *CLKIN*. The dummy of the frequency divider and the divider itself are designed to have nearly the same overall delay time. The frequency divider generates a non-inverted and inverted clock signal *HCLK* and $\overline{HCLK}$ with the half frequency of signals *DCLK* and $\overline{DCLK}$. The frequency divider has been

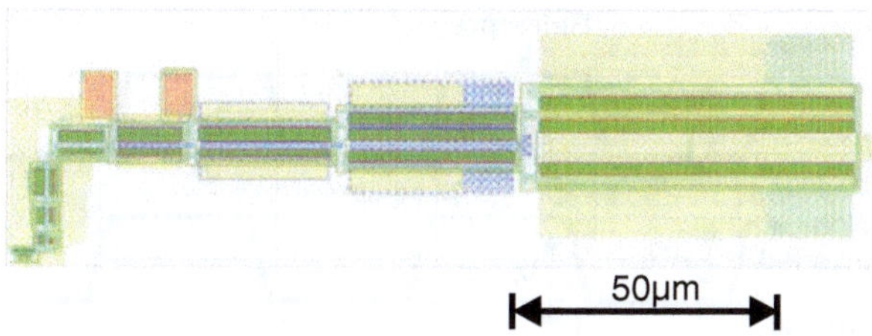

Fig. 5.40 Layout of the output driver (size: $50 \times 165\,\mu m^2$)

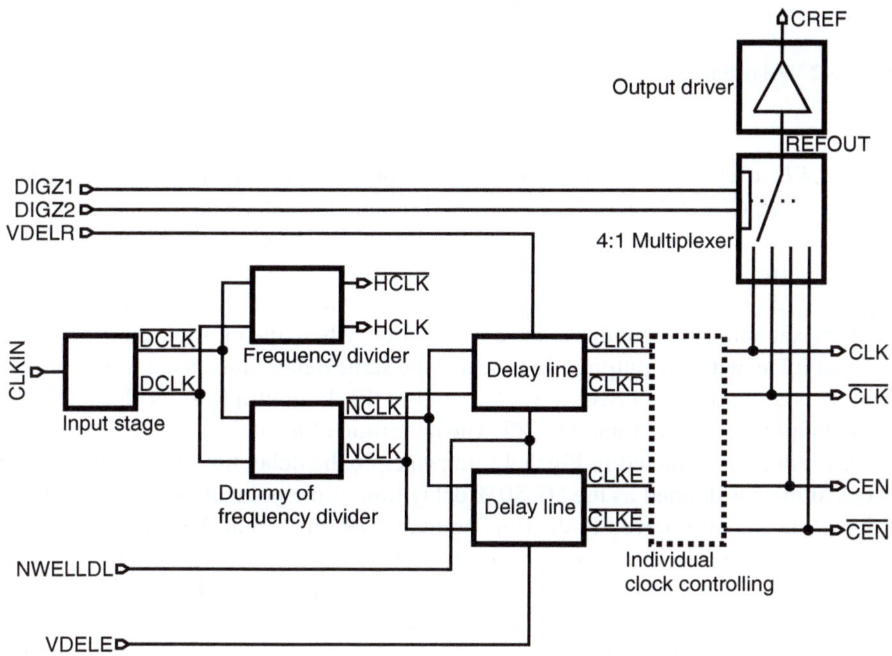

Fig. 5.41 Block diagram of the functional block, which delivers clock signals to the comparator

added to have the possibility of additional future on-chip measurement options of a comparator, e.g. investigating turning on and of at the maximum possible frequency if a power down function has been added to the comparator. The dummy of the frequency divider has two functions. First of all it reduces a probable appeared delay time mismatch between signals *DCLK* and $\overline{DCLK}$ and secondly it has always nearly the same delay time as the frequency divider. The output clock signals *NCLK* and $\overline{NCLK}$ therefore have the same frequency as signals *DCLK* and $\overline{DCLK}$, but their overlapping and delay time difference have been reduced significantly with the help of the dummy frequency divider. Clock signals *NCLK* and $\overline{NCLK}$ are applied to two voltage-controlled delay lines, where with bias voltages at the input pads *VDELR*, *VDELE* and *NWELLDL* the delay time difference between clock pairs *CLKR*, $\overline{CLKR}$

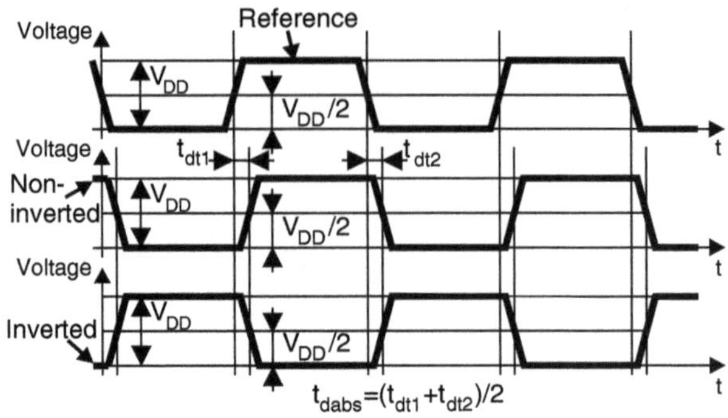

Fig. 5.42 Delay time definition

and *CLKE*, $\overline{CLKE}$ can be adjusted. Furthermore the overall sampling time point, when *CLKR*, $\overline{CLKR}$, *CLKE* and $\overline{CLKE}$ shall sample the data, which is applied to the inputs of the comparator can be altered in a fine way. The circuitry of the individual clock controlling block depends on the implemented comparator, to which clock signals *CLK*, $\overline{CLK}$, *CEN* or $\overline{CEN}$ are applied to. To monitor these clock signals, a 4:1 multiplexer and an output driver have been added, where the multiplexer is designed so, that it has with the output driver nearly the same delay time as the transfer stage with the output driver of Sect. 5.3. The input of the 4:1 multiplexer is chosen by the digital input pins *DIGZ*1 and *DIGZ*2. The definition of the delay time, which is used in this section is depicted in Fig. 5.42. In principle the delay time t_{Iabs} between two clock signals is defined as the 50–50 % delay time (time point at 50 % of V_{DD} at the edge of the signal referred to the time point at 50 % of V_{DD} at the appropriate edge of the reference signal), where the trigger voltage is $V_{DD}/2$ and where the mean of the individual delay of the rising- and falling ramp (t_{dt1}, t_{dt2}) is built. In most cases $t_{Iabs} \approx t_{dt1} \approx t_{dt2}$ is valid.

The schematic of the input stage is depicted in Fig. 5.43. The sine wave with a distinct clock frequency, which is applied to the input pad *CLKIN*, is typically biased with $V_{DD}/2$. Changing this bias would change as a consequence the duty cycle of clock signal pairs *CLK*, $\overline{CLK}$ and *CEN*, $\overline{CEN}$. The buffer, which consists of the two inverters $N0$, $P0$ and $N1$, $P1$, converts the sine wave to a digital, rectangular clock. The non-inverted clock *DCLK* is generated with the help of additional four inverters and the inverted clock $\overline{DCLK}$ is built out of three additional inverters, where the open transmission gate $N3$, $P3$ compensates the delay time of inverter $N2$, $P2$.

The circuit of the frequency divider is shown in Fig. 5.44, where for the design a rail-to-rail output swing (CMOS logic voltage level) was intended. The core is a latch, which consists of the two cross-coupled inverters $N10$, $P10$ and $N11$, $P11$. Two inverter chains, where each consists of three inverters (transistors $N18$–$N23$ and $P18$–$P23$), are used to drive the lines *HCLK* and $\overline{HCLK}$, because the switching speed of the latch depends on the load capacitances at its cross-coupled nodes. Transistors $N12$–$N15$ and $P12$–$P15$ form four transmission gates. If *DCLK* is *low*, transistors

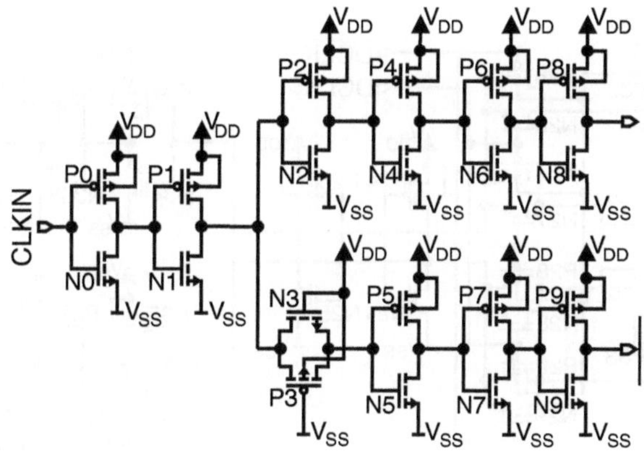

Fig. 5.43 The circuit of the input stage

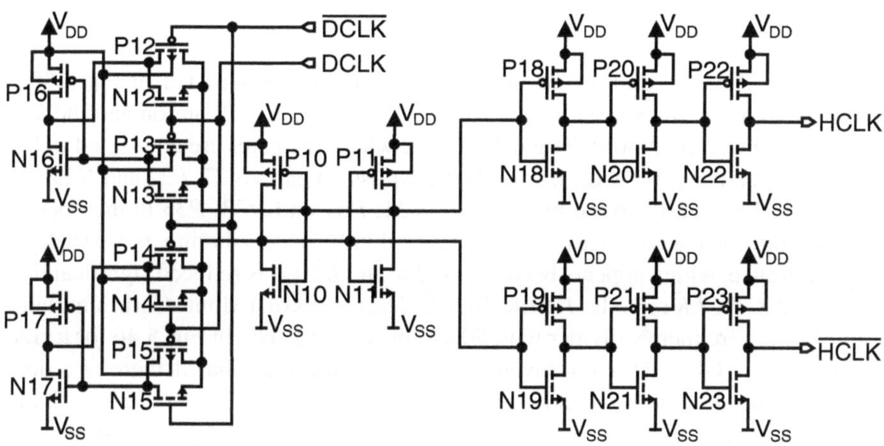

Fig. 5.44 The circuit of the frequency divider

$N12$, $P12$ and $N14$, $P14$ are switched off and $N13$, $P13$ and $N15$, $P15$ are switched on and both cross-coupled nodes of the latch are connected to the gates of inverters $N16$, $P16$ and $N17$, $P17$. If *DCLK* is *high*, transistors $N12$, $P12$ and $N14$, $P14$ are switched on and $N13$, $P13$ and $N15$, $P15$ are off and the previous state of the latch is stored dynamically at the gates of inverters $P16$, $N16$ and $P17$, $N17$. These inverters drive now the cross-coupled nodes of the latch, which changes state. If once again *DCLK* switches to *low*, the state of the latch hold because of $N12$, $P12$, $N14$, $P14$ are switched off and the transmission gates $N13$, $P13$ and $N14$, $P14$ connect the cross-coupled nodes to the gates of inverters $N16$, $P16$ and $N17$, $P17$ to the appropriate cross-coupled nodes of the latch. For such a frequency-divider structure it is easy to design a dummy structure, which toggles at the same frequency of the input clock and has nearly the same delay time.

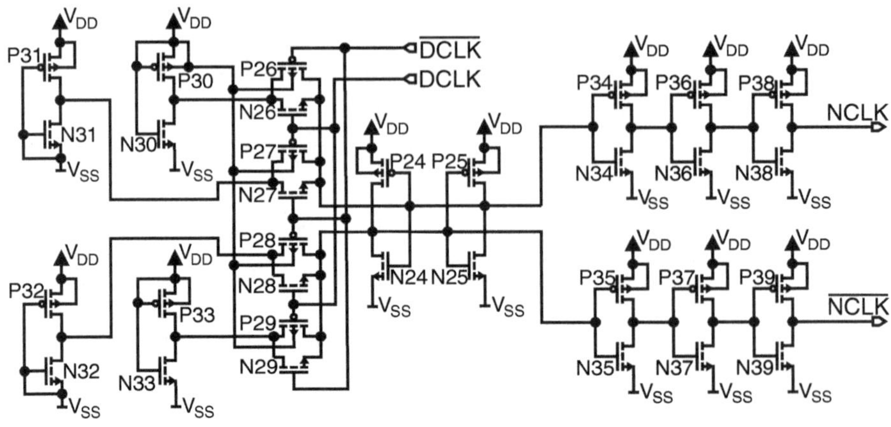

Fig. 5.45 The circuit of the dummy of the frequency divider

The circuit of the dummy of the frequency divider is depicted in Fig. 5.45. If *DCLK* is *high*, the transmission gates $N26$, $P26$ and $N28$, $P28$ are on and the latch is driven by inverters $N30$, $P30$ and $N32$, $P32$ where the gates are connected to V_{DD} and V_{SS} respectively. If *DCLK* is *low*, the transmission gates $N27$, $P27$ and $N29$, $P29$ are on and the latch is driven by inverters $N31$, $P31$ and $N33$, $P33$ in the opposite way. An additional advantage of the frequency divider and its dummy is, that a delay-time difference, which appears between *DCLK* and $\overline{DCLK}$ is reduced significantly at nodes *NCLK*, $\overline{NCLK}$ or nodes *HCLK*, $\overline{HCLK}$. The results of a Monte Carlo simulation at the dummy frequency divider with 50 samples can be seen in Fig. 5.46 for a clock frequency of 2 GHz. This improvement is due to the fact, that if there is a delay time difference or overlap regions between *DCLK* and $\overline{DCLK}$, there are time points, when *DCLK* and $\overline{DCLK}$ are both at V_{DD} or V_{SS}. Then all n-MOS transistors of the transmission gates are on in the case of V_{DD} or all p-MOS transistors are on in the case of V_{SS}. This only delays the switching of the latch, but due to the cross-coupling of two inverters to a latch, the delay time difference itself is reduced. Remarkable is also, that the standard deviations of the delay time difference between *DCLK* and $\overline{DCLK}$ and between *NCLK* and $\overline{NCLK}$ have nearly the same values of around 1.5 ps. The functionality of the frequency divider and the dummy of the frequency divider was verified at similar structures in [14].

The schematic of the voltage-controlled delay line with output signals *CLKR* and $\overline{CLKR}$ (controlled with bias voltage *VDELR*) is shown in Fig. 5.47. For the other delay line with outputs *CLKE* and $\overline{CLKE}$ (controlled with bias voltage *VDELE*) an identical schematic is used (see also block diagram in Fig. 5.41). A delay line consists of a series of delay cells, where one cell consists of an inverter and a p-MOS transistor, which is switched as a voltage-controlled capacitor in inversion mode (e.g. inverter $N40$, $P40$ with p-MOS transistor $P78$ used as inversion mode capacitor). Such voltage-controlled p-MOS capacitors are called p-MOS varactors, where the function is described in Chap. 4. The gate of such a p-MOS transistors is connected

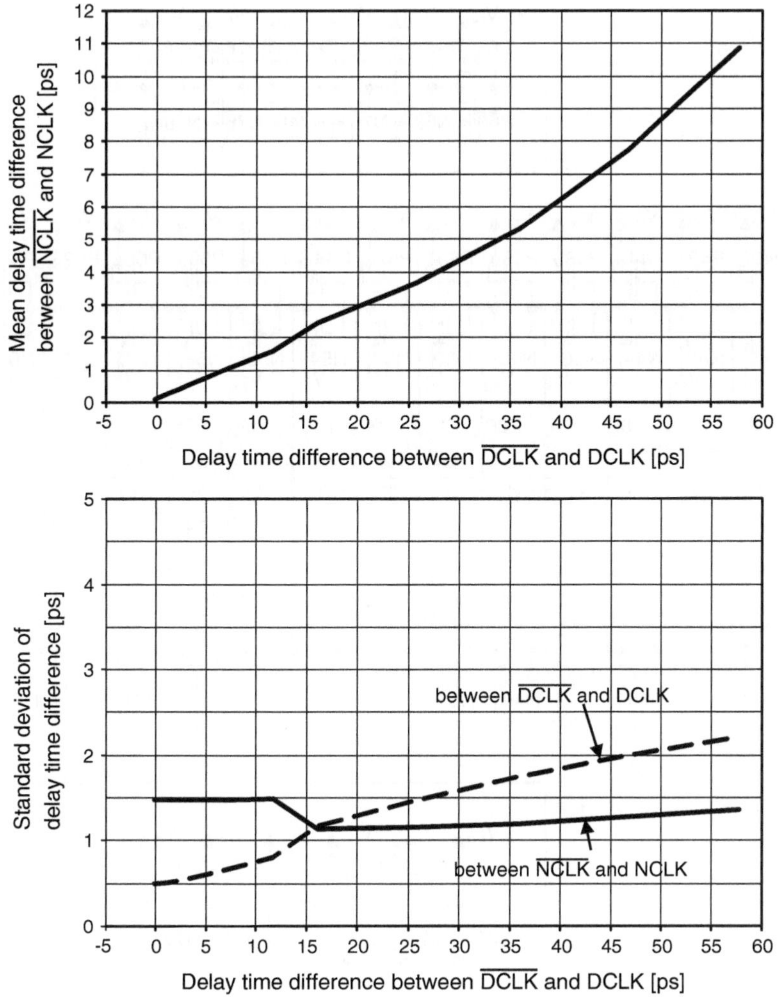

Fig. 5.46 Improvement of the delay time difference between *NCLK* and $\overline{NCLK}$ due to the dummy frequency divider in dependence on the delay time difference between *DCLK* and $\overline{DCLK}$ (Monte Carlo simulation of 50 samples at a 2 GHz clock frequency): To cause such a delay time difference, the gate lengths of transistors *N*3, *P*3 have been varied in simulation (see also Fig. 5.43)

to the output of an appropriate inverter, the drain and source are connected together and biased with a DC voltage at *VDELR* and the n-well is biased with pin *NWELLDL*, which is usually set to V_{DD}. The second delay line with outputs *CLKE* and $\overline{CLKE}$ is biased separately with a DC voltage at *VDELE*, which there has the same function as *VDLER* in the first delay line. The n-wells of both delay lines are connected together and biased with *NWELLDL*. To guarantee appropriate fast rise and fall times of the digital clock edges at an output of an inverter with a p-MOS varactor as load (so that

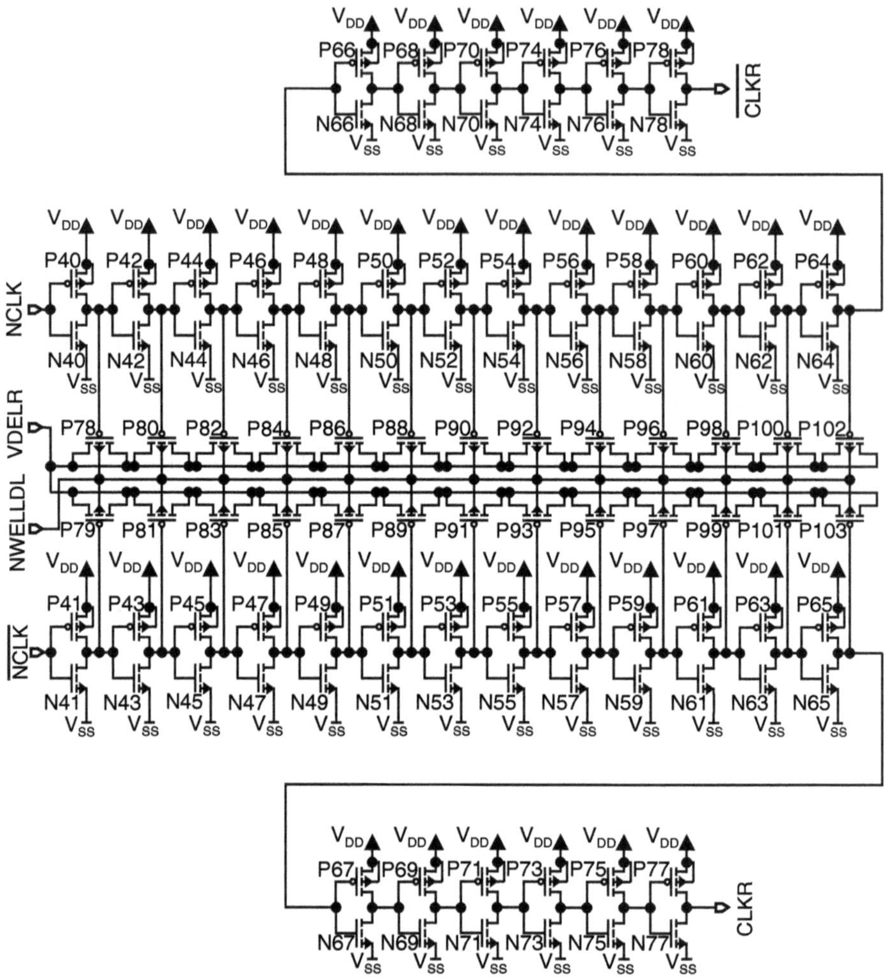

Fig. 5.47 Circuit of the voltage-controlled delay line with output signals *CLKR* and $\overline{CLKR}$

the overall delay line can be also used for high clock frequencies), 13 delay cells (a delay cell consists of an inverter with a p-MOS varactor as load capacitance) are connected in series to achieve a higher tuning range. So each p-MOS varactor can be designed smaller to reduce the load capacitance of the appropriate inverter. Hence the absolute delay time t_{Iabs} of the overall delay line is then larger, but this has no influence to the overall function of clocking a comparator. A sufficient high tuning range of the delay time between the clock signals (Δt_{Iabs}) and the input signals of a comparator is important to e.g. adjust the optimal sampling time delay between clock and input signals of the comparator. After the delay cells, one inverter chain of 6 inverters is added for each clock line to drive the parasitic capacitances of the clock lines, which are connected to the outputs *CLKR* and $\overline{CLKR}$ (*CLKE* and $\overline{CLKE}$ for the second delay line respectively).

A transient simulation of different clock signals of Fig. 5.41, where the bias voltage at pin *VDELE* = 0.6 V, at pin *NWELLDL* = 1.5 V and where *VDELR* is varied from 0 to 1.5 V is shown in Fig. 5.48. In Fig. 5.49 a Monte Carlo simulation (50 runs at a 2 GHz clock frequency) of the mean delay time difference $\mu(\Delta t_{Iabs})$ of clock signals *CLKR*, $\overline{CLKR}$, *CLKE* and $\overline{CLKE}$ related the absolute delay time t_{Iabs0}, which is the mean delay time for $\overline{CLKE}$ at *VDELE* = 0 V and *NWELLDL* = 1.5 V and where $\Delta t_{Iabs} = t_{Iabs} - t_{Iabs0}$, is shown. The standard deviation $\sigma(\Delta t_{Iabs})$ varied between 3 and 4 ps.

To monitor the clock signals *CLK*, $\overline{CLK}$, *CEN* and $\overline{CEN}$, a 4:1 multiplexer with an output driver, which has an identical circuit as the ones used after the transfer stages (see Fig. 5.27 in Sect. 5.3) to drive a 50 Ω measurement system (output pad *CREF*) is implemented. The multiplexer is designed, that it has together with the output driver nearly the same delay time as the transfer stage with the output driver of Sect. 5.3. The

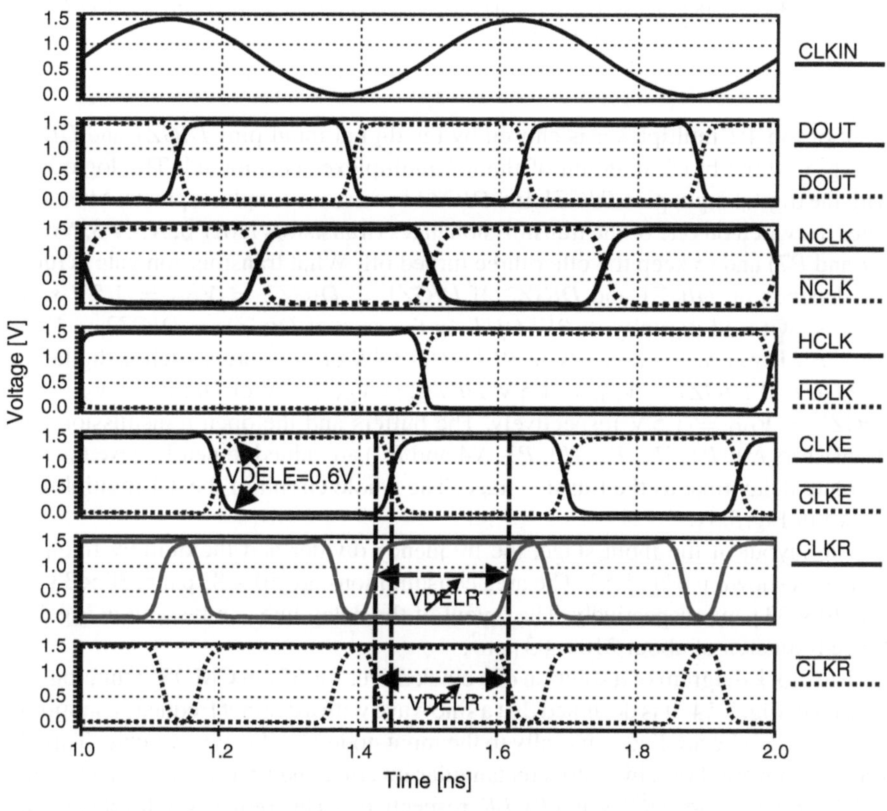

Fig. 5.48 Transient simulation of different clock signals of Fig. 5.41, where at *VDELE* = 0.6 V is applied, *NWELLDL* = 1.5V and *VDELR* is varied from 0 to 1.5 V

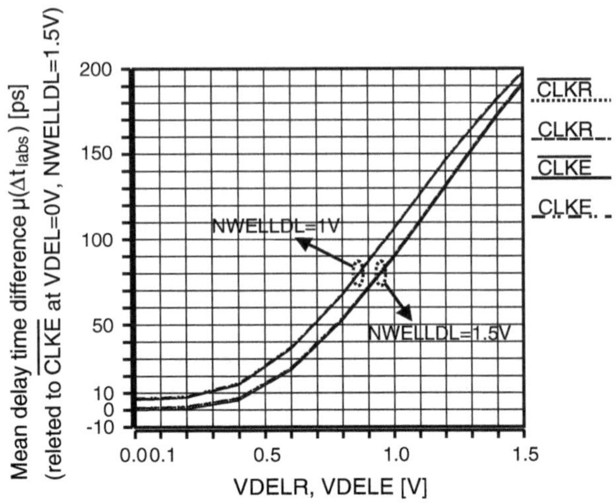

Fig. 5.49 Monte Carlo simulation (50 runs) of the mean delay time difference $\mu(\Delta t_{Iabs}) = \mu(t_{Iabs} - t_{Iabs0})$ of different clock signals related to the absolute delay time t_{Iabs0} for $\overline{CLKE}$ at $VDELE = 0\,\text{V}$ and $NWELLDL = 1.5\,\text{V}$; $\sigma(\Delta t_{Iabs}) \approx 3$–$4\,\text{ps}$

input of the 4:1 multiplexer is chosen by the digital input pins *DIGZ*1 and *DIGZ*2. In Fig. 5.50 the block diagram of the 4:1 multiplexer is depicted. The logic block with the digital input pins *DIGZ*1 and *DIGZ*2 consists of simple, typical CMOS logic gates to switch on one of the transmission gates (transistors *N*0, *P*0, *N*5, *P*5, *N*6, *P*6, *N*7 and *P*7) and to keep the other three turned off. What transmission gate is turned on, depends on *DIGZ*1 and *DIGZ*2. If $DIGZ1 = DIGZ2 = V_{DD} = 1.5\,\text{V}$ signal *CEN* is switched digitally to *REFOUT*. In the case of $DIGZ1 = DIGZ2 = V_{SS} = 0\,\text{V}$, $\overline{CEN}$ is redirected digitally to *REFOUT*. *CLK* or $\overline{CLK}$ are switched digitally to *REFOUT*, if $DIGZ1 = V_{DD} = 1.5\,\text{V}$, $DIGZ2 = V_{SS} = 0\,\text{V}$ or $DIGZ1 = V_{SS} = 0\,\text{V}$, $DIGZ2 = V_{DD} = 1.5\,\text{V}$ respectively. The buffers and the open transmission gates (transistors *N*1, *P*1, *N*2, *P*2, *N*3, *P*3, *N*4 and *P*4) are added to nearly have the same absolute delay time as the transfer stage. The whole circuit of the 4:1 multiplexer is shown in Fig. 5.51.

The layout of the input stage, the frequency divider and the dummy frequency divider is shown in Fig. 5.53. The area consumptions are $50 \times 83\,\mu\text{m}^2$, $30 \times 33\,\mu\text{m}^2$ and $30 \times 33\,\mu\text{m}^2$ respectively. The layout of the delay line can be seen in Fig. 5.52. It needs an area of about $210 \times 83\,\mu\text{m}^2$.

The block diagram of a solution of a clock driver for 65 nm CMOS technology can be seen in Fig. 5.54. It is designed that a sine wave with a distinct frequency, amplitude and offset can be applied externally to the input node *CLKIN* of a test chip. This sine wave is converted on-chip into a rectangular inverted and non-inverted clock, which is delivered at nodes $\overline{DCLK}$ and *DCLK*, respectively. The logical low-level is V_{SS} and the logical high-level is the supply voltage level of the comparator V_{Co}. A delay line is implemented to fine-adjust the time point when the comparator begins to decide.

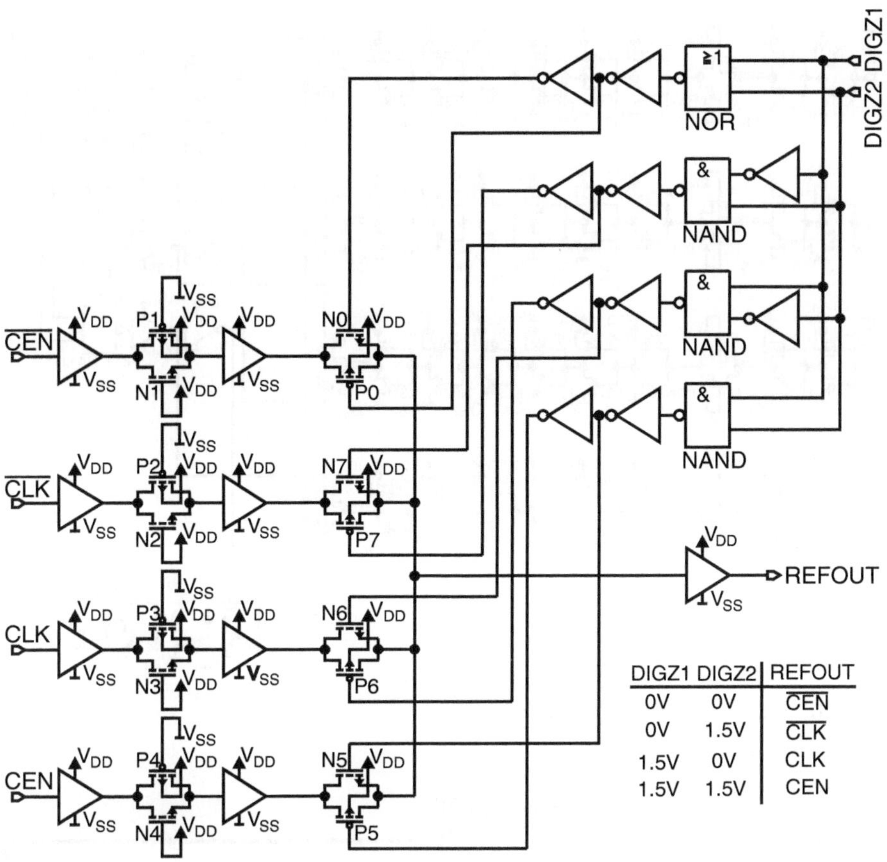

Fig. 5.50 Block diagram of the 4:1 multiplexer

This is useful for BER measurements to optimize the sampling point of data in a BER analyzer station. Cross coupled inverters (latches) are added to the clock driver to minimize the delay between *DCLK* and $\overline{DCLK}$. The duty cycle of the digital clock can be adjusted with the offset of the sine wave at node *CLKIN*. Transmission gate N40, P40 is added to compensate the delay of one inverter. A detailed schematic is shown in Fig. 5.55. The part with the delay line is depicted in the middle. The size of the capacitors (n-FET in n-well varactors) is controlled with a bias voltage at node *DEL* and thus the delay time. The last inverters, which drives nodes *DCLK* and $\overline{DCLK}$ are supplied with the supply voltage level V_{Co} of the comparator. The layout of the clock driver in 65 nm CMOS is shown in Fig. 5.56. It consumes a chip area of $130 \times 201\,\mu m^2$.

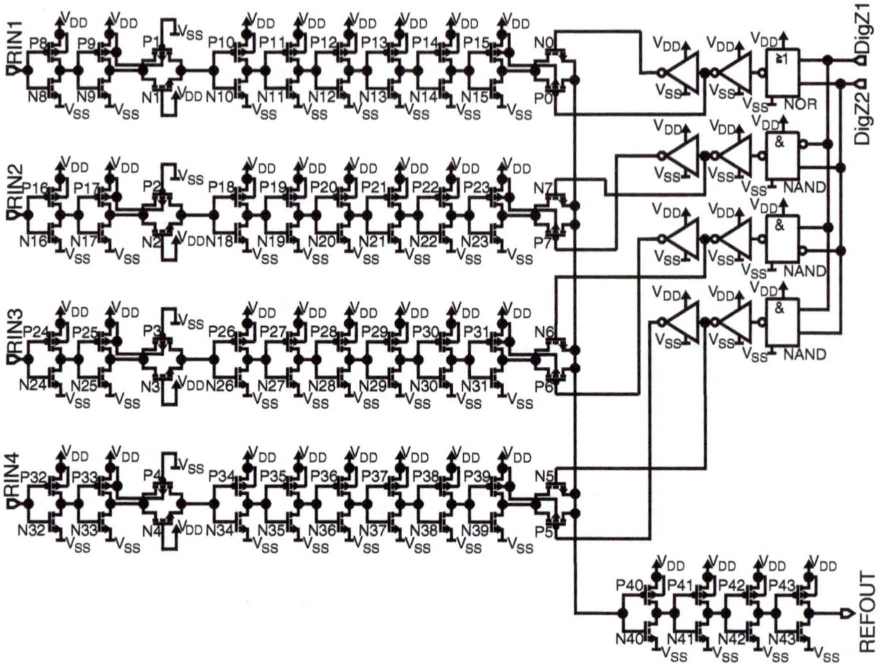

Fig. 5.51 Schematic of the 4:1 multiplexer

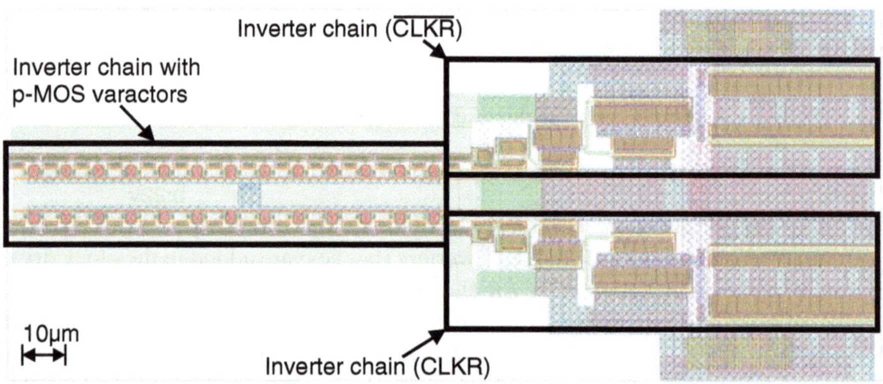

Fig. 5.52 Layout of the delay lines

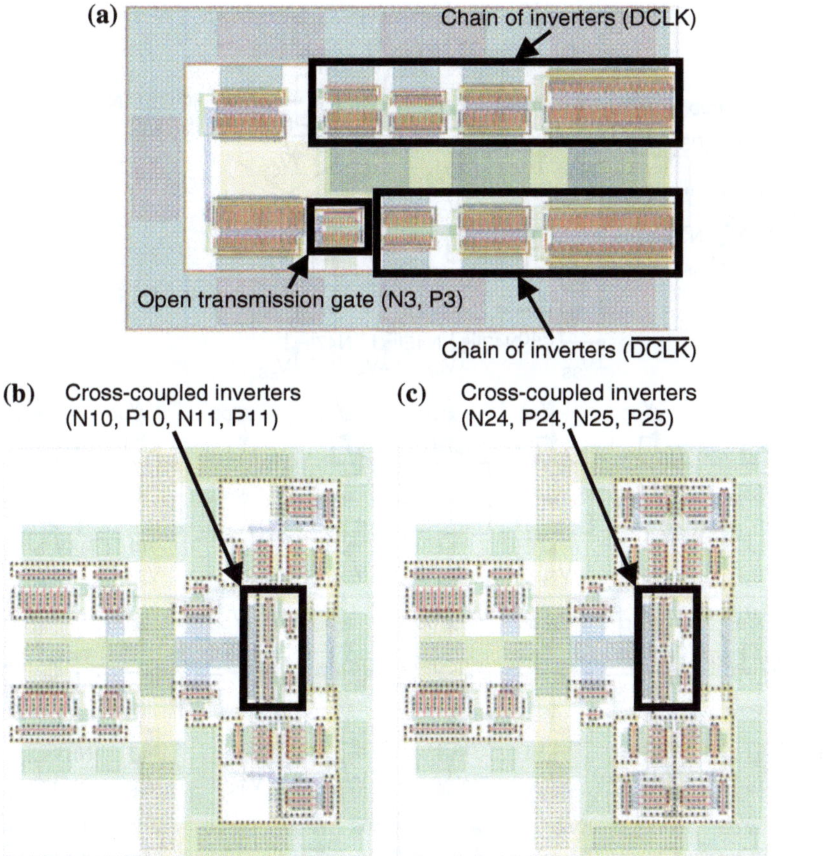

Fig. 5.53 Layouts of different blocks. **a** The input stage, **b** The frequency divider, **c** The dummy of the frequency divider

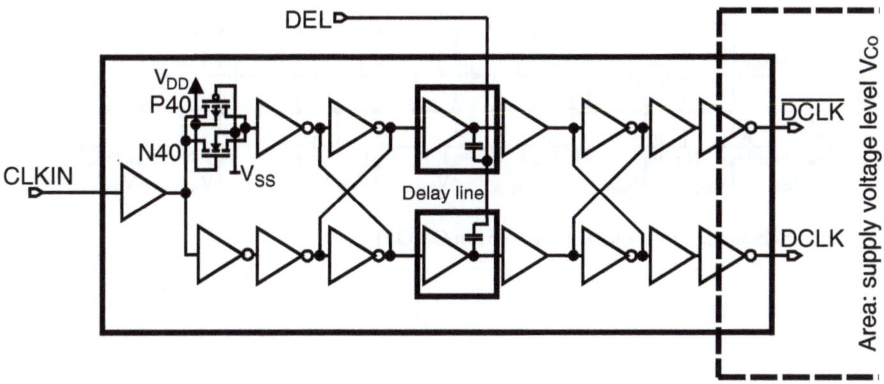

Fig. 5.54 Block diagram of the clock driver in 65 nm CMOS techology

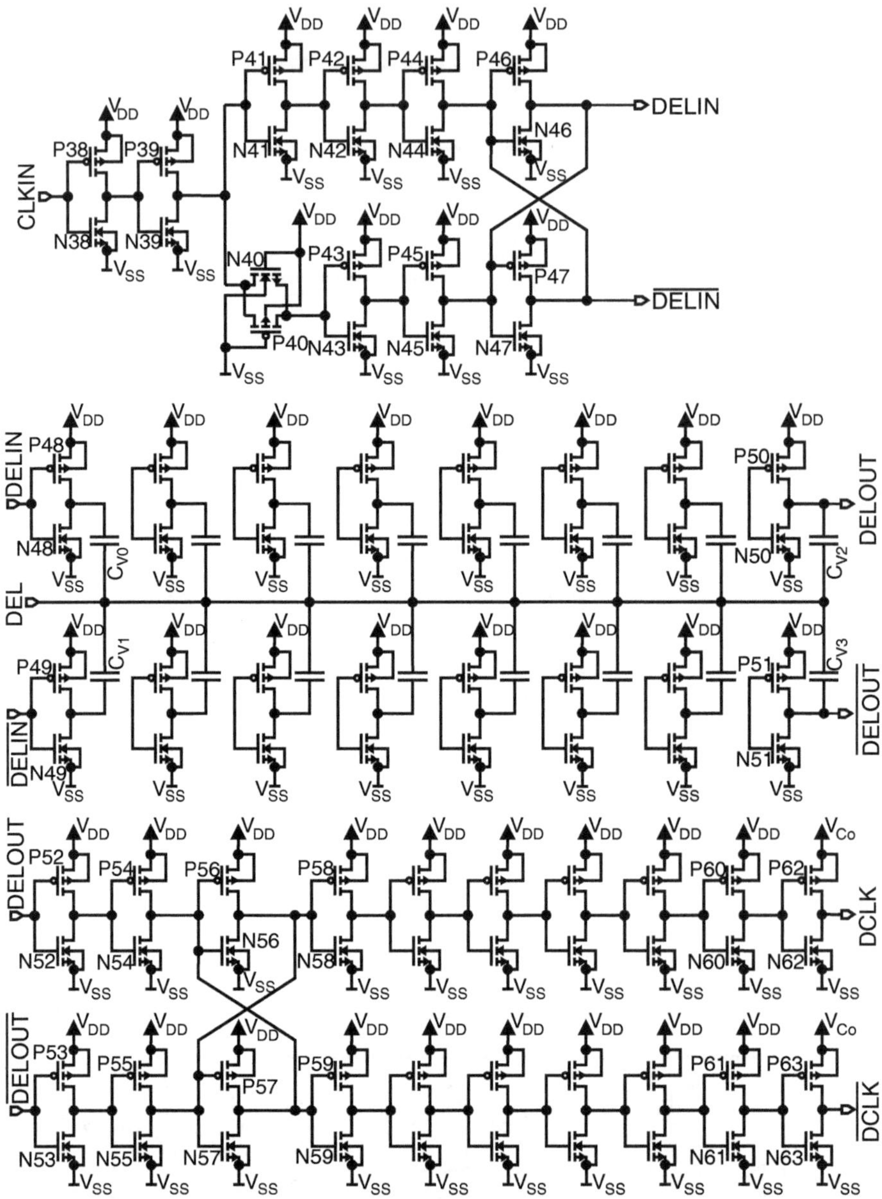

Fig. 5.55 Schematic of the clock driver (65 nm CMOS)

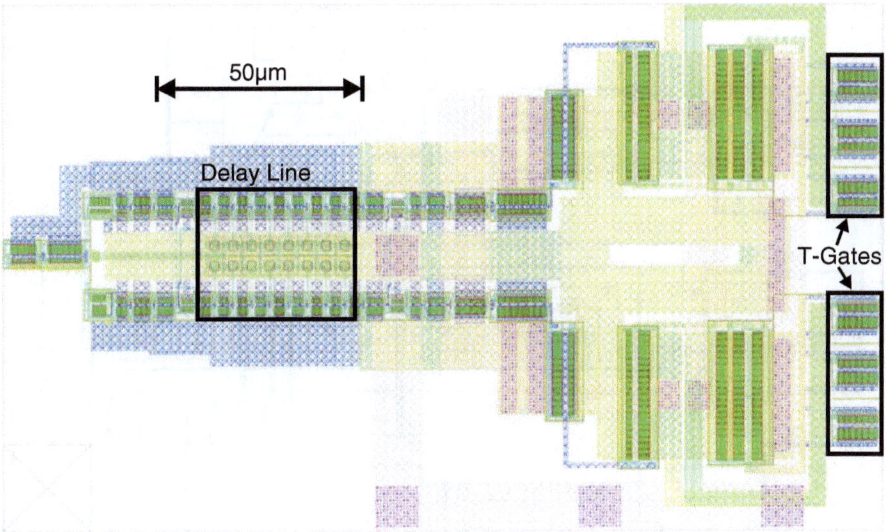

Fig. 5.56 Layout of the clock driver (size: $130 \times 201\ \mu m^2$)

5.5 Measurement Setup

The used measurement setup to characterize a test chip with a comparator is depicted in Fig. 5.57 and the shortcuts are depicted in Table 5.2.

In the center of Fig. 5.57 the test board with the test chip is depicted. The test board is supplied by several supply-, bias- and reference voltages, which can be adjusted with a 32 channel, 16 bit digital-to-analog converter (DAC) on a microcontroller board (MC-board) via a COM interface (RS232) by a personal computer (PC). With the help of the MC-board it is also possible to select one of the DC-voltages, which is wanted to be measured by the digital multimeter DMM (Keithley 2000 multimeter), with the PC by selecting one channel of an analog 32:1 multiplexer (MUX) on the MC-board. The DMM is also controlled via the COM interface from the PC to read out measurement values. If additionally a diode for temperature measurements is added on the test chip, also the actual temperature on-chip (*Tdut*) can be measured by the MC-board (section TEMP), read out by the PC, and additionally watched on a liquid crystal display (LCD). Temperature sensor *Tamb*, which is a diode connected pnp-transistor 2N3906, is added so that also the ambient temperature can be measured. For control purposes also the internal temperature *Tloc* of the temperature monitoring chip (section TEMP) is measured. *Tdut*, *Tloc* and *Tamb* are shown on the LCD. The clock, which is a biased sine-wave, for the test board is generated with the help of a vector network analyzer 10 MHz–20 GHz ZVM from Rhode& Schwarz (VNA). The frequency can be set at the VNA and the bias is applied from the MC-board to the bias input of the VNA, so that changing the bias and as a consequence the duty cycle after the on-chip clock driver is also controllable with the PC. With

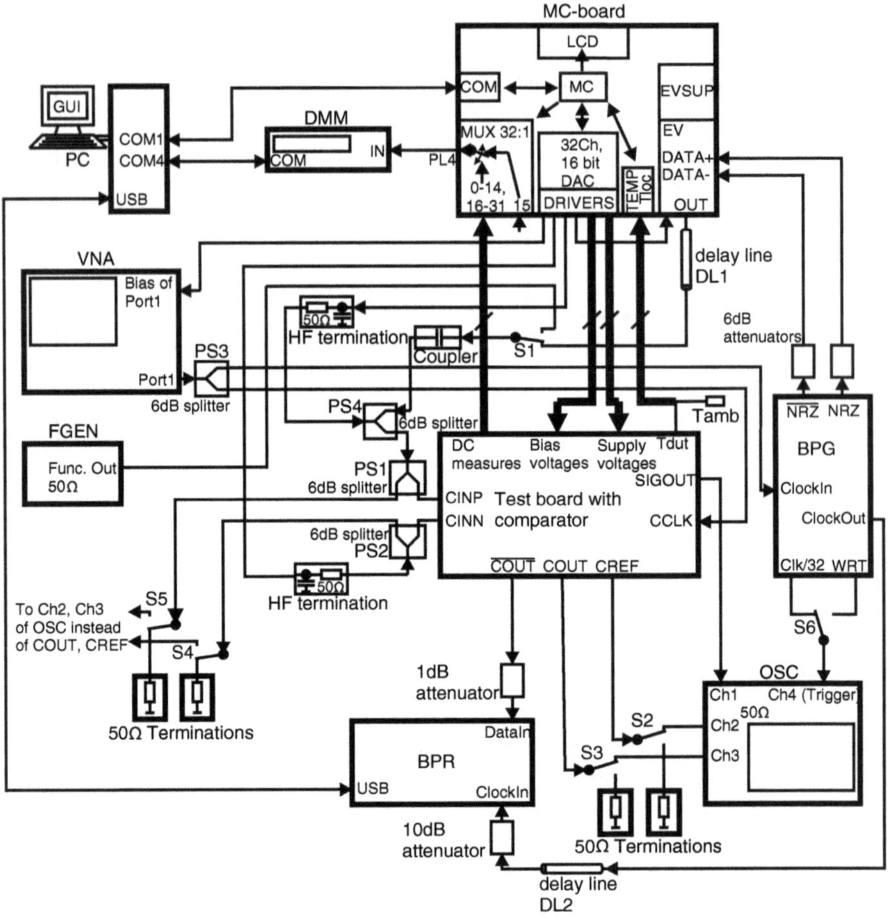

Fig. 5.57 Block diagram of the used measurement setup to characterize a test chip with a comparator

the help of the 6 dB-splitter PS3, the clock is applied to the bit pattern generator BPG (BMG12Gig, Sympuls Aachen) at *ClockIn* (AC coupled) and to the test board, where the test chip with the comparator is mounted, at input *CCLK*. The BPG generates a distinct bit pattern (output connectors *NRZ* and $\overline{NRZ}$, NRZ means no-return-to-zero), which is fed into a MAX3930 evaluation kit (EV) at pins (*DATA+* and *DATA−*) to generate better ramps and smaller rise and fall-times of the bit pattern. Furthermore the amplitude of the single-ended data at pin *OUT* can be controlled via MC-board with the PC by using an adjustable bias voltage output. The power supply of EV is generated with EVSUP. Delay line DL1 is inserted to get an optimal time delay in comparison to the clock at the sampling time-point of the comparator. The switches S1–S5 are in reality SMA cables, which were connected in different ways, which depends on the intended measurement. So plugging on or off a SMA cable for

Table 5.2 Description of Fig. 5.57

Shortcut	Description
BPG	Bit pattern generator, BMG12Gig, Sympuls Aachen
BPR	Bit pattern receiver, SBF10Gig, Sympuls Aachen
COM, COM1, COM4	COM-ports (communication equipment, RS-232)
DAC 32Ch 16 bit	32 channel, 16 bit digital-to-analog converter MAX5733B
DL1, DL2	Delay lines, semi-rigid coaxial cables with distinct lengths and SMA connectors
DMM	Digital multimeter, Keithley 2000 multimeter
DRIVERS	Supply drivers, reference drivers and reference current sources
EV	MAX3930 evaluation kit
EVSUP	Supply board for EV
FGEN	Function generator, Wavetek Model 275
GUI	Graphical user interface, programmed with GUI development equipment (GUIDE) in MATLAB
LCD	Liquid crystal display, JM162A
MC	Microcontroller PIC18F452-I/P from Microchip
MC-board	Self-constructed microcontroller board
MUX 32:1	32:1 Multiplexer with analog switches, consists of two chips MAX396CQI
OSC	Oscilloscope, digital storage oscilloscope Tektronix TDS6124
PC	Personal computer
PS1, PS2, PS3, PS4	6 dB power-splitter
S1, S2, S3, S4, S5, S6	Show possibilities of measurement arrangements with SMA-cables
TEMP	Temperature monitor chip MAX6655MEE
Tamb	pnp-transistor 2N3906, connected as diode to measure the ambient temperature
Tdut	On-chip parasitic pnp-transistor, connected as diode to measure on-chip temperature of the device under test (DUT)
Tloc	Internal measured temperature of the temperature monitor chip
USB	Universal serial bus
VNA	Vector network analyzer, 10 MHz–20 GHz ZVM from Rhode&Schwarz

different measurements is described schematically by the switches in Fig. 5.57. Switch S1 describes the possibility of either to apply a bit pattern from output *OUT* of EV or a signal (triangular-, rectangular- or sine-wave) from the function generator FGEN (Wavetek Model 275). The coupler, the HF termination and the 6 dB-splitter PS4 builds a Bias-T, where the DC voltage is fed in via the HF termination. For the Bias-T a 6 dB-splitter is used instead of a solution with a coil or instead of a Bias-T, where via a high resistor (due to a good 50 Ω matching) the DC voltage is applied, to achieve firstly a high transmission bandwidth, where resonances of a coil are avoided and secondly to need not to apply such a high DC voltage as in the case, where the

DC voltage is fed in via a high resistor (voltage divider high resistor / 50 Ω). The disadvantage of the implemented solution with P4 is, that the amplitude of the signal, which is coupled in, is reduced by a factor of two and also the bias voltage at input *CINP* of the test board with the chip is reduced by the series resistances. But for measurements of a comparator these disadvantages do not make any sense, because firstly the amplitudes and additionally the bias voltage at input *CINP* can be adjusted more accurately via the MC-board by the PC due to voltage attenuation and secondly the output voltages of the drivers of the MC-board can be adjusted between 0 and 10 V, which is enough for the range of 0–1.5 V for the test chips. The reference voltage, which is applied to input *CINN* uses also a 50 Ω termination to guarantee symmetry of the load, seen backwards from the test board. The 6 dB-splitters PS1 and PS2 have been added to have the possibility of watching the input signals and bias voltages for control reasons with the oscilloscope OSC (Digital storage oscilloscope, Tektronix TDS6124). During measurements the connections from PS1 or PS2 to the oscilloscope were terminated with 50 Ω (switches S4 and S5) to avoid influences of eventually appearing small kick-back disturbances of the oscilloscope. If for measurements higher input amplitudes of the bit pattern are needed, then the power splitters PS1 and PS2 may be removed to win 6 dB. The output signals *COUT* (non-inverted digital output of the comparator) and *CREF* (Monitoring output for digital on-chip clock) may be also watched by the oscilloscope. In the case of applying a bit pattern from BPG via EV, Coupler, PS4 and PS1 to *CINP*, the oscilloscope is triggered by the bit pattern frame (output WRT of BPG) or by clock frequency/32 (output Clk/32 of BPG).

To measure the bit-error-rate (BER), output $\overline{COUT}$ is connected to input *DataIn* of a bit pattern receiver BPR (SBF10Gig, Sympuls Aachen). BPR is clocked by BPG and the delay line DL2 is inserted to get the optimal delay-time between the data at DataIn and the clock at ClockIn for sampling. BPR is controlled via USB by the PC to read out the result of a BER measurement. Delay lines DL1 and DL2 are made out of semi-rigid 50 Ω coaxial cables, where SMA-connectors have been mounted. The different delay times were optimized by a set of such delay lines, when the clock frequency was changed and in some test chips additionally voltage-controlled delay lines for fine adjusting were added.

Controlling the MC-Board, the DMM and the BPR were done with the help of a programmed graphical user interface (GUI) from the PC. Also a microcontroller software was programmed to have an easy hand-shake protocol for the communication via COM interface. Figure 5.58 shows a photo of the measurement setup.

5.6 Microcontroller Board

The schematic of the microcontroller board (MC-board) is shown in Fig. 5.59. It consists of a microcontroller PIC18F452-I/P from Microchip (IC1) [15], which in principle forwards and adapts the commands and data, which are sent by the PC via the serial port COM (COM1 of PC to COM of MC-board), to the appropriate

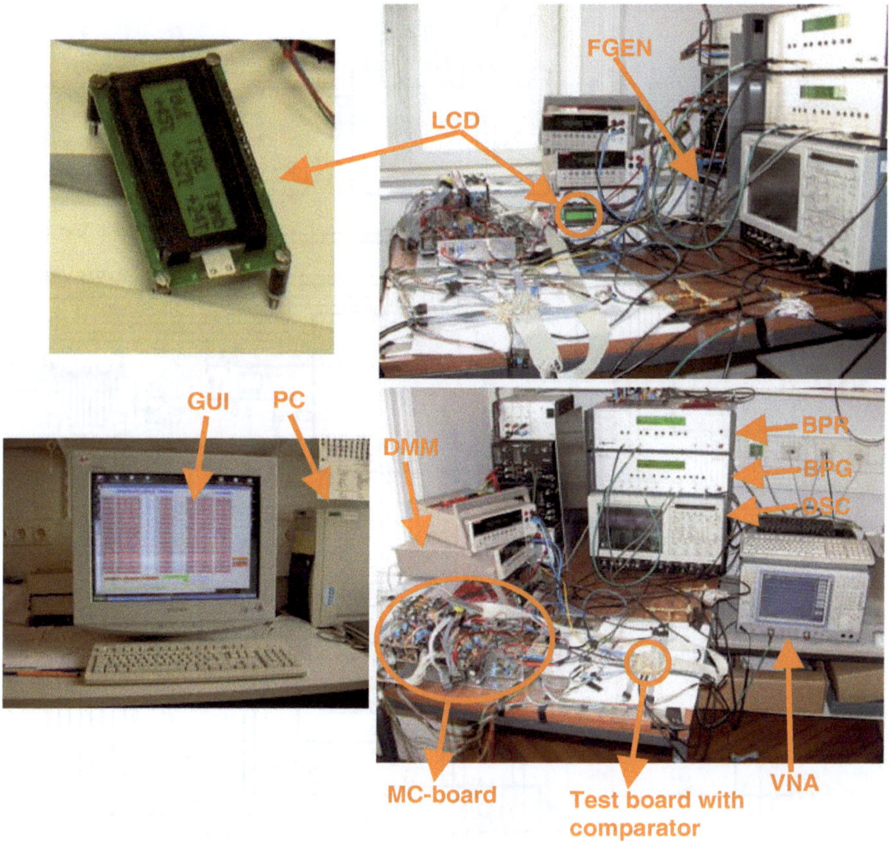

Fig. 5.58 Photo of the used measurement setup to characterize a test chip with a comparator

bus systems to the chips, e.g. Serial Peripheral Interface (SPI) to DAC MAX5733B (IC2) [16] or e.g. the System Management Bus (SMB) to temperature monitoring chip MAX6655MEE (IC5) [11]. Furthermore the data, received by these chips are adapted to COM interface and sent back to the PC. The internal clock frequency of the microcontroller consists of four non-overlapping quadrature clocks of 4 MHz, which are generated from a ceramic 16 MHz resonator MUR CSA 16.00 MXZ from Murata (XTAL1 in Fig. 5.59).

The microcontroller can be programmed at pins $\overline{MCLR}/Vpp$, $RB6/PGC$ and $RB7/PGD$, where $RB5/PGM$ is connected to DGND. To change the controller to programming mode, switch SW1 is switched to the position so that the voltage VD+10 V is applied to pin $\overline{MCLR}/Vpp$. Then pin $RB6/PGC$ works as clock line and pin $RB7/PGD$ as data line to program IC1 serial with the new program. Furthermore the outputs of part 1 (pin names with 1 at the beginning) of the bus driver 74HCT244, IC9, [17] are switched to high impedance, because $1\overline{OE}$ is set to VD + 5 V by switch

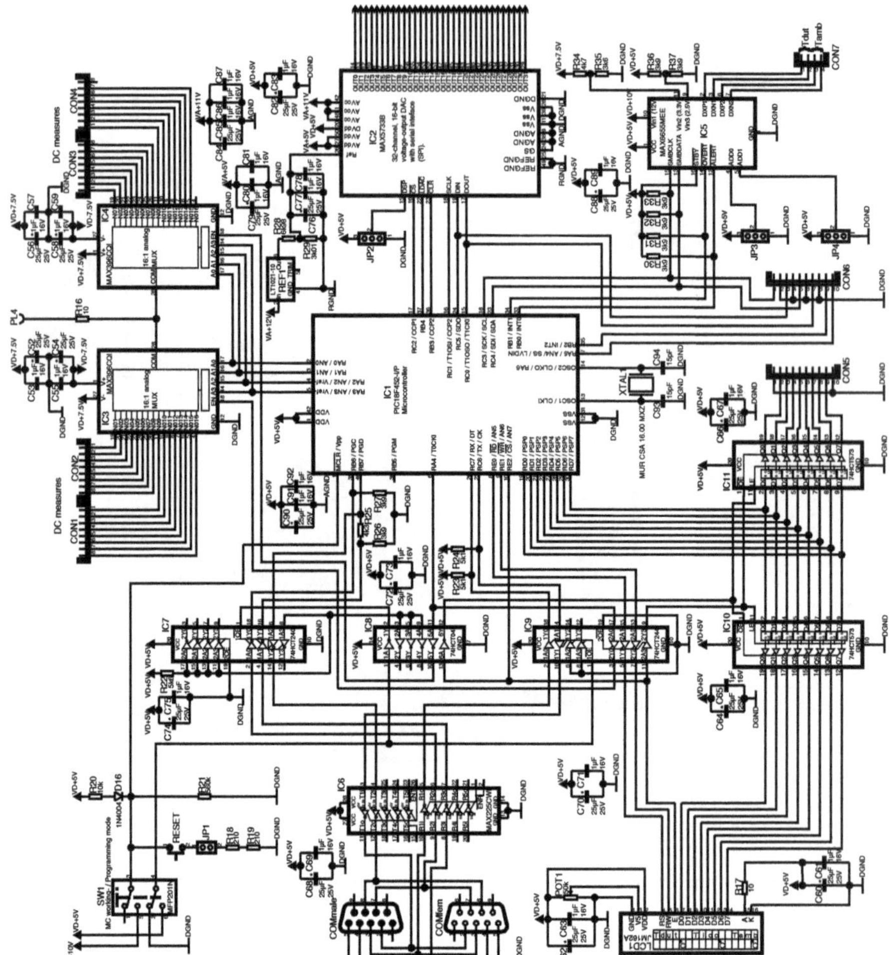

Fig. 5.59 Schematic of the microcontroller board (part 1), Schematic of the microcontroller board (part 2)

SW1 too so that the regular transmission pins 2 (TX) and 3 (RX) of the COM port are cut off from IC1. On the other hand part 1 of the inverted bus driver 74HCT240, IC7, [18] is enabled by setting $1\overline{OE}$ to DGND with the help of an inverter of the hex inverter chip 74HCT04, IC8, [19]. So pins *RB*6/*PGC* and *RB*7/*PGD* of IC1 are connected with pin 7 (RTS) and pin 8 (CTS) of the COM Port respectively. Pin 4 (DTR) is also connected via working resistor *R*25 to *RB*7/*PGD*. This is because *RB*7/*PGD* is a bidirectional data line during programming. When *RB*7/*PGD* is switched to driver mode, then *R*25 adds up as an additional pull up or pull down resistor (depends on logic state at pin 1*Y*1 of IC7) and data flow via input pin 1*A*2 of IC7 to pin 8 (CTS) of the COM port. In the other case, when *RB*7/*PGD* is switched

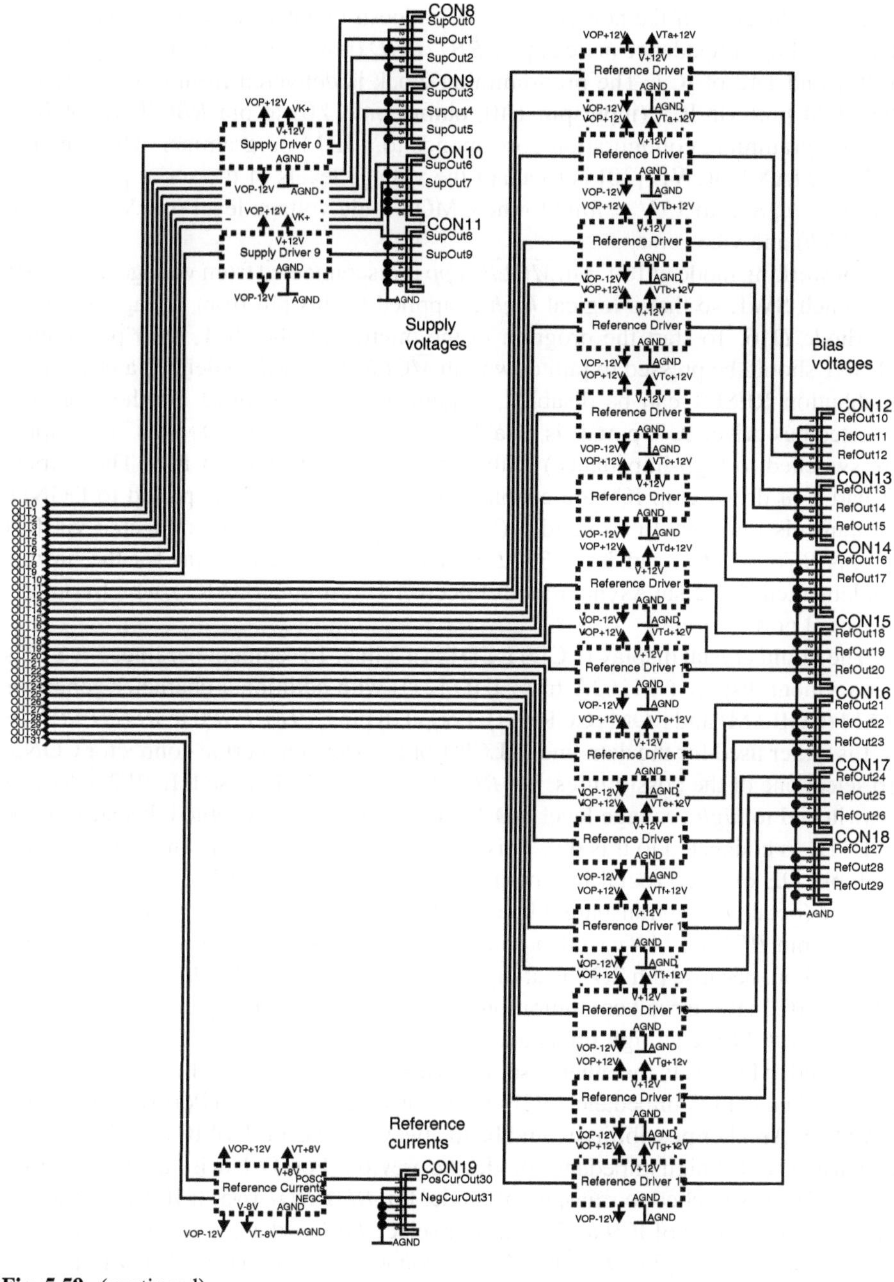

Fig. 5.59 (continued)

to receive the code of the new program, *R*25 doesn't contribute a high voltage drop due to the high-impedance node at pins $RB7/PGD$ (now input with high-impedance) of IC1 and 1*A*2 of IC7. The programming clock is delivered from pin 7 (RTS) of the COM port via IC7 (input pin 1*A*0, output pin 1*Y*0) to pin $RB6/PGC$ of IC1. This programming configuration corresponds to a JDM programmer. The function of IC6 (MAX225CWI) [20] is to adapt the voltage levels of the serial port (−3 to −15 V = *high*, 3 to 15 V = *low*) to the CMOS logic voltage level of 0 V (*low*) and 5 V (*high*) and vice versa.

For working mode of IC1, pin $\overline{MCLR}/Vpp$ is disconnected from voltage VD +10 V by switch SW1, so that a logical *high* is applied to this pin from voltage VD+5 V via diode *D*16. To start the program in the microcontroller IC1, reset pushbutton RESET should be pressed to pull down pin $\overline{MCLR}/Vpp$, which defines a reset state. Pushbutton RESET can be disabled, if jumper JP1 is removed. Furthermore the inverted bus driver IC7, part 1, is disabled by setting $1\overline{OE}$ to VD+5 V (The outputs are switched to high-impedance) with the help of an inverter of IC8. The outputs of part 1 of driver chip IC9 are enabled, because $1\overline{OE}$ of IC9 is pulled to DGND. So transmission pins 2 (TX) and 3 (RX) of the COM port are connected via IC9 to pins $RC6/TX/CK$ and $RC7/RX/DT$ of IC1, which are the pins for the, in IC1 implemented Universal Asynchronous Receiver Transmitter (UART), which controls the serial port.

The liquid crystal display (LCD) LCD1 is a 2 line, 16 segment per line, 5 × 8 dots per segment display (JM162A from JHD [21]) with additional character generator RAM (CGRAM) and ROM (CGROM). Port D (8 pins, $RD0/PSP0$ to $RD7/PSP7$) of IC1 is either used for the data line to LCD1 or as extension port at connector CON5. To select one of the possibilities, pin $RE2/\overline{CS}/AN7$ of IC1 is used. If $RE2/\overline{CS}/AN7$ is switched to *high* (voltage level VD + 5 V), then LCD1 is enabled, because pin *E* of LCD1 is connected to this pin. This node is also connected via an inverter in IC8 with pin $\overline{OE}$ of IC10 (octal D-type transparent latch 74HCT573 [22]) and pin *LE* of IC11 (octal D-type transparent latch 74HCT573 [22] like IC10). Additionally the node at pin $RE2/\overline{CS}/AN7$ is also connected via a buffer of IC9, part 2 (is always switched on, because pin $2\overline{OE}$ is at digital ground DGND) with $\overline{OE}$ of IC11 and pin *LE* of IC10. In this circuit configuration IC10 is switched to transparent mode and the outputs of IC11 are at high impedance state, when $RE2/\overline{CS}/AN7$ is *high*. Port D is connected to LCD1. In the other case, when $RE2/\overline{CS}/AN7$ is *low*, the outputs of IC10 are switched to high impedance, IC11 is in transparent mode and Port D is connected to CON5. Another possibility would be to always connect $\overline{OE}$ of IC11 to DGND, so that instead of the high impedance mode, the previous Port D data is latched at CON5, when IC11 is switched to transparent mode. Pin $RE0/\overline{RD}/AN5$ of IC1 is connected via a driver in part 2 of IC9 to *RS* (register select, *high* = data, *low* = instruction) of LCD1. Furthermore pin $RE1/\overline{WR}/AN6$ is connected via a driver of IC9 to pin *RW* (read/write, *low* = write data to LCD, *high* = read data from LCD) of LCD1. So Port E is used to control LCD1 and to select the extension port at connector CON5 and Port D delivers 8 bit data.

Connectors CON1, CON2, CON3 and CON4 are used to connect there various DC voltages from the test board with the comparator chip, which are wanted to be measured. One of them is selected by IC1 (pins *RA0/AN0*, *RA1/AN1*, *RA2/AN2/Vref−*, *RA3/AN3/Vref+*, *RA4/TOCKI*) with the help of two 16:1 analog multiplexer (MUX) chips IC3 and IC4 (MAX396CQI, [23]). Pin *RA4/TOCKI* is the extension of the 4 bit address (*RA0/AN0*, *RA1/AN1*, *RA2/AN2/Vref−*, *RA3/AN3/Vref+*), which is additionally needed to select a distinct channel, because two multiplexer chips are used. *RA4/TOCKI* is connected via an inverter in IC8 to *EN* of IC4 and via a line driver in part 2 of IC9 to *EN* of IC3. The selected analog channel is connected via resistor *R*10 to plug PL4, where the digital multimeter (DMM), which has an input resistance of about 10 GΩ, is plugged in to measure the voltage.

The temperature sensors are handled with the monitoring chip IC5 (MAX6655MEE [11]). It is controlled by IC1 via a System Management Bus (SMB), which is compatible to an Inter-IC bus (I2C). The address for SMB communication for IC5 is fixed with jumpers JP3 and JP4. Both pins *ADD*0 and *ADD*1 are connected with these jumpers to *DGND* to define the 7 bit device address 0b0011000(0) of IC5. Pin $\overline{STBY}$ is connected to *high* so that IC5 is not in standby mode. At IC5, output pin $\overline{OVERT}$ (open drain) become *low* when a measured temperature is higher than a programmed value and output pin $\overline{ALERT}$ (open drain) is switched to *low* by IC5, if e.g. a measured temperature or voltage is higher or lower than a programmed value or if e.g. a remote diode for temperature measurements is fault. Pin $\overline{OVERT}$ is connected with *RB1/INT1* and pin $\overline{ALERT}$ is connected with *RB0/INT0* of IC1. Communication with IC5 is done by the internal I2C bus muster of IC1 between pins *RC3/SCK/SCL* of IC1 and pin *SMBCLK* (clock line) of IC5 and between *RC4/SDI/SDA* of IC1 and *SMBDATA* of IC5 (data line). Between pins *DXP1* and *DXN*1 the on-chip diode connected pnp-transistor (provided by this CMOS process but with a very poor β gain) of the comparator test chip is plugged on (connector CON7) to measure *Tdut*. The ambient temperature sensor (pnp-transistor 2N3906, which is diode connected) for *Tamb* is also plugged to connector CON7 between pins *DXP2* and *DXN*2 of IC5. Inputs *Vin*1–*I*3 are for monitoring of different supply voltages, which is also provided by IC5.

To generate digitally adjustable analog DC voltages, the 16 bit, 32 channel, voltage output DAC MAX5733B [16] (IC2) has been added to the MC-board. Controlled is IC2 via a serial peripheral interface (SPI) by IC1, where the SPI is programmed at pins *RC1/T1OSI/CCO2* (clock line), *RC5/SCO* (data, which is sent to IC2) and *RC0/T1OSO/T1CKI* (data, which is received from IC2). With jumper JP2 the logical level at pin $\overline{DSP}$ of IC2 can be selected (*high* = VD + 5V, *low* = DGND). JP2 was set that $\overline{DSP}$ is *low*, so that a stand-alone SPI device is defined. Hence a logic *high* would indicate, that IC2 is one of more SPI devices in a Daisy chain or parallel connected. To select and activate the SPI of IC2, pin *RC2/CCP*1 of IC1 must be set to *low*, which is connected with $\overline{CS}$ of IC2. Pin *RB*4 of IC1 is connected with pin $\overline{LDAC}$ of IC2, where a *low* updates simultaneously the value for every output, which are stored in the appropriate input registers. Hence this pin was always set to *high* (normal operation), because this can be also done with a software command via SPI. Pin *RB*3 of IC1 is connected with pin $\overline{CLR}$ of IC2, where a *low* sets all output channels including the implemented offset DAC to 0 V (0x0000h). Pin $\overline{CLR}$ was also set to *high*, because this command can be also done with a software command via SPI.

The reference voltage is generated with the 10 V reference source REF1 (LT1021-10 [24]), which is reduced to 3 V by precision resistors $R28$ and $R29$. These 3 V is the demanded reference voltage, which is applied to pin *Ref* of IC2. St each output channel (pins $OUT0$ to $OUT31$) a voltage between 0 and 10 V with a resolution of 16 bit can be generated, which is controlled by the PC via the COM port of the MC-board with the microcontroller IC1, which is programmed to control the SPI to DAC IC2.

To also have the possibility of using extension devices, which needs a SPI or a I2C bus, CON6 has been added, where pin 1 and pin 5 are the lines $RC3/SCK/SCL$ and $RC4/SDI/SDA$ of IC1 (I2C bus) and pin 3 and pin 7 are the bus lines $RC5/SDO$ and $RC0/T1OSO/T1CKI$ of the programmed SPI of IC1. Pin 9 and pin 10 are the port outputs $RB2/INT2$ and $RA5/AN5/\overline{SS}/LVDIN$ which can be used universally, e.g. as interrupt input or as an additional, programmed clock line or data input or as a chip select line.

$OUT0$ to $OUT9$ of IC2 are connected to circuit blocks Supply Driver 0 to Supply Driver 9 and are used as reference voltages to generate the final output voltages *SupOut*0 to *SupOut*9 (connectors CON8, CON9, CON10 and CON11), which have the same voltage value as $OUT0$ to $OUT9$ respectively, but with the difference, that more current can be delivered for supply purposes. The circuit blocks Reference Driver 0 to Reference Driver 13 (*RefOut*0 to *RefOut*13, connectors CON12 to CON18) also have the same voltage levels as $OUT10$ to $OUT29$, but are also designed to deliver more current, but not as much as Supply Drivers 0–9. $OUT30$ and $OUT31$ delivers the reference voltage to adjust a positive current source (*PosCurOut*30 at CON19), where the current flows out of CON19 and a negative current source (*NegCurOut*31 at CON19), where the current flows into CON19. The schematics of Supply Driver 0 and Reference Driver 0 are depicted in Fig. 5.60a, b. Supply Driver 1–9 and Reference Driver 1–18 have the same circuit schematic as Supply Driver 0 and Reference Driver 0 respectively. In Fig. 5.60a the operational amplifier $OPV1$ and the transistor $T1$ regulates the voltage *SupOut*0, so that it is equal to the voltage level of $OUT0$. Every Supply Driver consists of an own 12 V standard voltage regulator (VR17, MC7812CT) due to the possibility of a higher output current. Resistor $R40$ and $T2$ are added to limit the output current. If a voltage drop of 0.7 V appears at resistor $R40$, then transistor $T2$ turns on and reduces the base current of transistor $T1$. Jumper JP5 is added to have the possibility of plugging on a multimeter to measure the current. The circuit of the Reference Driver in Fig. 5.60b has a similar function as the circuit in Fig. 5.60a, but without an own voltage regulator MC7812CT. Here a group of three or two Reference Drivers is supplied by one MC7812CT (supply voltages $VTa + 12$ V to $Vtg + 12$ V). In Fig. 5.61 the circuits in block Reference Currents are depicted.

The currents are delivered at connector CON19. The positive output current, delivered to pin *PosCurOut*30, (see Fig. 5.61a) is regulated by the operational amplifier $OPV3$ with pnp-transistor $T5$ to $(8V - OUT30)/R46$, (VT + 8 V = 8 V), where with changing $R46$ the current range, which can be adjusted by the PC, is defined.

The voltage VT + 8V is chosen to be +8 V to have an appropriate offset in comparison to $OUT30 = 0$ V to 10 V to reach also zero current flow at the output due to

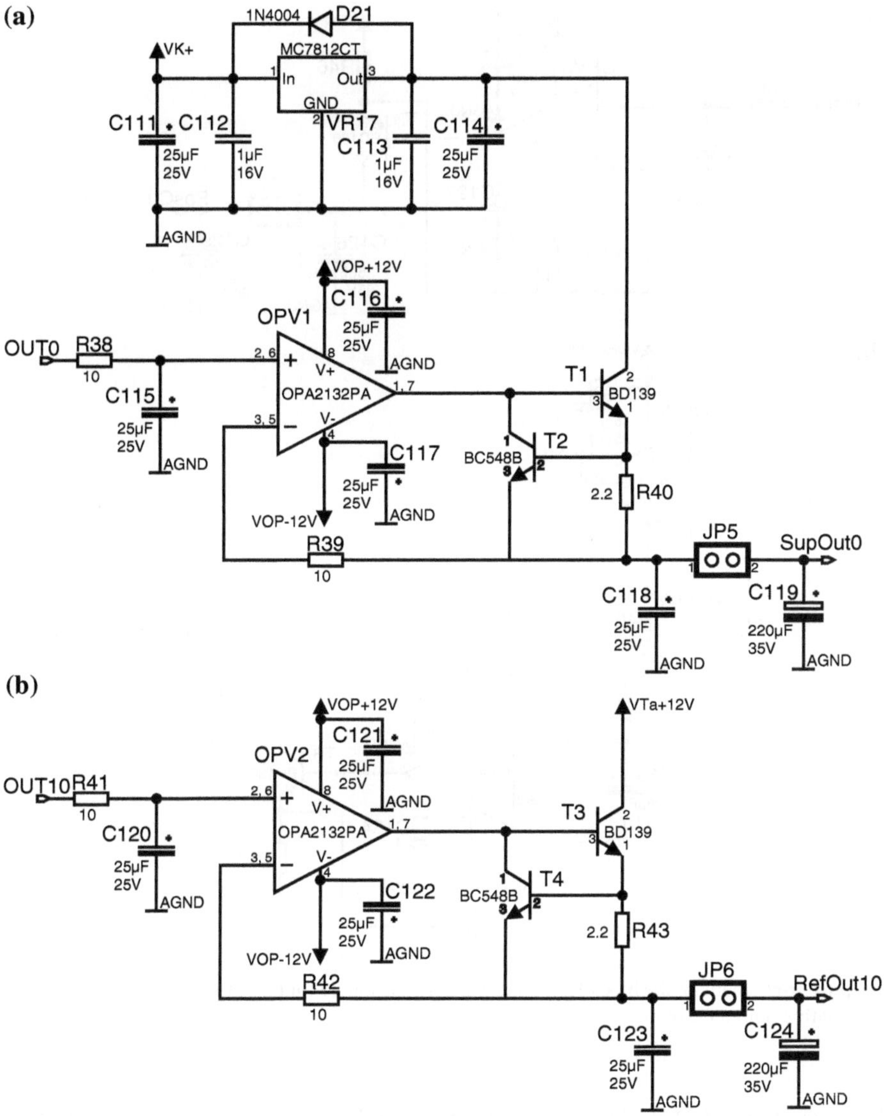

Fig. 5.60 Schematics of the output voltage drivers. **a** Supply Driver 0, **b** Reference Driver 0

tolerances of electronic components. Jumper JP7 is added to have the possibility of plugging on a multimeter to measure the current. The negative output current, delivered to pin *NegCurOut*31, (see Fig. 5.61b) is regulated in the same way as depicted in Fig. 5.61a with the difference, that the sign of the voltage at *OUT31* is changed to negative with the help of the operational amplifier *OPV4*, the supply voltage VT−8V is now −8 V and transistor *T6* is now npn. So a current $(-OUT31 - 8V)/R51$ is

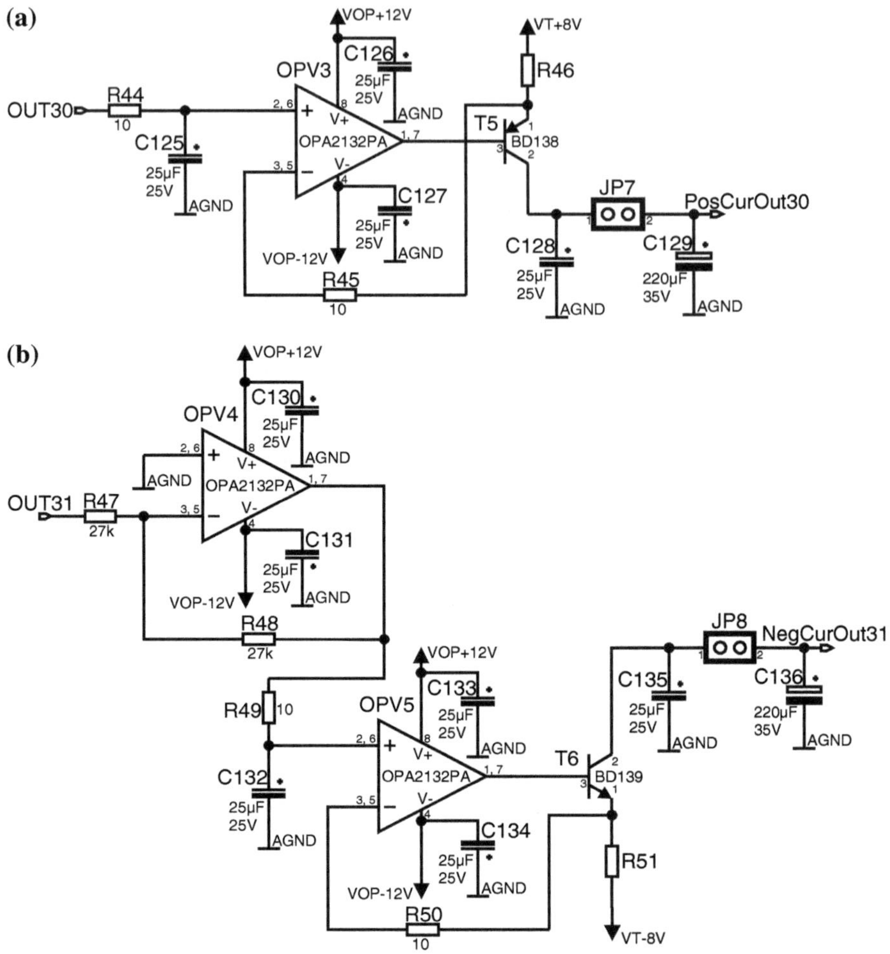

Fig. 5.61 Schematic of the block Reference Currents. **a** source for positive output current, **b** source for negative output current

defined with $OUT31 = 0\,\text{V}$ to $10\,\text{V}$ adjustable and regulated by operational amplifier *OPV5*.

The generation of the different supply voltages are shown in Fig. 5.62 for completeness, where an explanation is omitted, because only standard circuits are used. The input voltages $\text{VK}+ = 15\,\text{V}$ and $\text{VK}- = -15\,\text{V}$ are delivered from a standard power supply for a electronic laboratory.

The bit pattern, which is generated by the BPG (see Fig. 5.57) and delivered at outputs *NRZ* and $\overline{NRZ}$ is fed into a MAX3930 evaluation kit [25] (EV) at pins (*DATA*+ and *DATA*−) to generate better ramps and lower rise times of the bit pattern.

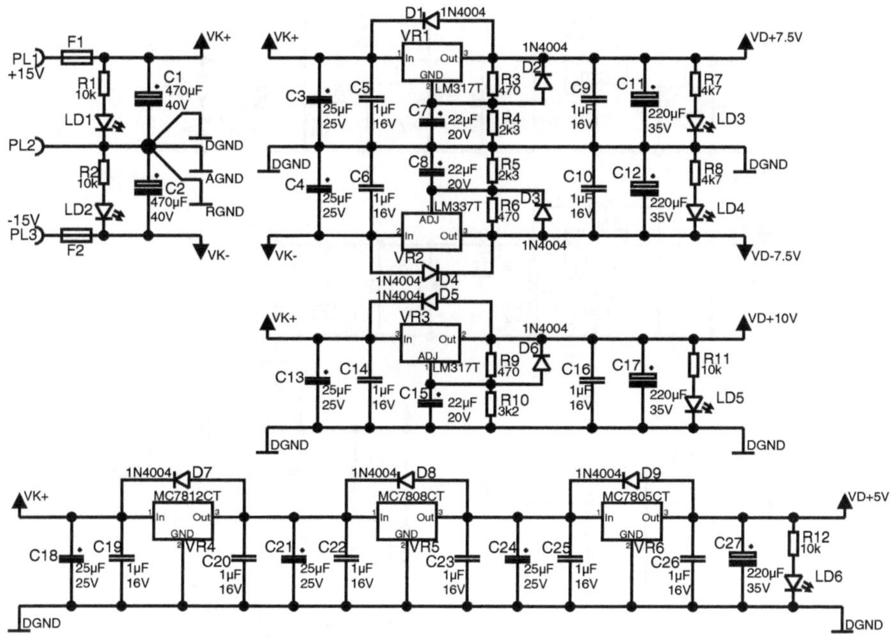

Fig. 5.62 Schematic of the generation of different power supplies for the microcontroller board (part 1), Schematic of the generation of different power supplies for the microcontroller board (part 2)

Furthermore the amplitude of the single-ended data at pin *OUT* of EV can be adjusted via reference voltage output *RefOut29* of the MC-board with the PC.

A schematic of the evaluation kit with the extension, that the amplitude of the bit pattern at *OUT* can be controlled with the voltage at *RefOut*29, is shown in Fig. 5.63. Pins *DATA+* and *DATA−* are the high-frequency, differential inputs of the bit patter, generated by the BPG. With potentiometer *P*12 the bias current at the output *OUT* of EV can be adjusted and with *P*1 the pulse width can be set to achieve e.g. a duty cycle of 50 %. Jumper Ju1 enables/disables data retiming. If Ju1 is shunt data retiming is enabled, where additionally to the data at *DATA+* and *DATA−* a clock has to be applied to pins *CLK+* and *CLK−*. For the measurement setup Ju1 was removed for direct data transmission, where only pins *DATA+* and *DATA−* were used. Jumper Ju2 is also shunt so that the output *OUT* is in normal operation and not disabled (Ju2 would be open). To control the amplitude of the bit pattern at output *OUT*, potentiometer *P5* is removed and instead the voltage divider *R*73, *R*74 and *R*75 is connected as shown in Fig. 5.63. The reference voltage at *RefOut*29 (0–10 V), which is defined by the DAC IC2 (see also Fig. 5.59) is subtracted by 5 V (−5 V supply voltage of EV, which is inverted at the inverting amplifier *OPV*8, *R*67, *R*68) at the subtracter circuit *OPV*9, *R*69, *R*70, *R*71, *R*72, so that a voltage between −5 and 5 V

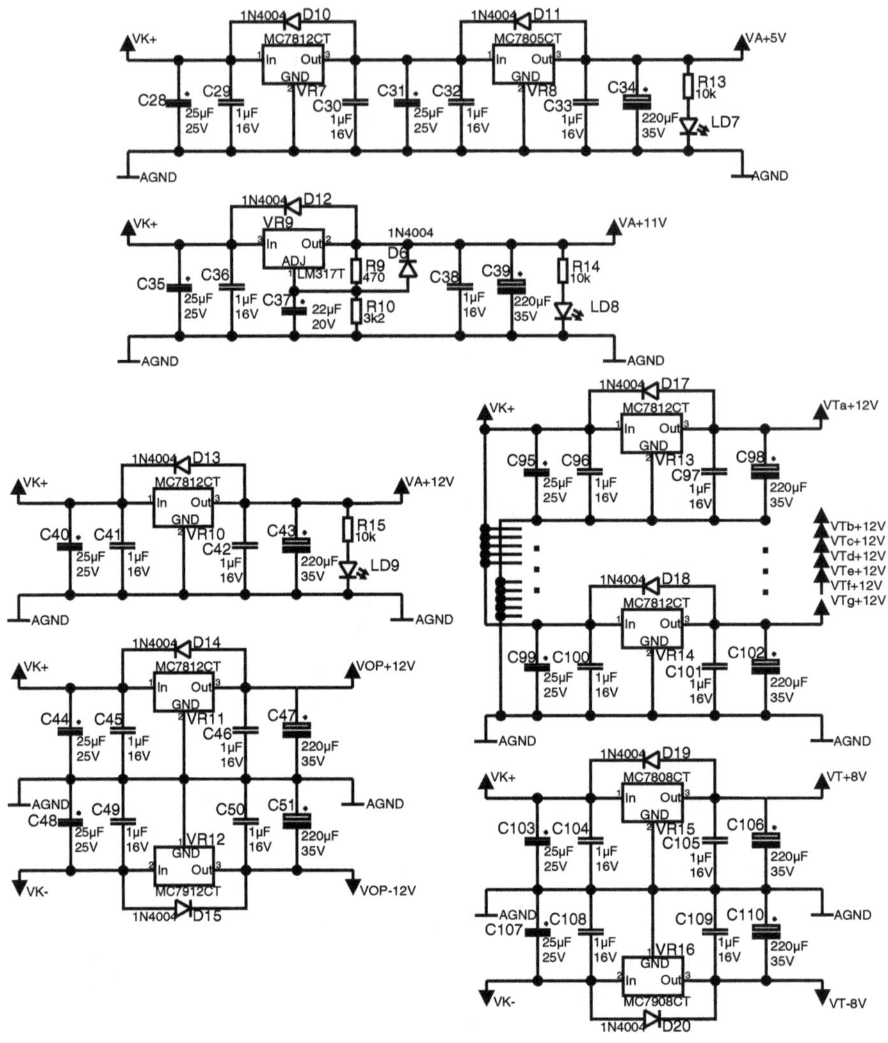

Fig. 5.62 (continued)

is applied to the voltage divider $R73$, $R74$ and $R75$. There the voltage -5 and 5 V is reduced to fit between -5 and -3.75 V, because the removed potentiometer $P5$ was mounted also between -5 and -3.75 V. This voltage defines the amplitude at OUT.

The power supply voltages (-5 V and -2 V) of EV are generated with the circuit block EVSUP and are connected at connector CON22 to EV. The schematic of EVSUP is depicted in Fig. 5.64.

With the reference voltage $REF2$ (TL1431, [26]) and resistors $R58$, $R59$ and $R60$ a voltage of -8 V (regulated by $VR21$, MC7908CT) is used to regulate a constant voltage of -5 V as depicted in Fig. 5.64. With potentiometers POT2 and POT3 a

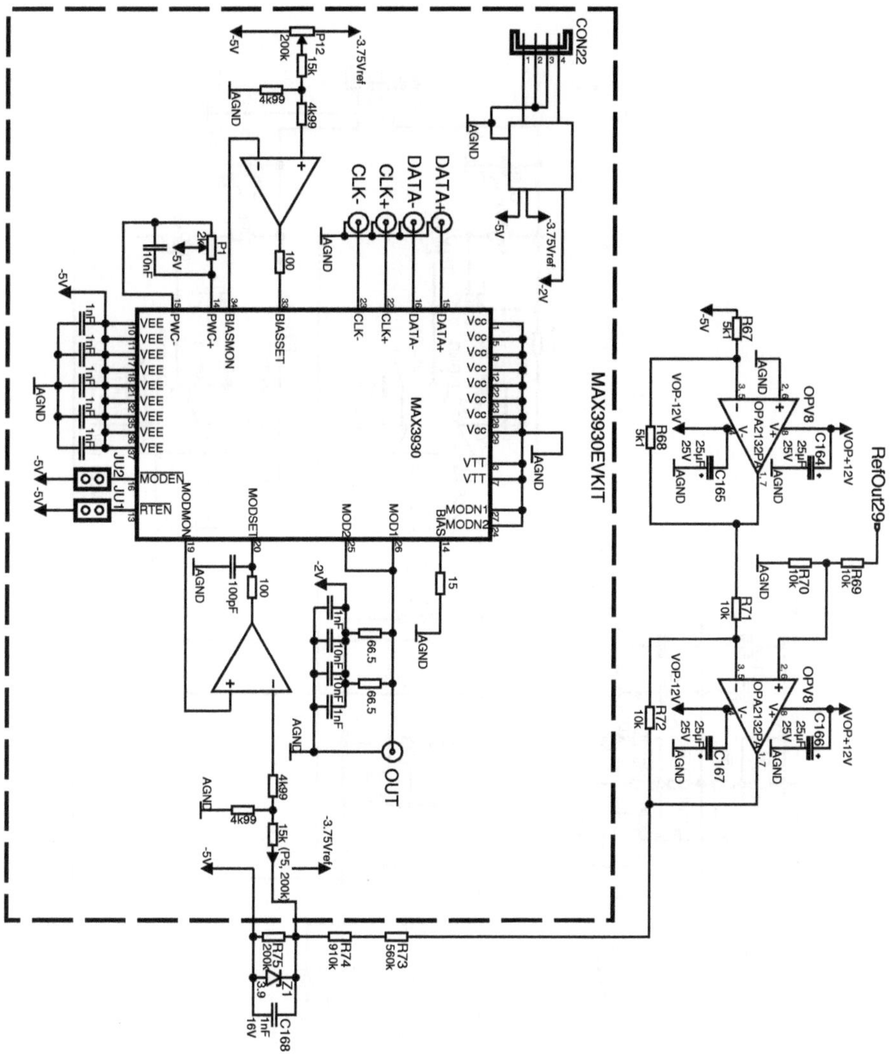

Fig. 5.63 Simplified schematic of the MAX3930 evaluation kit (MAX3930EVKIT circuit in dashed rectangle) and the extension to control the bit pattern amplitude via PC and microcontroller board instead of the orignal mounted potentiometer *P*5

voltage between −5 and 0 V can be adjusted. These reference voltages are driven in the same way as in the circuit block Reference Driver (see also Fig. 5.60b), which is already described. So at connector CON21 the demanded and relatively exact supply voltages of −5 and −2 V, which are connected to connector CON22 of the MAX3930EVKIT can be generated and used for EV.

A photo of the microcontroller board (MC-board) with the MAX3930EVKIT (EV) and its supply board (EVSUP) is shown in Fig. 5.65 (see also Fig. 5.59).

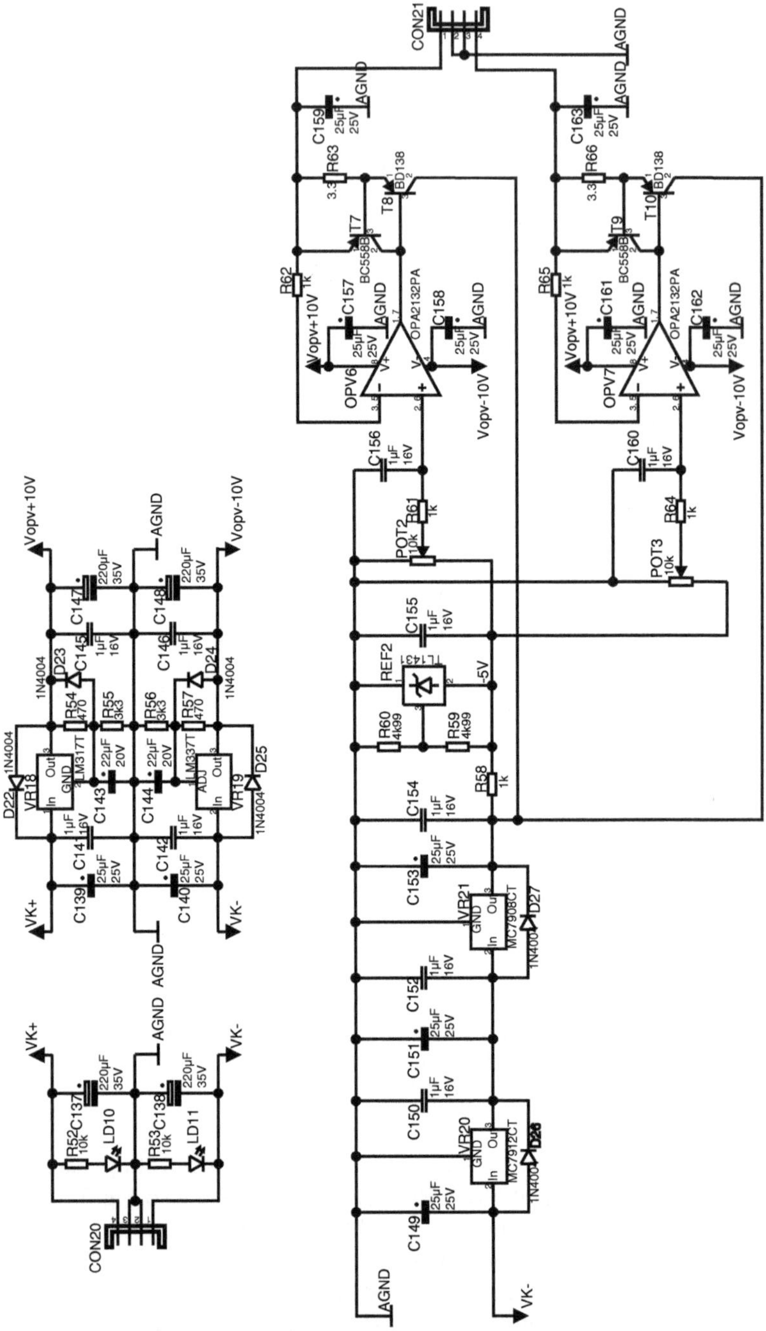

Fig. 5.64 Schematic of the supply voltage board EVSUP for EV

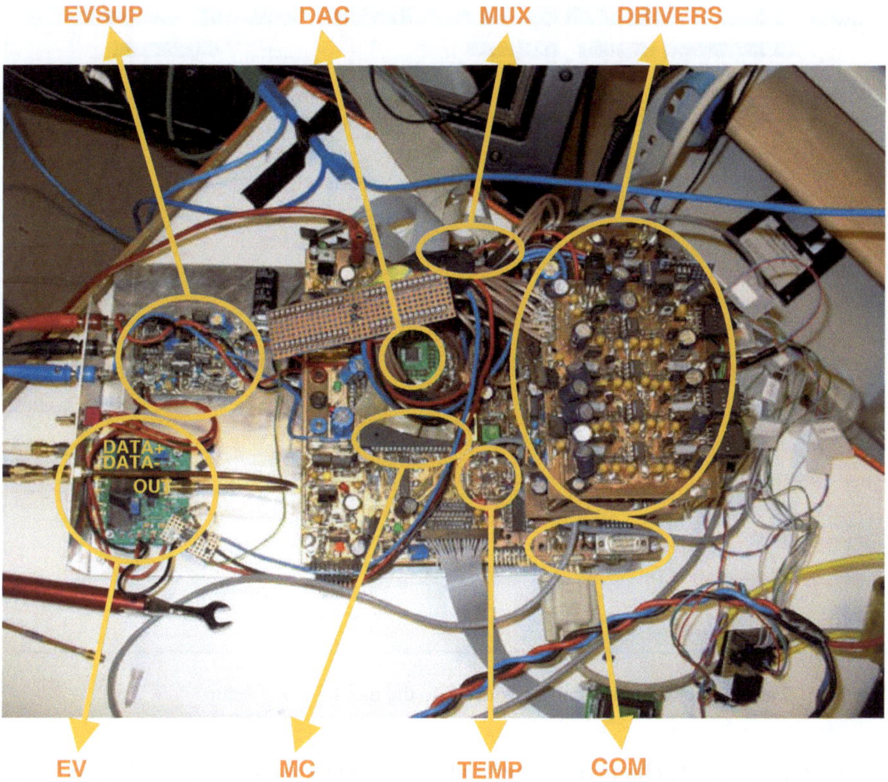

Fig. 5.65 Photo of the microcontroller board with the MAX3930EVKIT (EV) and its supply board (EVSUP)

The graphical user interface (GUI) on the PC was programmed with help of the graphical user interface development equipment (GUIDE) of MATLAB with callback functions, which indicates whether e.g. a button has been pressed, or whether e.g. a value or string has been entered in a command line. For the following description of GUI see also Fig. 5.57. In Fig. 5.66 a screen shot of the programmed GUI is shown.

The different DC voltage outputs of the microcontroller-board (MC-board) are marked with OUT0 to OUT31, which corresponds to *SupOUT*0 to *SupOut*9, *RefOut*10 to *RefOut*29, *PosCurOut*30 and *NegCurOut*31 of Fig. 5.59. For every OUTx (x= 0 to 31) a decimal value can be entered in a command line, which is the 16 bit value of the appropriate DAC (IC2) channel. Beside, with the push buttons 'Up' and 'Dn' this value can be increment or decrement by 1 bit respectively. In the program each maximum allowed value, which can be entered in the appropriate field, is stored in a variable. The program controls, that no higher value can be entered to have an additional safety feature. A text field beside the push buttons shows the corresponding, estimated voltage value, which will appear at the channel output of the DAC. With pushbutton 'LoadDAC Outx' the entered bit amount is loaded into

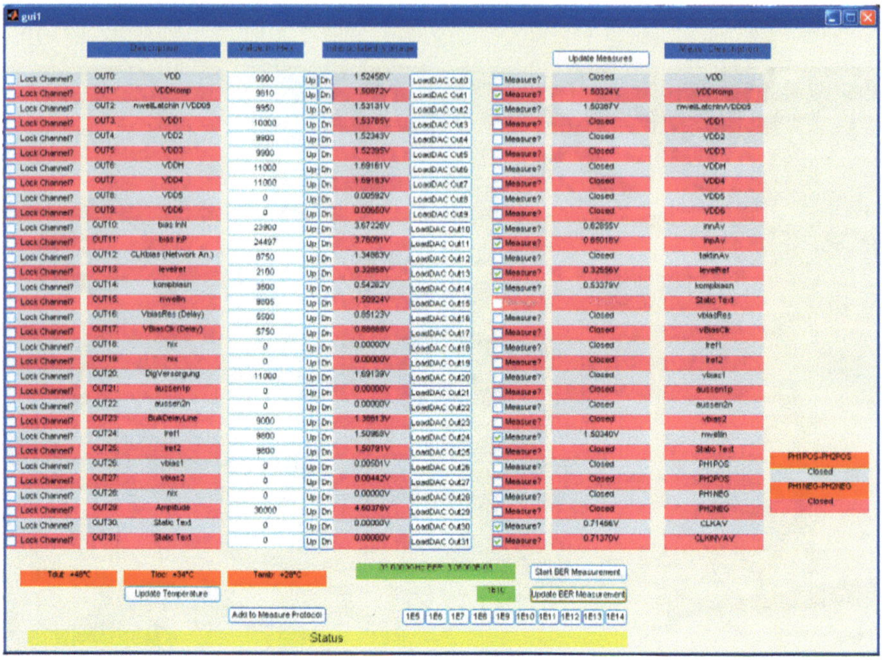

Fig. 5.66 The GUI for communication from PC to the measurement setup

the appropriate DAC channel register and the output of this channel is changed to the new DC voltage level corresponding to the bit value entered. Check box 'Lock Channel' locks the appropriate command line for being changed erroneously because of stress and inattention, which occur during measurements, which needs a long time.

With pushbutton 'Update Temperature' the measured temperatures *Tdut*, *Tloc* and *Tamb* are read out from the monitoring chip IC5 and displayed on the GUI.

Of what DC voltage, which is connected to a analog switch of the 32:1 MUX, (connectors CON1, CON2, CON3 and CON4 in Fig. 5.59) will be measured is selected by a check box 'Measure?'. If the pushbutton 'Update Measures' is pressed, then every analog switch, which is checked at 'Measure' is closed in turn and the appropriate DC voltage value is measured by the digital multimeter DMM, which is connected to plug PL4. The measured voltage is read out by the PC via the COM-port of the DMM and displayed on the GUI right of the appropriate check box 'Measure?' in a text field. A short description of the measured DC voltage can be found in the text field right of the measurement result.

The bit pattern receiver (BPR) is also controlled by the PC via a USB port. With push buttons '1E5' to '1E14' the count length can be selected. The actual selection is shown in the small text field above. The push button 'Start BER Measurement' starts a new BER measurement at BPR. If the push button 'Update BER Measurement' is pressed, then the measured value of the BER measurement and the clock frequency is read out of BPR and displayed in the text field left beside the button 'Start BER Measurement'.

The long text field at the bottom of the GUI displays errors and instructions.

With the push button 'Add to Measure Protocol' the checked DC voltage measurements, the actual temperatures *Tdut* and *Tamb*, the measured BER, the clock frequency and the saving time point is added to a file 'measures.txt'. When GUI has been started and the push button 'Add to Measure Protocol' is pressed the first time, also the date is added before the first measurement results. Below an example of the file 'measures.txt' is shown.

Year, 2013, month, 8, day, 18

VDDKomp, 1.50235, nwellLatchIn/VDD05, 1.50239, innAv, 0.79411, InpAv, 0.79993, levelRef, 0.05152, kompbiasn, 0.71473 , nwellIn, 1.50206, CLKAV, 0.67113, CLKINVAV, 0.67431, BER:, 8.60700E-07, Ampl, 50000, Freq, 05.0001, Tdut, +47C, Tamb, +25C, Time, 9-39-22

VDDKomp, 1.50234, nwellLatchIn/VDD05, 1.50239, innAv, 0.79411, InpAv, 0.79993, levelRef, 0.05151, kompbiasn, 0.71473 , nwellIn, 1.50202, CLKAV, 0.67112, CLKINVAV, 0.67432, BER:, 8.59000E-07, Ampl, 50000, Freq , 05.0001, Tdut, +47C, Tamb, +25C, Time, 9-55-13.

References

1. P.R. Gray, P.J. Hurst, S.H. Lews, R.G. Meyer, *Analysis and Design of Analog Integrated Circuits* (Wiley, New York, 2001)
2. K. Tanno, H. Matsumoto, O. Ishizuka, Zheng Tang, Simple CMOS voltage follower with resistive load driveability. IEEE Trans. Circuits Syst. II **46**(2), 172–177 (1999)
3. V. Kasemsuwan, P. Boonyaporn, A. Thanachayanont, ±1.5 V high performance CMOS rail to rail voltage follower. IEEE Asia-Pacific Conf. **2**, 425–428 (2002)
4. M. Milkovic, A high-speed local feedback unity-gain amplifier. IEEE Proceedings of the 35th Midwest Symposium on Circuits and Systems **2**, 904–907 (1992)
5. G. Palmisano, P. Palumbo, S. Pennisis, High-performance and simple CMOS unity-gain amplifier. IEEE Trans. Circuits Syst. I **47**(3), 406–410 (2000)
6. B. Goll, H. Zimmermann, A Five-Stage 10 GHz Voltage Buffer in 120 nm CMOS Technology, in *12th International Conference of Mixed Design of Integrated Circuits and Systems*, pp. 385–389 (2005)
7. T. Voo, C. Toumazou, High-speed current mirror resistive compensation technique. IET Electron. Lett. **31**(4), 248–250 (1995)
8. F. Schlögl, H. Zimmermann, in *1.5 GHz OPAMP in 120 nm digital CMOS, IEEE European Solid-State Circuits Conference*, pp. 239–242 (2004)
9. M. Spinola-Durante, *Development of an on-chip circuit for delay measurement of a comparator in 120 nm CMOS technology*. Master-Thesis, Vienna University of Technology, Supervisors: H. Zimmermann, B. Goll, G. Ripamonti (2005)
10. Temperature Sensors, Application Note 1057, www.maxim-ic.com. Accessed 17 April 2002
11. MAX6655/MAX6656 Dual Remote/Local Temperature Sensors and Four-Channel Voltage Monitors, Datasheet, 19–2117, Rev 0, 7/01, MAXIM Integrated Products (2001), www.maxim-ic.com
12. B. Goll, M. Spinola Durante, H. Zimmermann, in *An on-chip technique to measure the delay time of a comparatorin 120 nm CMOS technology*, Austrochip, pp. 71–75 (2005)

13. B. Goll, M. Spinola Durante, H. Zimmermann, A Measurement Technique to obtain the delay time of a comparator in 120 nm CMOS, in *13th International Conference of Mixed Design of Integrated Circuits and Systems (MIXDES)*, pp. 563–568 (2006)
14. B. Goll, H. Zimmermann, in *Simple Creation of Half and Full Frequency, Inverted and Non-Inverted Clock Signals with Maximum 10 ps Delay Time Differences in 120 nm CMOS*, Austrochip, pp. 143-148 (2006)
15. PIC18FXX2 28/40-pin High Performance, Enhanced FLASH Microcontrollers with 10-Bit A/D, Datasheet, Microchip Technology Inc. (2002), www.microchip.com
16. MAX5732-MAX5735 32-Channel, 16-Bit, Voltage-Output DACs with Serial Interface, Datasheet, 19–3148, Rev 6, 4/05, MAXIM Integrated Products (2005), www.maxim-ic.com
17. 74HCT244 Octal buffer/line driver; 3-state, Datasheet, Philips Semiconductors (1990), www.philips.com
18. 74HCT240 Octal buffer/line driver; 3-state; inverting, Datasheet, Philips Semiconductors (1990), www.philips.com
19. 74HC/HCT04 Hex inverter, Datasheet, Philips Semiconductors (1993), www.philips.com
20. MAX202E-MAX213E, MAX232E/MAX241E ±15 kV ESD-Protected, +5 V RS-232 Tranceivers, Datasheet, 19–0175, Rev 6, 3/05, MAXIM Integrated Products (2000), www.maxim-ic.com
21. JM162A or 162A, Datasheet, Shenzhen Jinghua Display (JHD) (2005), www.china-lcd.com
22. 74HC/HCT573 Octal D-type transparent latch, Datasheet, Philips Semiconductors (1990), www.philips.com
23. MAX396/MAX397 Precision 16-Channel/Dual 8-Channel, Low-Voltage, CMOS Analog Multiplexers, Datasheet, 19–0404, Rev 1, 9/96, MAXIM Integrated Products (1996), www.maxim-ic.com
24. LT1021 Precision Reference, Datasheet, Linear Technologies (1995), www.linear-tech.com
25. MAX3930 Evaluation Kit, Datasheet, 19–2070, Rev 0, 6/01, MAXIM Integrated Products (2001), www.maxim-ic.com
26. TL1431 Programmable Voltage Reference, Datasheet, former SGS-Thomson Microelectronics, now STMicroelectronics (1998), www.st.com

Chapter 6
Comparators in 120 nm CMOS

This chapter presents five comparators in 120 nm CMOS technology. Circuit details and measurement results are described for each of them. Starting from 1.5 GHz decision rate the performance is improved up to 6 GHz. One comparator is appropriate for low-voltage operation down to 0.5 V supply voltage and shows then an ultra-low power consumption of only 18 μW. Another comparator demonstrates an advanced sensitivity tuning technique.

6.1 A 660 μW, 1.5 GHz Comparator in 120 nm CMOS

In this section a test chip with a pre-amplifier in connection with a comparator (CMOS latch) is described. In principle pre-amplifiers are added to a clocked, regenerative comparator to enhance its sensitivity.

6.1.1 Circuit Description

The block diagram of the test chip with the comparator [1] is depicted in Fig. 6.1. The circuit is designed in a 0.12 μm CMOS process with a nominal supply voltage of 1.5 V. The comparator itself is a switched latch. To enhance sensitivity, a pre-amplifier is placed in front of the comparator. During reset phase the previous decision of the comparator is held in the transfer stage. At pins *DOUTP* and *DOUTN* the common-mode level at the inputs *INP* and *INN* of the comparator can be measured for control purposes. In Fig. 6.2a the pre-amplifier is depicted. It is a simple n-MOS differential amplifier with two input transistors $N0$ and $N1$ and with two poly-silicon resistors R_0 and R_1 for the load. It is designed to have an small-signal amplification of approximately 9. The output pins of the pre-amplifier are connected to the inputs *INP* and *INN* of the comparator. The common-mode level at the outputs of the pre-amplifier is adjustable at pin *BIAS* with a bias voltage.

B. Goll and H. Zimmermann, *Comparators in Nanometer CMOS Technology*,
Springer Series in Advanced Microelectronics 50, DOI 10.1007/978-3-662-44482-5_6

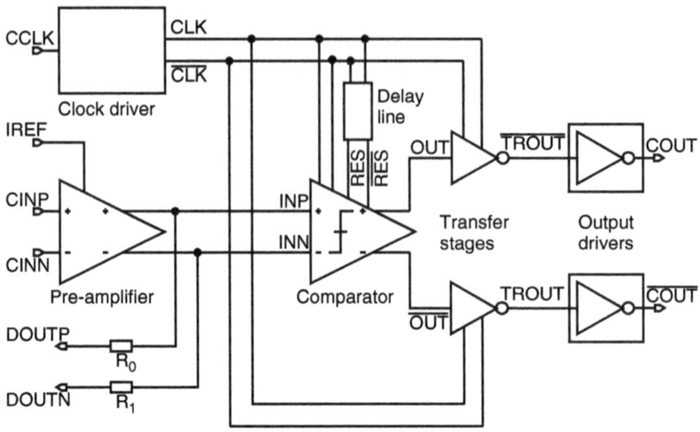

Fig. 6.1 Block diagram of the test chip

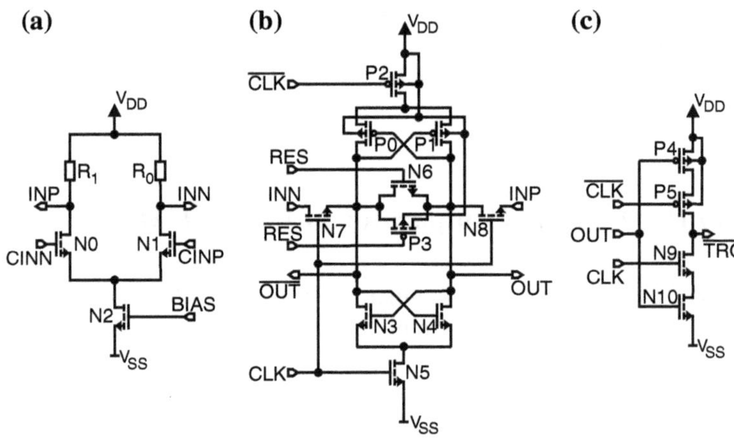

Fig. 6.2 Circuit of the pre-amplifier, the comparator and the latch [1]. **a** Pre-amplifier. **b** Comparator. **c** Transfer stage (clocked CMOS inverter)

The comparator is shown in Fig. 6.2b. It is only supplied with current during the regeneration phase by the switches $N5$ and $P2$. Transistors $N3$, $P0$, $N4$ and $P1$ build two cross-coupled inverters for positive feedback. During a whole clock cycle (signal *CLK*) there are a regenerative and a reset phase. When *RES* is *low* (voltage level V_{SS}) the transmission gate $N6$, $P3$ resets the latch to the metastable point. Using a transmission gate reduces the voltage drop via the reset transistors during reset phase. Such a drop would result in a slight hysteresis in the characteristic of the comparator. It occurs if there is not enough time for pulling both output nodes to the same voltage level of the metastable state so that an unwanted higher voltage drop appears at the transmission gate. Furthermore the influence of charges, introduced by the reset switches is reduced due to the complementary transistors. Mismatch of injected charge would affect the decision of the latch. When *RES* switches to *high* (voltage level V_{DD}) and transistors $N5$, $P2$ are turned on by *CLK* and $\overline{CLK}$, the transmission gate is turned off and the voltages *INP* and *INN* are introduced to the comparator via transistors $N7$ and $N8$. Regeneration starts and the output *OUT* is pulled to a voltage near to V_{DD} (near to V_{SS}) and $\overline{OUT}$ near to V_{SS} (near to V_{DD}), if *INP* is higher (smaller) than respectively. Full rail-to-rail swing of the outputs of the comparator is not possible, because the pre-amplifier is connected via transistors $N7$ and $N8$ during regeneration. The influence of injected charges from $N7$ and $N8$ can be neglected, because these transistors are much smaller than the reset transistors. On the other hand transistors $N7$ and $N8$ are large enough so that the degeneration of the amplification due to their on-resistances can be also neglected.

Each output ($OUT, \overline{OUT}$) of the comparator is connected to a transfer stage, which is depicted in Fig. 6.2c. During the regeneration phase (*CLK high* and *RES low*) transistors $P5$ and $N9$ are turned on and transistors $P4$ and $N10$ build an inverter. When the latch is in the reset-phase (*CLK low* and *RES* = *high*) the output of the transfer stage is switched to high-impedance and the state of the comparator is stored dynamically at the input capacitance of the output driver (nodes *TROUT*, $\overline{TROUT}$). Each output driver itself consists of a chain of five inverters. In Fig. 6.3a the clock driver is depicted. From a biased sine-wave, which is applied to pin *CCLK* a non-inverted clock *CLK* and an inverted clock $\overline{CLK}$ is generated with the help of inverters. Transmission gate $P7, N12$ is added to compensate the delay time difference between *CLK* and $\overline{CLK}$.

In Fig. 6.1 it can be seen, that the reset signals *RES* and $\overline{RES}$ are delayed in relation to the clock. So parasitic charge injections from switches $N7$ and $N8$ are shorted by the reset switch $P3$ and $N6$ at the beginning and a slight pre-charge time is generated before $P3$ and $N6$ are released so that the comparator can compare. The delay line consists of a chain of three inverters for each signal and has got a delay of approximately 40 ps (see Fig. 6.3b). Another positive effect is, that if the

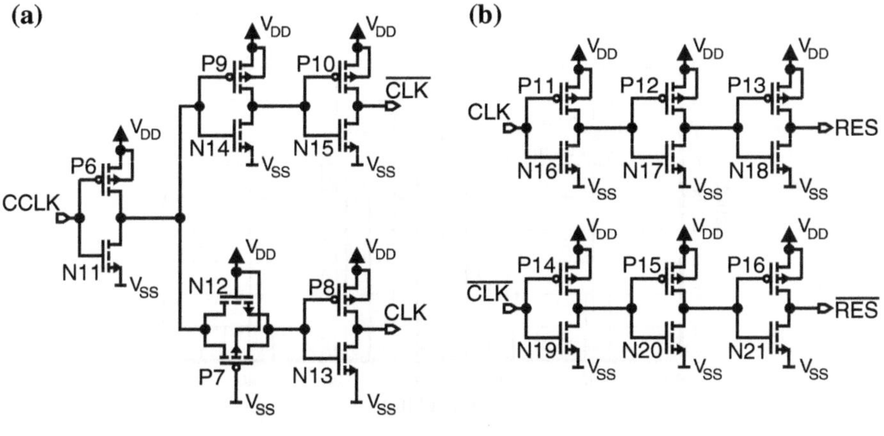

Fig. 6.3 Circuit of the clock driver and the delay line. **a** Clock driver. **b** Delay line

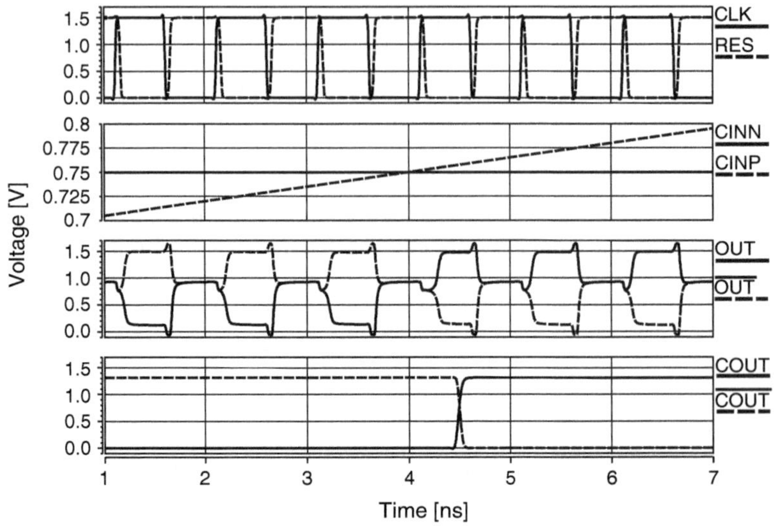

Fig. 6.4 Simulated waveforms

reset transistors are switched on delayed to the disabling of transistors $N5$ and $P2$, parasitic charge injection of $N5$ and $P2$ helps to enhance the final output voltage difference $OUT - \overline{OUT}$. So if e.g. OUT is near V_{DD} and $\overline{OUT}$ is near V_{SS} charge injection from $P2$ raises OUT through open transistor $P1$ and injection from $N5$ decreases $\overline{OUT}$ through open transistor $N3$.

Figure 6.4 shows simulated waveforms. There the delay between *CLK* and *RES* can be seen. In dependence on the sign of *CINP* – *CINN* (mismatch and noise neglected) the output node OUT switches towards a valid logical level near V_{DD} in the positive case and a valid logical level near V_{SS} in the negative case. In Fig. 6.5 the histogram of the offset of the comparator from a Monte Carlo simulation of

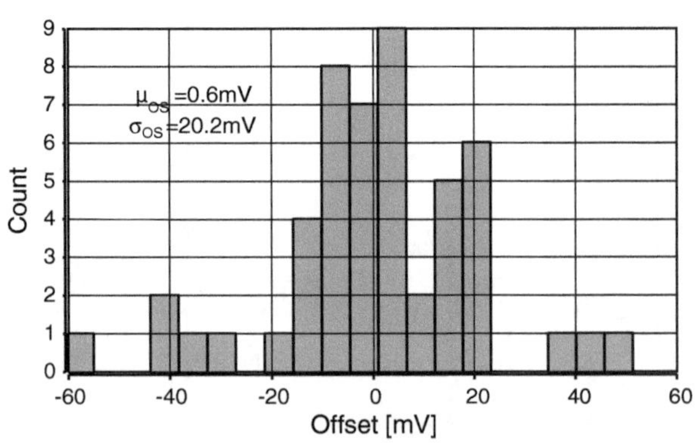

Fig. 6.5 Monte Carlo simulation of the offset (pre-amplifier and comparator)

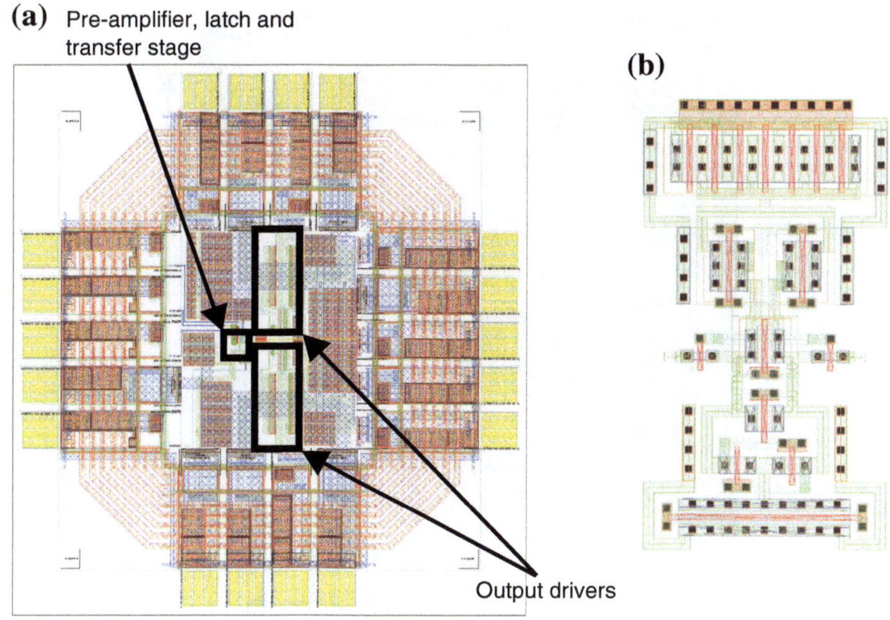

Fig. 6.6 Layout plots. **a** Text chip. **b** The latch

50 samples is depicted. For such a simulation a time duration of several hours on application servers were needed. The input common-mode voltage of the comparator was 0.75 V and the clock frequency was 1 GHz. The simulated standard deviation of the offset is $\sigma_{OS} = 20.2\,\text{mV}$.

Figure 6.6a shows a layout plot of the test chip with the comparator circuit and in Fig. 6.6b the layout plot of the latch can be seen. Instead the die photo does not show any details due to the passivation layer and the planarisation and fill structures of the metal layers, as it can be seen in Fig. 6.7. The area of the whole test chip amounts to $787\times857\,\mu\text{m}^2$. Thereof $16\times34\,\mu\text{m}^2$ is dedicated to the pre-amplifier and $10\times5\,\mu\text{m}^2$ is the size of the latch. A transfer stage needs a place of $4\times4.5\,\mu\text{m}^2$.

6.1.2 Measurement Results

The fabricated chip was bonded directly to a test board to reduce the influence of the inductances of a package and its long bond wires. Photos of the chip mounted on the test board are shown in Fig. 6.8. The wires, connected with an additional small print, where a ribbon cable with connectors is mounted, are for DC measurements via the microcontroller-board to the PC.

On the test board, 50 Ω microstrip lines were designed for high-frequency input and output signals of the chip (pins *CINP*, *CINN*, *CCLK*, *COUT* and $\overline{COUT}$). The

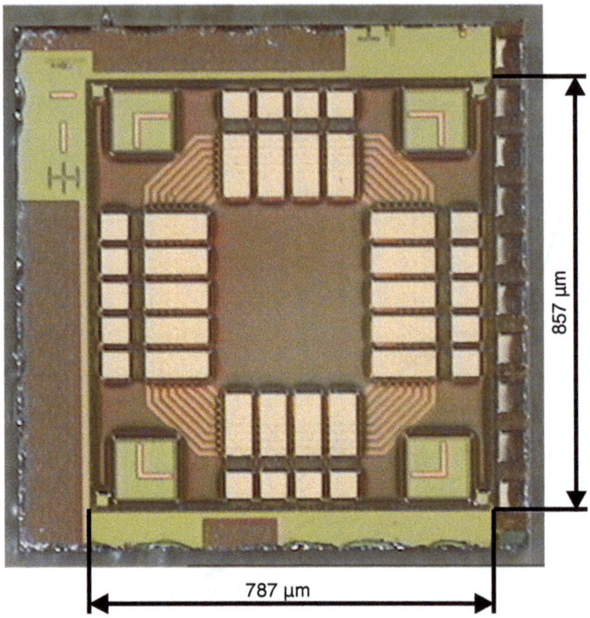

Fig. 6.7 Microphotograph of the test chip

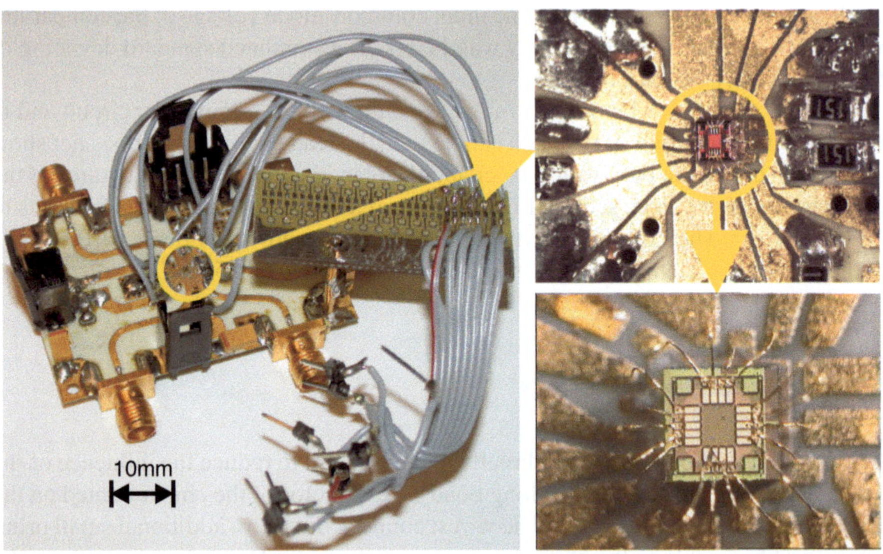

Fig. 6.8 Photos of the test board

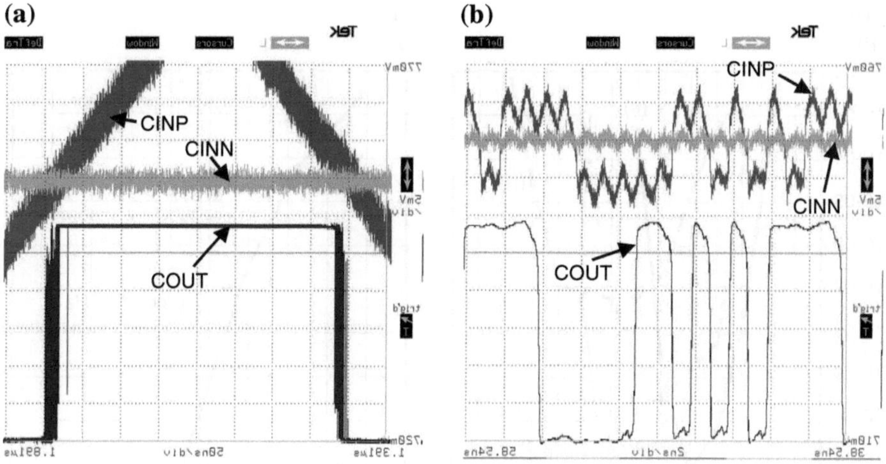

Fig. 6.9 Oscilloscope measurements [1]. **a** Oscilloscope picture of the 1.5 GHz clocked comparator, where a saw-tooth wave-form with 800 kHz is applied (50 ns/div, *CINN* and *CINP*: 5 mV/div, *COUT*: 200 mV/div). **b** Oscilloscope picture of the comparator, clocked with 1GHz, when a test wave with max. 500MHz switching rate is applied to *CINP* (2 ns/div, *CINN* and *CINP*: 5 mV/div, *COUT*: 200 mV/div)

microstrip lines for the input signals are conducted as near as possible to the chip and terminated by a 50 Ω resistor on board. Off-chip series resistors were introduced between the 50 Ω termination and the bond wire to the chip pad to damp oscillations caused by the inductance of the bond wire and the capacitance of the pad. To furthermore reduce the influence of bond wires, the pre-amplifier, the latch and the transfer stage were supplied with a separate power supply pad. Also the output drivers were supplied by separate pads. The typical measuring setup is described in Sect. 5.5. To get a first overview of the behavior of the comparator a reference voltage of 0.75 V was connected to CINN and a saw-tooth with a frequency of 800 kHz and with variable offset was applied to *CINP*. The sampling frequency was 1.5 GHz. The oscilloscope picture can be seen in Fig. 6.9a. Due to noise the comparator switches randomly near the cross-over of *CINN* and *CINP*. Figure 6.9b shows the behavior of the comparator, when a test-wave with a maximum rate of a half of the clock frequency was applied. To determine the sensitivity, BER measurements were done. On the test chip at pin *CINP* a reference voltage was applied, which was superimposed by a pseudo-random-bit-sequence (PRBS) $2^{31}-1$. At *CINN* a bias voltage, which compensated the offset was applied (best BER measured at a distinct bias voltage at *CINN*). Also the duty cycle of the clock was adjusted to measure best BER. The results are depicted in Fig. 6.10. Amplitude of PRBS $2^{31}-1$ means, that data (ideal in rectangular form) is switching maximal ± amplitude around the bias level at *CINP* or it can also be stated in another way, that amplitude $= \max\{CINP - CINN + \text{offset}\}$. Due to the fact, that the delay time of the comparator depends e.g. on the input voltage difference or on the input common-mode voltage level of the pre-amplifier, the delay between external reference clock and received comparator data had to be

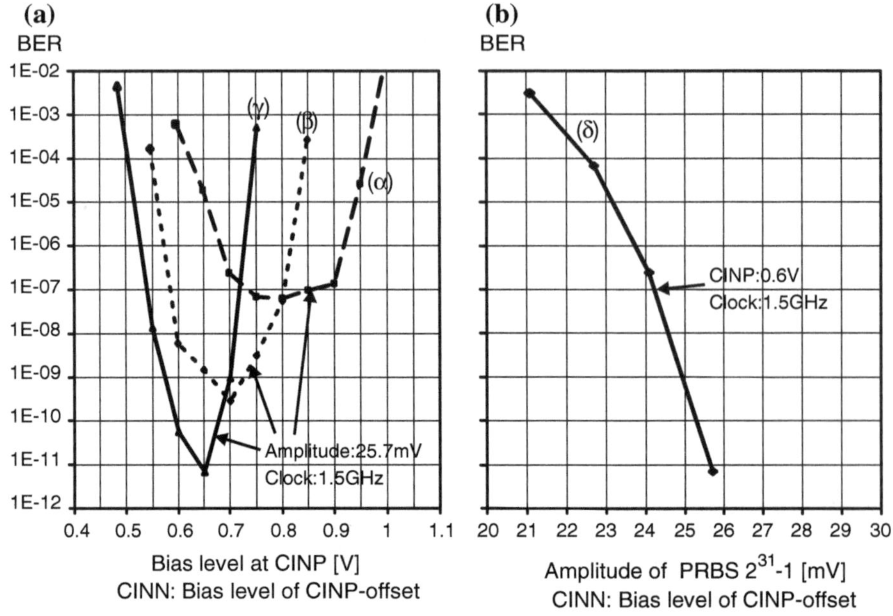

Fig. 6.10 BER measurements with a PRBS $2^{31}-1$. **a** BER versus bias level at CINP, clock: 1.5 GHz, *BIAS*: (α) 0.45 V, (β) 0.46 V, (γ) 0.47 V. **b** BER versus amplitude of $2^{31}-1$ clock: 1.5 GHz, *BIAS*: (δ) 0.47 V Bias level of *CINP*: 0.65 V

optimized to get best BER by adjusting the delay line in the bit pattern receiver for every measurement. In Fig. 6.10a at a clock frequency of 1.5 GHz the BER was measured versus the bias level at *CINP*, where the amplitude of PRBS was held constant at 25.7 mV at different reference voltages of $BIAS = (\alpha)$ 0.45 V, (β) 0.46 V, (γ) 0.47 V. When raising the bias voltage, the operating point with the optimal BER moves to lower input common-mode levels and BER becomes better. For applications an lower input common-mode level of about 0.65 V causes no design problems and so for a BER of 10^{-9} an minimal input amplitude of 25 mV at 1.5 GHz is needed (see Fig. 6.10b). Typically an offset of 5 mV was measured, but here no exact statistics of measurement values was done.

A reason, that the BER depends on *CINP* is, that the current sources $N0$, $N1$ and $N2$ of the pre-amplifier are not ideal (output resistance) and therefore the output common-mode voltage of the pre-amplifier $(INP + INN)/2$ depends also on the input common-mode voltage $(CINP + CINN)/2$. A lower $(CINP + CINN)/2$ causes a higher $(INP + INN)/2$, because the tail current through $N2$ also depends on the input voltages, it decreases and a smaller voltage drops occur at resistors R_0 and R_1 as a consequence. The optimal (lowest) BER occurs, when the output common-mode voltage of the pre-amplifier is equal to the output common-mode voltage of the latch during reset $(OUT + \overline{OUT})/2$, which is the switching threshold of inverters $P0$, $P3$ or $P1$, $N4$. Then the series on-resistances of switches $N7$

and $N8$ causes minimal voltage drops and don't reduce the amplification of the pre-amplifier. Therefore, the optimal point of BER is shifted to lower input bias levels *CINP* if *BIAS* is raised, because more current flows through the pre-amplifier and $(CINP + CINN)/2$ has to be lowered more to reach optimal $(INP + INN)/2$. On the other hand the small-signal amplification of the pre-amplifier tends to increase, if more tail current flows and the sensitivity becomes better (BER decreases) at the optimal point of operation.

The power consumption was determined at a clock frequency of 1.5 GHz under the condition that at *CINP* a bias voltage of 0.9 V, which was superimposed by a PRBS $2^{31} - 1$ with a sufficiently high amplitude of 30 mV was applied. At *CINN* a bias voltage, which compensated the offset and at *BIAS* 0.53 V were connected. A current consumption of 440 μA was measured for the pre-amplifier, the comparator and the transfer stage with the help of a separated supply voltage pad. Also at higher voltages at *BIAS* and operating points at lower input common-mode voltage levels, a current of about 440 μA was determined. With a supply voltage of 1.5 V this gives 660 μW power consumption.

6.2 A 360 μW, 2 GHz Comparator in 120 nm CMOS with Delayed Reset

In this section, a comparator with the complementary structure [2] compared to that presented in [3] is described. Additionally a delayed reset signal is used, where an increase of the output voltage difference is obtained with the help of parasitic charge injection. Because of the complementary structure, also the possibility to adjust the bias voltage of the separated n-wells of p-MOS transistors in the CMOS process used is given to alter the threshold voltages. The advantage of these modifications is, that they can easily be realized. In the used CMOS technology a p-MOS transistor also has a slightly better noise characteristics compared to its n-MOS counterpart.

6.2.1 Circuit Description

In Fig. 6.11 the schematic of the comparator is shown [2]. The capacitor C_A represents the parasitic capacitances at node A. C_L is the overall load capacitance at the output nodes OUT and $\overline{OUT}$. At pin *NWELL* the n-wells of the p-MOS transistors $P1$ to $P4$ are biased. In principle, if $\overline{CLK}$ is *high* (voltage level V_{DD}) the comparator is in the reset state and if $\overline{CLK}$ is *low* (voltage level V_{SS}) the input voltages at *CINP* and *CINN* are compared. During reset, transistor $P0$ is off and transistors $N2$ and $N3$ pull down the output nodes OUT and $\overline{OUT}$ to V_{SS}. When $P0$ is turned on and the reset switches $N2$ and $N3$ are released, transistors $P1$ and $P2$ start to charge the output nodes. In dependence on whether the analog voltage at *CINP* is higher

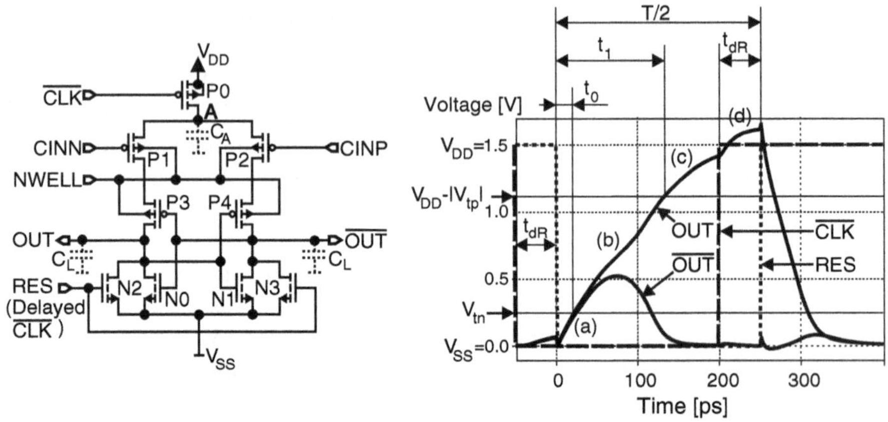

Fig. 6.11 Schematic of the comparator [2]. *a* Charging of output nodes. *b* Latch $N0$, $N1$, $P3$, $P4$ regenerates *c* $N0$, $P4$ are off and node OUT is charged, *d* charge injection of $P0$ raises the voltage level of node OUT because of a delay t_{dR} between $\overline{CLK}$ and RES

or lower than $CINN$ one output node, e.g. OUT in the case $CINP > CINN$, is charged faster (see Fig. 6.11a). This node reaches at time t_0 the threshold voltage $V_{tn} = V_{tn1} = V_{tn0}$ of $N1$, which is assumed to be the same as $N0$, before $\overline{OUT}$ and the latch $N0$, $N1$, $P3$, $P4$ is turned on and pulls OUT towards supply level V_{DD} and $\overline{OUT}$ towards ground level V_{SS} because positive feedback of the latch is established (regeneration of latch, see Fig. 6.11b). In [3, 4] an analytical expression for the time for charging the output nodes and the following regeneration time of the latch is presented. For the following explanations the case is assumed, that OUT goes *high* ($CINP > CINN$) and that for simplicity the regeneration time of the latch lasts till $N0$ and $P4$ turn off at about the same time (see Fig. 6.11a, b, small transition regions are neglected). Then the output-voltage difference $\Delta V_{OUT}(t) = OUT(t) - \overline{OUT}(t)$ reaches $\Delta V_{OUT}(t = t_1) \approx V_{DD} - |V_{tp}| - V_{tn}$ at the end of the regeneration time t_1, where $P0$ is assumed to have sufficient high W/L-ratio to pull node A very close to V_{DD} and input transistors $P1$ and $P2$ are sufficient large. $V_{tp} = V_{tp3} = V_{tp4}$ is the threshold voltage of transistors $P3$ and $P4$. The expression of [3] adapted for Fig. 6.11a, b leads then to 6.1.

$$t_1 = \frac{2C_L V_{tn}}{I_0} + \frac{C_L}{g_m} ln\left(\frac{I_0 \left(V_{DD} - V_{tn} - |V_{tp}|\right)}{2V_{tn} g_{mp1} \Delta V_{in}}\right) \tag{6.1}$$

I_0 is the current through $P0$, $g_{mp1} = g_{mp2}$ are the transconductances of $P1$ and $P2$, $g_m \approx g_{mn0} + g_{mp3}$ ($g_{mn1} = g_{mn0}$, $g_{mp4} = g_{mp3}$ is assumed) is the effective transconductance of the latch (transistors $N0$, $N1$, $P3$ and $P4$ with transconductances g_{mn0}, g_{mn1}, g_{mp3} and g_{mp4} respectively) and $\Delta V_0 = CINP - CINN$ is the input voltage difference, which is assumed, that it does not change during the comparison phase of the comparator. After $N0$ and $P4$ have turned off (case $CINP > CINN$,

$OUT \rightarrow V_{DD}$, $OUT \rightarrow V_{SS}$) OUT is charged via $P0$, $P1$ and $P3$ to V_{DD} and $\overline{OUT}$ is discharged via $N1$ to V_{SS} (see Fig. 6.11c). In this phase the output voltages can be approximated to the relations stated in 6.2 and 6.3 [2].

$$OUT(t) = V_{DD} - |V_{tp}|e^{\frac{-(t-t_1)}{C_L R_P}} \qquad t_1 < t < \frac{T}{2} \tag{6.2}$$

$$\overline{OUT}(t) = V_{tn}e^{\frac{-(t-t_1)}{C_L R_N}} \qquad t_1 < t < \frac{T}{2} \tag{6.3}$$

T is the clock period and R_N is a resistor, which approximates the speed of discharging node $\overline{OUT}$ to V_{SS} through transistor N_1, which is assumed to be in triode region. Resistor R_P characterizes the charging speed of node OUT to V_{DD} through transistors $P0$, $P1$ and $P3$, which can also be assumed to be in triode region, because of the voltage level of OUT is assumed to be sufficiently high. R_P depends also on the common mode level at the inputs ($CINP$, $CINN$) of the comparator.

If the signal RES at reset transistors $N2$ and $N3$ is delayed by a time t_{dR} to $\overline{CLK}$ (see Fig. 6.11d) then the reset transistors $N2$ and $N3$ turns on the time t_{dR} after $P0$ has turned off. The voltage at node OUT can be assumed to be $OUT(t = T/2 - t_{dR})$ (50 % duty cycle of RES and $\overline{CLK}$ assumed), when $P0$ turns off. Turning off $P0$ also injects a charge Q_{inj} to node A, which depends on the dimensions of this transistor, which as a consequence defines the overlap gate-drain capacitance, the channel charge and part of capacitance C_A of node A. A larger transistor $P0$ tends to inject more charge to node A, but it also has to be considered, that C_A also increases. Here a sufficient high W/L-ratio of $P0$ also has the advantage of an enhanced speed of the comparator, but a lower immunity against noise and mismatch [3]. The charge Q_{inj} adds momentarily the voltage $\Delta V_E = Q_{inj}/C_A$ to node A and redistributes over $P1$ and $P3$ between node A and OUT during time t_{dR} till the reset switches $N2$ and $N3$ are turned on. Redistribution of the charge Q_{inj} causes the voltage level at OUT to rise. Because $\overline{OUT}$ is near V_{SS} and transistor $P4$ is turned off, no charge flows to node $\overline{OUT}$ and the output voltage difference gets higher. When the parasitic capacitances in the intermediate nodes between $P1$, $P3$ and between $P2$, $P4$ are neglected and the initial voltage of node A before charge injection is assumed to be V_{DD} (sufficient high W/L-ratio of $P0$, node A is momentarily raised to $V_{DD} + \Delta V_E$), the output voltage difference $\Delta V_{OUT}(T/2, t_{dR}) = OUT(T/2, t_{dR}) - \overline{OUT}(T/2, t_{dR})$ at the time point, when reset starts, can be described analytically ($0 \leq t_{dR} \leq T/2 - t1$) with 6.4, 6.5 and 6.6 (charge redistribution between node A and OUT in the case $CINP > CINN$).

$$\begin{aligned} OUT(T/2, t_{dR}) = {} & \frac{C_A}{C_A + C_L}(V_{DD} + \Delta V_E)\left(1 - e^{\frac{-t_{dR}}{\tau_{Qdis}}}\right) \\ & + \frac{OUT(T/2 - t_{dR})}{C_A + C_L}\left(C_A e^{\frac{-t_{dR}}{\tau_{Qdis}}} + C_L\right) \quad 0 \leq t_{dR} \leq \frac{T}{2} - t_1 \end{aligned} \tag{6.4}$$

$$\overline{OUT}(T/2, t_{dR}) = \overline{OUT}(T/2) \quad 0 \leq t_{dR} \leq \frac{T}{2} - t_1 \tag{6.5}$$

$$\Delta u(T/2, t_{dR}) = OUT(T/2, t_{dR}) - \overline{OUT}(T/2) = \frac{C_A}{C_A + C_L}(V_{DD} + \Delta V_E)\left(1 - e^{\frac{-t_{dR}}{\tau_{Qdis}}}\right)$$
$$+ \frac{OUT(T/2 - t_{dR})}{C_A + C_L}\left(C_A e^{\frac{-t_{dR}}{\tau_{Qdis}}} + C_L\right) - \overline{OUT}(T/2) \qquad 0 \leq t_{dR} \leq \frac{T}{2} - t_1 \quad (6.6)$$

In 6.4 and 6.6 $\tau_{Qdis} = C_A C_L R_{P2}/(C_A + C_B)$ represents the time constant of charge distribution, where R_{P2} is the average resistance, which represents transistors $P1$ and $P3$ (assumed to be in triode region). In Fig. 6.12 simulated waveforms of the comparator are depicted, where the influence of a delay time t_{dR} between RES and $\overline{CLK}$ is shown. It can be seen, that the overall output voltage difference $OUT - \overline{OUT}$ gets higher and broader (see also Fig. 6.13), when a delay time t_{dR} is introduced, which helps that the decision is recognized better from a following logic gate to reduce occurrence of metastability errors at a higher clock frequency. After a maximum of the output voltage difference at a distinct time t_{dRmax} the output voltage difference gets smaller when t_{dR} is raised. A simulation of the output voltage difference

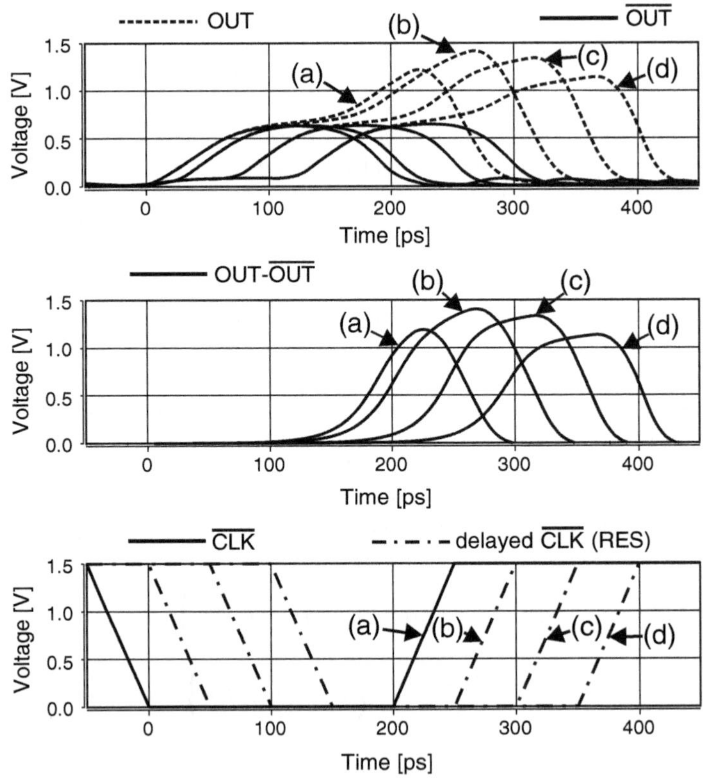

Fig. 6.12 Transient simulation of the comparator [2] (2 GHz clock), where RES is delayed with a $t_{dR} = 0\,ps$, b $t_{dR} = 50\,ps$, c $t_{dR} = 100\,ps$, d $t_{dR} = 150\,ps$ from $\overline{CLK}$ at $CINP = 0.85\,\text{V} + 1\,\text{mV}$, $CINN = 0.85\,\text{V}$

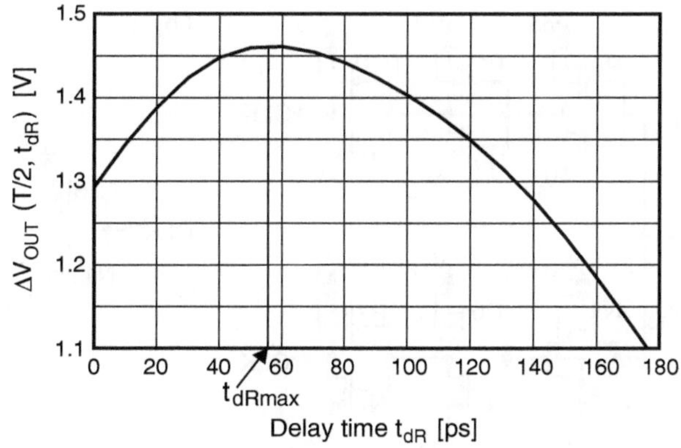

Fig. 6.13 Simulation of the output voltage difference $\Delta V_{OUT}(T/2, t_{dR}) = OUT(T/2, t_{dR}) - \overline{OUT}(T/2, t_{dR})$ versus t_{dR} at $CINP = 0.85\,\text{V} + 1\,\text{mV}$, $CINN = 0.85\,\text{V}$ and 2 GHz clock frequency [2]

$\Delta V_{OUT}(T/2, t_{dR}) = OUT(T/2, t_{dR}) - \overline{OUT}(T/2, t_{dR})$ at the beginning of the reset phase versus t_{dR} can be seen in Fig. 6.13, where a maximum exists at $t_{dRmax} > 0$. For the simulations, which are depicted in Figs. 6.12 and 6.13, extracted parasitic capacitances of the layout has been added and a clock edge of 50 ps has been taken into account. In this design at 2 GHz the output voltage difference was enhanced by about 0.17 V (13 % related to 1.3 V) due to simulation, if a delay time of about 50 ps is used. The delay time t_{dRmax} can be calculated numerically by equating the derivation of (6.6) with zero. The block diagram of the test chip with the comparator [2] is depicted in Fig. 6.14. The circuit is designed in a 0.12 μm CMOS process with a nominal supply voltage of 1.5 V. The circuit of the clock driver is shown in Fig. 6.15a. From a biased sine-wave, which is applied to pin *CCLK* a non-inverted clock *CLK* and an inverted clock $\overline{CLK}$ is generated with the help of inverters. Transmission gate

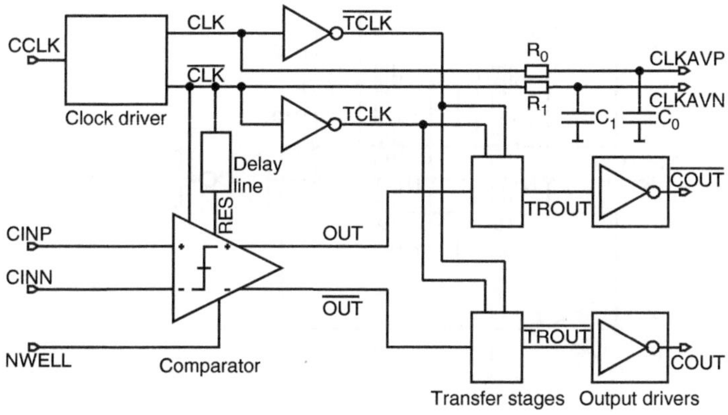

Fig. 6.14 Block diagram of the test chip

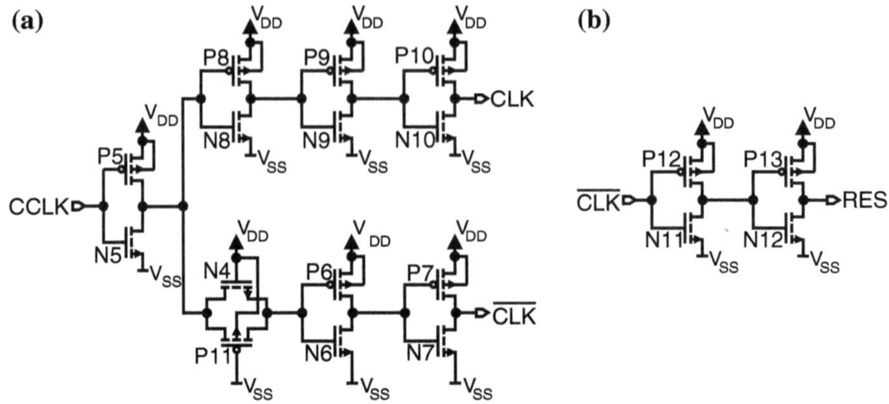

Fig. 6.15 Circuit of the clock driver and the delay line. **a** Clock driver. **b** Delay line

*N*4, *P*11 is added to compensate the delay time difference between *CLK* and $\overline{CLK}$. The schematic of the delay line is depicted in Fig. 6.15b. A delay time between *RES* and $\overline{CLK}$ of about 50 ps is introduced with the help of a delay line, which consists of a chain of two inverters. With modifying the bias voltage of the sine-wave and measuring pins *CLKAVP* and *CLKAVN*, the duty cycle could be adjusted. Via pin *NWELL* (see Figs. 6.11 and 6.14) a voltage for biasing the n-well can be applied from outside the chip. Each output *OUT* and $\overline{OUT}$ of the comparator is connected with the input of a transfer stage. Figure 6.16 shows the schematic of the transfer stage. The transfer stage contains two clocked inverters, which consist of transistors *N*13, *N*14, *P*14, *P*15 and transistors *N*15, *N*16, *P*16, *P*17 and act like a shift register. The transfer stage is added to hold the decision of the comparator during its reset phase and to have in principle an overall constant delay time between the clock *CCLK*

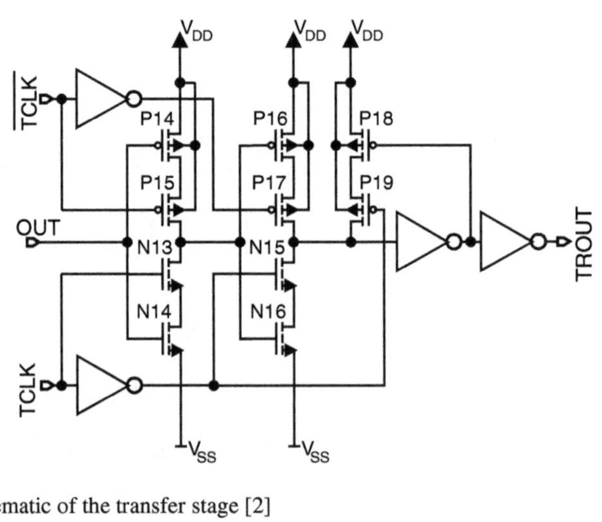

Fig. 6.16 Schematic of the transfer stage [2]

and the output data *COUT* or $\overline{COUT}$ of the test chip. Because of the decision time of the comparator depends on various factors, e.g. the input-voltage difference or the common-mode voltage at the inputs of the comparator (*CINP*, *CINN*), such a transfer stage simplifies BER measurements in that way, that for a next BER measurement with new conditions at the inputs the delay time between *CCLK* and *COUT* or $\overline{COUT}$ has not to be adjusted once again in the off-chip bit-pattern receiver. Signals *TCLK* and $\overline{TCLK}$ are delayed adequately by inverters, because the delay between signal *RES* and $\overline{CLK}$ has to be taken into account (see Fig. 6.14). In the transfer stage *TCLK* and $\overline{TCLK}$ are delayed once against compensate the delay time of the first clocked inverter (transistors $N13$, $N14$, $P14$, $P15$). When signal *RES* is *low* (voltage level V_{SS}), the comparator is in the decision phase. During this phase signal *TCLK* is *high* (voltage level V_{DD}), $\overline{TCLK}$ is *low*, transistors $N13$ and $P15$ are switched on and transistors $N15$ and $P17$ are switched off so that the second clocked inverter (transistors $N15, N16, P16, P17$) is in a high-impedance state. The input capacitance of the second clocked inverter is loaded with the decision of the comparator. The previous decision of the comparator, which has happened one clock period before, is hold at the output of the second clocked inverter via $P18$ by switching on $P19$. When the comparator changes to reset phase (*RES* is *high*), the first clocked inverter is switched to high-impedance state by turning off transistors $N13$, $P15$, the second clocked inverter is switched on by turning on transistors $N15$, $P17$ and transistor $P19$ is turned off. The decision of the comparator is stored dynamically at the input capacitance of the second clocked inverter so that the output of the second clocked inverter is loaded with the new decided logical value of the comparator. Each output *TROUT* and $\overline{TROUT}$ of the transfer stage is connected with an output driver (see Fig. 6.14). The output driver is simply a chain of five inverters to drive an off-chip 50 Ω measurement system. Of the comparator circuit, where also the extracted parasitic capacitances of the layout have been taken into account, a Monte Carlo simulation of 100 runs was performed for an input common-mode voltage of 0.85 V at *CINP* and *CINN* at a clock frequency of 1 GHz. The histogram is depicted in Fig. 6.17 and

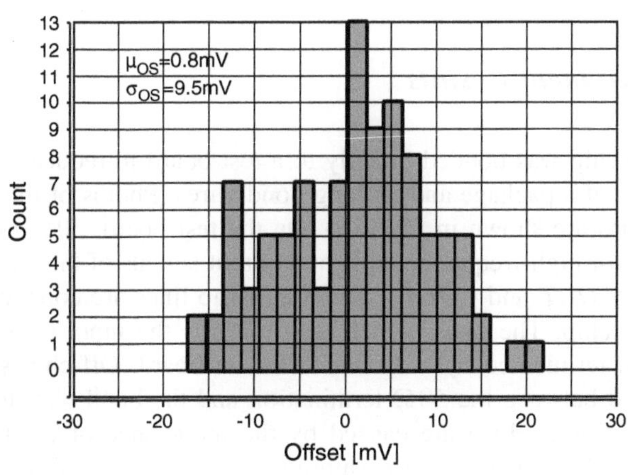

Fig. 6.17 Monte Carlo simulation of the comparator

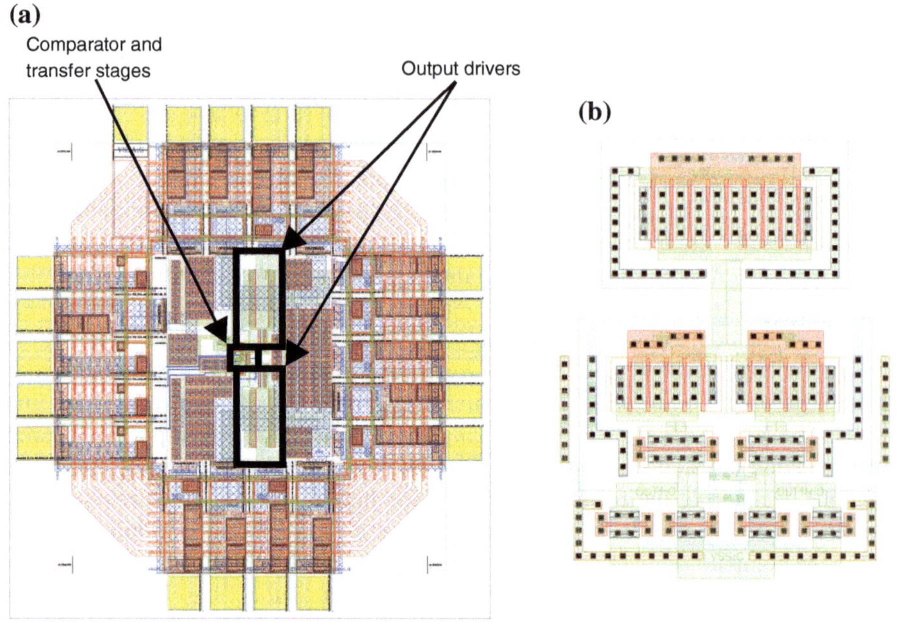

Fig. 6.18 Layout plots. **a** Test chip. **b** Comparator

the standard deviation of the offset was simulated to be $\sigma_{OS} = 9.5\,\text{mV}$. Figure 6.18a shows a layout plot of the test chip with the comparator and a plot of the comparator itself. Instead the die photo does not show any details due to the passivation layer and the planarisation and fill structures of the metal layers in this 0.12 µm CMOS technology, as it can be seen in Fig. 6.19. The area of the whole test chip amounts to $787 \times 857\,\mu\text{m}^2$. Thereof $11 \times 14\,\mu\text{m}^2$ is dedicated to the comparator and $10 \times 8\,\mu\text{m}^2$ is the size of a transfer stage.

6.2.2 Measurement Results

The fabricated chip was bonded directly to a test board to reduce the influence of the inductances of a package and its long bond wires. Photos of the chip mounted on the test board are shown in Fig. 6.20. On the test board, 50 Ω microstrip lines were designed for high-frequency input and output signals of the chip (pads *CINP*, *CINN*, *CCLK*, *COUT* and $\overline{COUT}$). The microstrip lines are conducted as near as possible to the chip. The lines, which lead towards the input pads *CINP*, *CINN* or *CCLK* were terminated by a 50 Ω resistor on board. Off-chip series resistors were introduced between the 50 Ω termination and the bond wire to the chip pad to damp oscillations, which are caused by the inductance of the bond wire and the capacitance of the pad. To furthermore reduce the influence of bond wires the

Fig. 6.19 Microphotograph of the test chip

Fig. 6.20 Photos of the test board

comparator and the output drivers were supplied by separate power supply pads. The typical measuring setup is described in Sect. 5.5. To get a first overview of the behavior of the comparator a reference voltage of 0.85 V was connected to CINN and a saw-tooth with a frequency of 800 kHz with variable offset was applied to *CINP*. The sampling frequency was 2 GHz and the bias voltage at pin *NWELL* was

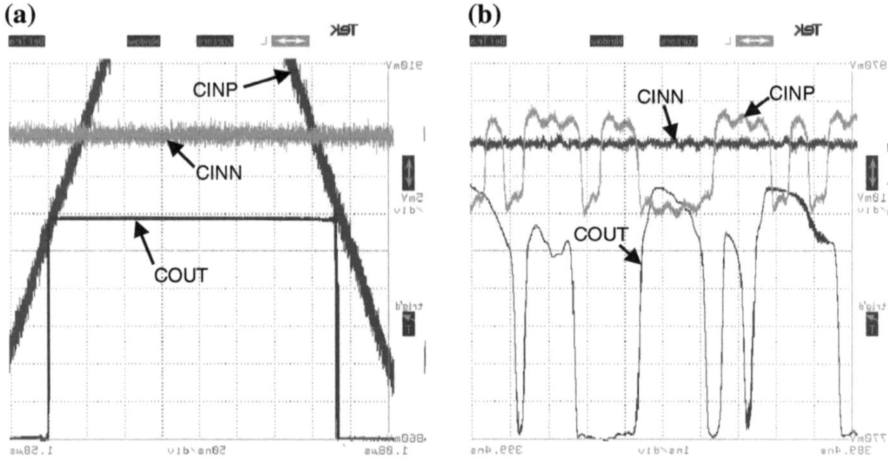

Fig. 6.21 Oscilloscope measurements [2]. **a** Oscilloscope picture of the 2 GHz clocked comparator, where a saw-tooth wave-form with 800 kHz is applied (50 ns/div, *CINN* and *CINP*: 5 mV/div, *COUT*: 200 mV/div). **b** Oscilloscope picture of the 2 GHz clocked comparator, when a test wave with max. 1 GHz switching rate is applied to *CINP* (1 ns/div, *CINN* and *CINP*: 10 mV/div, *COUT*: 200 mV/div)

1 V. The oscilloscope picture can be seen in Fig. 6.21a. Due to noise the comparator switches randomly near the cross-over of *CINN* and *CINP*. Figure 6.21b shows the behavior of the comparator, when a test-wave with a maximum rate of a half of the clock frequency was applied (*NWELL* = 1.5 V). To determine the sensitivity, BER measurements were done. On the test chip at *CINN* a reference voltage was applied. At pin *CINP* a bias voltage was applied, which compensated the offset and which was superimposed by a pseudo-random-bit-sequence (PRBS) $2^{31} - 1$. The bias voltage at *CINP* was chosen so that best BER was measured at a distinct voltage difference *CINP* − *CINN*, which can be seen as offset. The duty cycle of the clock *CLK* and $\overline{CLK}$ were adjusted by measuring the average voltages at pins *CLKAVP*, *CLKAVN* and and varying the bias voltage of the sine-wave at the input pin *CCLK* of the test chip. The results are depicted in Fig. 6.22. Amplitude of PRBS $2^{31} - 1$ means, that data (ideal in rectangular form) is switching maximal ± amplitude around the bias level at *CINP* or it can also be stated in another way, that amplitude = max(*CINP*−*CINN*+ offset). In Fig. 6.22a at a clock frequency of 1.5 GHz the BER was measured versus the bias level at *CINN*, where the amplitude of PRBS $2^{31} - 1$ was held constant at 9.5 mV and the bias voltage of *NWELL* was varied from (α) 1.5 V, (β) 1 V to (γ) 0.8V. For comparison on Fig. 6.22a also the case with an amplitude of 10 mV and *NWELL* = (α) 1.5 V is depicted.

In Fig. 6.22a it can be seen, that there exists a distinct reference voltage at *CINN* (similar to the input common-mode voltage (*CINN* + *CINP*)/2), where the BER is optimal. For *NWELL* = 1.5 V and a clock frequency of 1.5 GHz the BER is at *CINN* = 0.85 V optimal. If *CINN* is raised the input common-mode level gets nearer to the cut-off of input transistors *P*1 and *P*2. Here the BER approaches more and

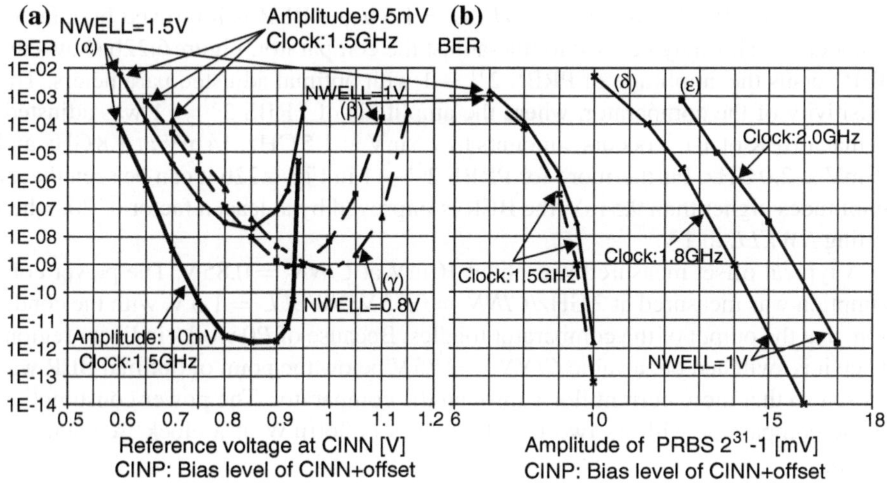

Fig. 6.22 BER measurements with a PRBS $2^{31}-1$ [2]. **a** BER versus bias level at *CINN*, clock: 1.5 GHz, *NWELL*: (α) 1.5 V, (β) 1 V, (γ) 0.8 V. **b** BER versus amplitude of $2^{31}-1$ *CINN*: (α) 0.95 V, (β) 0.85 V, (δ) 0.95 V, (ε) 0.9 V duty cycle: (α, β) 50 %, (δ, ε) 46 %

more 0.5, because at a constant amplitude charging nodes OUT and $\overline{OUT}$ at the beginning of the comparison phase of the comparator needs more time (see Fig. 6.11a) and the overall decision time t_1 lasts longer, even at a higher initial output voltage difference ($\Delta V_0 = \Delta V_{OUT}(t = t_0) = OUT(t = t_0) - \overline{OUT}(t = t_0)$) of the latch (transition from (a) to (b) in Fig. 6.11). This approaches more and more a metastability error, where also noise influence the decision (BER→0.5) of the comparator (see Sect. 2.4.2). Because of tail transistor $P0$ (see Fig. 6.11) works as a switch (linear region of $P0$) the tail current depends among other things also on the input common-mode voltage $(CINN + CINP)/2$. In [4, 5] an equation for the initial output voltage difference $\Delta V_0 = \Delta V_{OUT}(t = t_0) = OUT(t = t_0) - \overline{OUT}(t = t_0)$, which appears at the start of the regeneration of the latch ($N0, N1, P3, P4$) is stated. In other words OUT or $\overline{OUT}$ has just reached V_{tn} of $N1$ or $N0$ respectively and the latch is switched on and begins to regenerate with an initial voltage difference ΔV_0. Adapted to the comparator in Fig. 6.11a the equation of ΔV_0 is depicted in 6.7 [4, 5].

$$\Delta V_0 = \Delta V_{OUT}(t = t_0) = 2V_{tn}\frac{g_{mp1}\Delta V_{in}}{I_0}, \quad t_0 = \frac{2C_L V_{tn}}{I_0} \tag{6.7}$$

This initial voltage difference ΔV_0 can be seen as the amplified input voltage difference ΔV_{in}, which causes the latch to decide. As mentioned above $P0$ works, when it is switched on, in the linear region. Therefore the tail current I_0 increases, when the input common-mode voltage $(CINN + CINP)/2$ is lowered. This causes ΔV_0 to decrease and as a consequence the immunity against noise and mismatch also decreases and the measured BER becomes worse. By adjusting the bias voltage at *NWELL* (see Fig. 6.22a) the optimal point of BER can be shifted and for

NWELL = 0.8 V, 1 V instead of *NWELL* = 1.5 V the BER is improved by a factor of about 20. This may be used to fine-adjust the comparator. Figure 6.22b shows the BER versus the amplitude of PRBS $2^{31} - 1$ with optimal adjusted parameters. The sensitivity of the comparator, where the amplitude of PRBS $2^{31} - 1$ was adjusted, so that a BER of 10^{-9} occurs, amounts to 9.5 mV at 1.5 GHz, 14 mV at 1.8 GHz and 16 mV at 2.0 GHz. Furthermore for PRBS $2^{31} - 1$ in Fig. 6.22b it can be seen that at amplitudes higher than 9.5 mV, the BER is improved by at least a factor of 20, when setting *NWELL* to 1 V.

A typical offset measured was about 10 mV at *CINN* = 0.85 V. The power consumption was measured at 2 GHz/*CINN* = 0.75 V/*NWELL* = 1.5 V, with the condition, that the output of the comparator toggles. Because of *P*0 works in linear region, when it is switched on, a value *CINN* = 0.75 V below the point of optimal BER was chosen so that more current flows through the comparator. The power consumption of the comparator with its two transfer stages is 360 μW at a clock frequency of 2 GHz.

6.3 A 812 μW, 4 GHz Comparator in 120 nm CMOS with Adjustable Sensitivity

In this section a comparator with the capability of a high decision speed is described. The circuit was designed to avoid a static power consumption so that only a current flow during the decision action of the comparator takes place (dynamic power consumption). Furthermore a technique to adjust and improve the sensitivity of the comparator, which is designed for a distinct maximum clock frequency, in the case, that a lower clock frequency is applied, is described. An advantage of this technique is also a reduction of the dynamic power consumption while the sensitivity is improved.

6.3.1 Circuit Description

For first analytical considerations a basic underlying circuit of the comparators from the literature is extracted and shown in Fig. 6.23. During reset phase of the comparator, when the clock signals $\overline{EN}$ and $\overline{CLK}$ are at V_{DD} = 1.5 V (*high*), transistors *P*0 and *P*1 are switched off and transistors *N*2 and *N*3 pull both output nodes *OUT* and $\overline{OUT}$ to V_{SS} = 0 V (*low*). Advantageous here is, that during reset phase the output nodes define a logical level and therefore no voltage level anywhere between V_{DD} and V_{SS} occurs, which may cause an undesired current flow in a following CMOS logic gate. When clock signal nodes $\overline{EN}$ and $\overline{CLK}$ change to *low*, the reset switches *N*2 and *N*3 are released and transistors *P*0 and *P*1 are switched on. Due to the fact that immediately after reset *OUT* and $\overline{OUT}$ are still at V_{SS}, *P*4 and *P*5 are now switched on, but work in saturation mode, because their drain-source voltages have to be considered

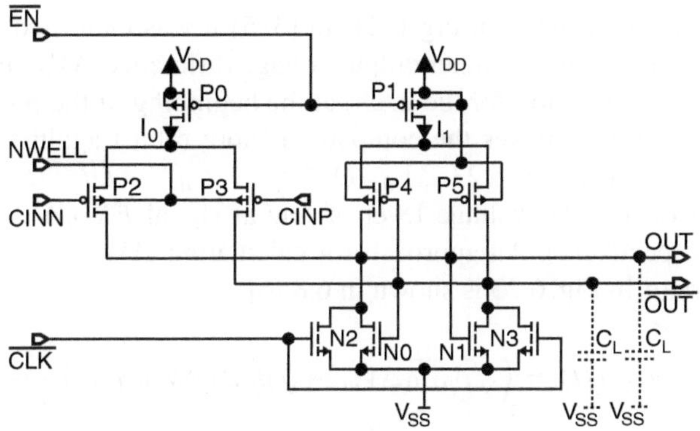

Fig. 6.23 Typical comparator structure from the literature

to be sufficiently large. Transistors $P2$ and $P4$ charge output node OUT and $P3$ and $P5$ charge $\overline{OUT}$. Transistors $P2$ and $P3$ are chosen to be p-MOS transistors in this comparator circuit, because in the used technology they have a slightly better noise characteristics than the n-MOS counterpart, but have also a lower transconductance, which degenerates amplification of the input voltage difference $CINP - CINN$. The drain current difference between $P2$ and $P3$, $\Delta I_0 = \sqrt{\beta_{p2} I_0} \Delta V_{in}$, is determined by the input voltage difference $\Delta V_{in} = CINP - CINN$ while the drain current difference between $P4$ and $P5$, $\Delta I_1 = \sqrt{\beta_{p4} I_1} \Delta V_{OUT}(t)$, is determined by the output voltage difference $\Delta V_{OUT}(t) = OUT - \overline{OUT}$. Here it is assumed, that the transconductance parameter β_{p2} of transistor $P2$ is equal to that of $P3$ ($\beta_{p2} = \beta_{p3}$) and β_{p4} of $P4$ equals that of $P5$ ($\beta_{p4} = \beta_{p5}$). When one of the output nodes reaches the threshold voltage V_{tn0} of transistor $N0$ or V_{tn1} of $N1$ ($V_{tn} = V_{tn0} = V_{tn1}$ is assumed) after a time duration t_{Init} (see times t_{Inita} and t_{Initb} in Fig. 6.24) and turns the appropriate transistor ($N0$ or $N1$) on, the whole latch ($N0$, $N1$, $P4$, $P5$) begins to switch due to positive feedback, because now a current flow between V_{DD} and V_{SS} is possible. A simulation

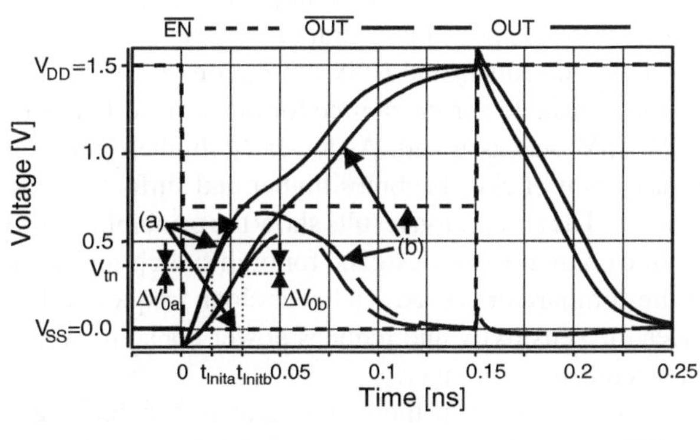

Fig. 6.24 Simulated signals of the comparator in Fig. 6.23 [6], where at $\overline{EN}$ the voltage level of *low* is *a* 0 V and *b* 0.6 V

of the waveforms is depicted in Fig. 6.24. In [3–5] it was identified at a latch-type sense amplifier circuit, that a higher output voltage difference ΔV_0 (amplified ΔV_{in}, which is the initial condition for the latch) at the beginning of the positive feedback of the latch at time t_{Init} makes the comparator more robust against mismatch and noise. For the circuit in Fig. 6.23 $\Delta V_0 = OUT(t = t_{Init}) - \overline{OUT}(t = t_{Init})$ can be enhanced with raising the voltage level of *low* at signal $\overline{EN}$ (ΔV_{0a} and ΔV_{0b} in Fig. 6.24a, b respectively). An approach for calculating $\Delta V_0 = \Delta V_{OUT}(t = t_{Init})$ for the comparator in Fig. 6.23 is shown in 6.8 [6].

$$\Delta V_{OUT}(t) = \left(\sqrt{\beta_{p2} I_0}\Delta V_{in} + \sqrt{\beta_{p4} I_1}\Delta V_{OUT}(t)\right)\frac{t}{C_L} \tag{6.8}$$

With $t_{Init} \approx 2C_L V_{tn}/(I_0 + I_1)$, $\Delta B_0 = \Delta V_{OUT}(t = t_{Init})$ can be calculated, where the result is depicted in 6.9.

$$\Delta V_0 = \Delta V_{OUT}(t = t_{Init}) = \frac{2V_{tn}\sqrt{\beta_{p2} I_0}}{I_0 + I_1 - 2V_{tn}\sqrt{\beta_{p4} I_1}}\Delta V_{in} \tag{6.9}$$

It can be seen, that if I_1 is reduced, ΔV_0 is enhanced. When varying I_0 at a fixed I_1 there exists a maximum of ΔV_0 at a current I_{0opt}, which is calculated in 6.10.

$$I_{0opt} = I_1 - 2V_{tn}\sqrt{\beta_{p4} I_1} \tag{6.10}$$

So if the currents I_0 through $P0$ and especially I_1 through $P1$ are reduced, ΔV_0 becomes larger, which makes the comparator more robust against noise and thus enhancing the sensitivity, but as a trade-off also the decision time of the comparator tends to last longer. A solution to somewhat control and reduce the drain currents of $P0$ and $P1$ is also shown in Fig. 6.24, where the lower voltage level of the rectangular clock signal at $\overline{EN}$ is raised from $V_{SS} = 0$ V (Fig. 6.24a)–0.6 V (Fig. 6.24b), where then $\Delta V_{0a} < \Delta V_{0b}$, but in the other way $t_{Inita} < t_{Initb}$. It should be also stated that a higher ΔV_0 causes the latch to switch faster and so a slight tendency to somewhat compensate the overall delay time of the comparator over a short tuning range is given [3–5]. Furthermore the power consumption of the comparator is also reduced because of lower drain currents I_0 and I_1. A simulation of ΔV_0 is shown in Fig. 6.25. For this simulations a variable current source for I_0 was inserted instead of transistor $P0$ and ΔV_{in} = 50 mV was assumed. At lower I_1 (higher gate voltage at $P1$, see Fig. 6.25b) the maximum of ΔV_0 becomes higher and shifts towards a smaller I_0 as depicted in Fig. 6.25b. The simulation results show the principle behavior as described in 6.8–6.10. So if the comparator is designed for high switching speeds the sensitivity degrades, when the comparator is used at a lower clock frequency. Reducing currents I_0 and I_1 enhances the sensitivity and reduces power consumption, but the decision of the comparator tends to last longer.

As a consequence the circuit of the comparator of Fig. 6.23 has been improved to the proposed circuit, which is shown in Fig. 6.26 [6]. When Fig. 6.26 is compared with Fig. 6.23, transistors $P6$ and $P7$ are added. In the reset phase, when transistors

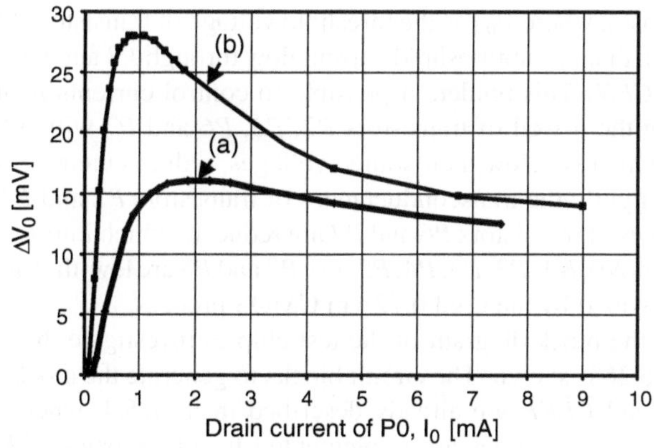

Fig. 6.25 Simulation of the output voltage difference $\Delta V_0 = \Delta V_{OUT}(t = t_{Init})$, at the beginning of positive feedback of the latch [6], where the lower voltage level at $\overline{EN}$ is *a* 0 V (standard clocking condition at $\overline{EN}$) and *b* 0.6 V

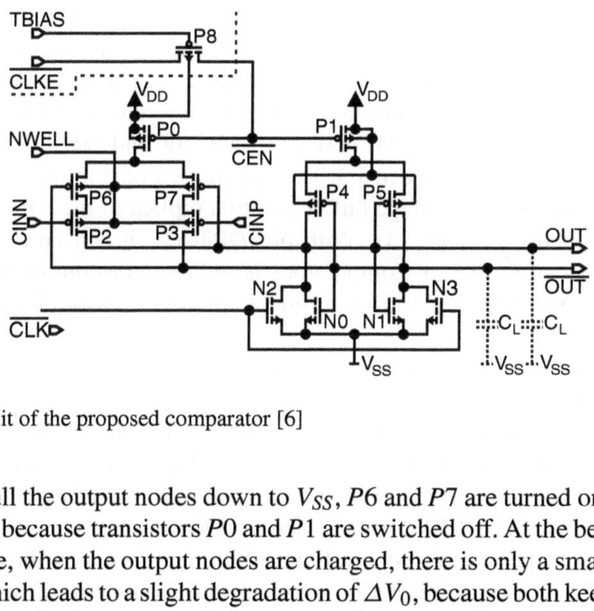

Fig. 6.26 Circuit of the proposed comparator [6]

$N2$ and $N3$ pull the output nodes down to V_{SS}, $P6$ and $P7$ are turned on, but no static current flows, because transistors $P0$ and $P1$ are switched off. At the beginning of the decision phase, when the output nodes are charged, there is only a small influence of $P6$ and $P7$, which leads to a slight degradation of ΔV_0, because both keep conducting. When the latch starts positive feedback and pulls e.g. OUT towards V_{DD} and $\overline{OUT}$ towards V_{SS}, $P7$ turns off and $P6$ stays on. So the current path from V_{DD} to node $\overline{OUT}$, which switches to *low*, is disconnected. Therefore $\overline{OUT}$, which is discharged from $N1$, can reach V_{SS}. Transistors $P6$ and $P4$ are on and OUT can reach V_{DD}, the output voltage difference reaches V_{DD} when the decision of the comparator is done and no static current flows. So the power consumption is only defined by a dynamic current. To have the possibility to adjust ΔV_0, transistor $P8$ is added in series to the clock signal $\overline{CLKE}$. With an appropriate bias voltage *TBIAS* at the gate of $P8$, signal $\overline{CEN}$ switches between V_{DD} (during reset phase) and $TBIAS + |V_{tp8}|$ (during comparison phase), where V_{tp8} is the threshold voltage of transistors $P8$ and the ideal case is assumed, that no subthreshold current flow through $P8$ and no leakage current exists at node $\overline{CEN}$. This renders it possible to control currents I_0 and I_1 to some extend. Biasing the n-well of transistors $P2$, $P3$, $P6$ and $P7$ with a bias voltage at pin NWELL, which is below their source voltages, reduces their threshold voltages and enhances slightly the transconductances of transistors $P2$ and $P3$. Furthermore the on-resistances of transistors $P6$ and $P7$ are reduced, which improves also slightly ΔV_0. Transistors $N0, N1, P2, P3, P4, P5, P6, P7$ and $P8$ are low-threshold transistors, which were provided by the used 0.12 μm CMOS process.

In Fig. 6.27 the block diagram of the test chip to investigate the proposed comparator of Fig. 6.26 is shown. The circuit blocks to generate the clock signals *CLKR*, $\overline{CLKR}$, *CLKE* and $\overline{CLKE}$ are already described in Sect. 5.4, where two separated voltage-controlled delay lines are implemented for clock pairs *CLKR*, $\overline{CLKR}$ and *CLKE* and $\overline{CLKE}$, which are controlled with bias voltages at pins *VDELR*, *VDELE* respectively and a bias voltage for the n-wells, which is applied at pin *NWELLDL*. The 4:1 multiplexer and the driver for the reference output *CREF* to monitor the internal clock signals are also described in Sect. 5.4.

The test chip consists of two comparators (Comparator1 and Comparator2), where each has the same circuit, which is shown in Fig. 6.26. Comparator1 is connected with its outputs *OUT* and $\overline{OUT}$ to transfer stages, which are used for BER and delay time measurements. The circuit of the transfer stage and the output drivers with the additional outputs *DM0AV DM1AV DM2AV* and *DM3AV* for delay time measurement are described in Sect. 5.3. The outputs of Comparator2 (*OUTA*, $\overline{OUTA}$) are connected to an analog voltage buffer and to a replica of the buffer to guarantee a symmetrical load (see Sect. 5.1). With pin *DIGBUF* it can be chosen, whether the output *OUTA* or $\overline{OUTA}$ of Comparator2 is redirected to output *SIGOUT* to be measured. The functional description of the temperature sensor diode can be found in Sect. 5.2.

With the help of pad *NWELL* a bias voltage for the n-wells of transistors $P2$, $P3$, $P6$ and $P7$ of the comparators can be applied.

Comparator1 is clocked with $\overline{CLK}$ and $\overline{CEN}$ and comparator2 with *CLK* and *CEN*, which are in principle the clock signals $\overline{CLKR}$, $\overline{CLKE}$ and *CLKR*, *CLKE* respectively, but where each of it is lead through a transmission gate. The transmission gates between $\overline{CLKR}$ and $\overline{CLK}$ and between *CLKR* and *CLK* are always on and are only added to compensate the delay time differences between the clock signals. $\overline{CLK}$ and *CLK* are used to reset Comparator1 and Comparator2 respectively and are the standard clock, which are also connected to the transmission gates. With the digital input *DIGC* it can be chosen, whether the transmission gates between $\overline{CLKE}$ and $\overline{CEN}$ and between *CLKE* and *CEN* are always switched on (standard clocking of comparators) or only a p-MOS transistor is switched in between instead of the transmission gates (the n-MOS transistor is switched off). Then at the gate of both p-MOS transistors a bias voltage *TBIAS* is applied from a pad to the test chip so that $\overline{CLKE}$ and *CLKE* switches between $V_{DD} = 1.5$ V (reset phase) and about $TBIAS + |V_{tp}|$ (comparison phase), where $V_{tp} = V_{tp8}$ is the threshold voltage of both p-MOS transistor. These two possibilities of clocking the comparators were

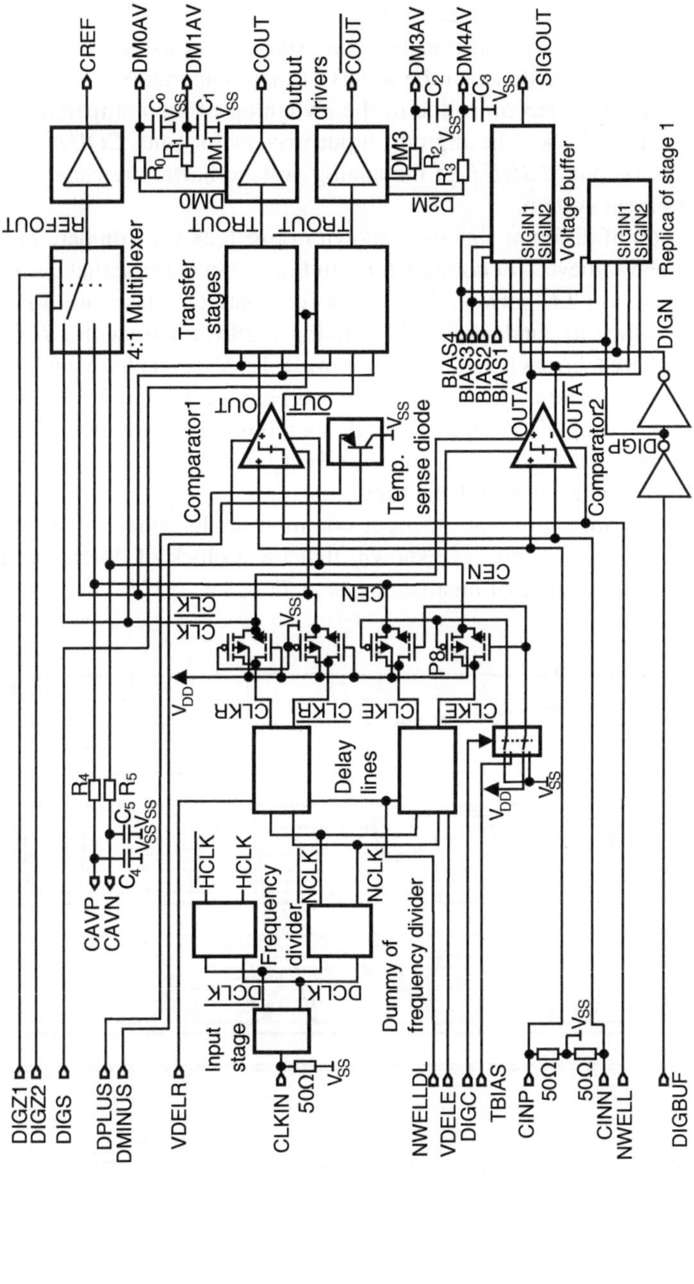

Fig. 6.27 Block diagram of the test chip to investigate the proposed comparator

implemented to compare the case of standard clocking with the case of raising the lower voltage level of signals *CEN* and $\overline{CEN}$. With the help of output pins *CAVP* and *CAVN* the duty cycle of clock signals *CEN* and $\overline{CEN}$ can be measured. The duty cycle can be adjusted by varying the bias voltage at pad *CLKIN*.

The high-frequency inputs *CLKIN* for the clock, *CINP* and *CINN* for the comparators of the test chip are terminated with 50 Ω poly-resistors to achieve a good matching. This low impedance at the inputs of the comparators has also the effect, that no considerable distortions due to the switching of the comparator (kickback noise) are coupled back. The digital output drivers for pins *COUT*, $\overline{COUT}$ and *CREF* and the output *SIGOUT* of the analog voltage buffer are capable to drive a 50 Ω measurement system.

A simulation of different signals is shown in Fig. 6.28 for Comparator1. Similar results would be achieved for comparator2. In Fig. 6.28a the case that $\overline{CEN}$ switches in the same way as $\overline{CLK}$ between $V_{SS} = 0\,\text{V}$ (comparison phase) and $V_{DD} = 1.5\,\text{V}$ (reset phase) is shown (standard clocking). In Fig. 6.28b the case with *TBIAS* = 0.3 V can be seen, where $\overline{CEN}$ switches between $V_{DD} = 1.5\,\text{V}$ (reset phase) and about 0.7 V (comparison phase) and $\overline{CLK}$ switches as usual between $V_{DD} = 1.5\,\text{V}$ (reset phase) and $V_{SS} = 0\,\text{V}$ (comparison phase). For the demonstration simulation in Fig. 6.28, *TBIAS* was chosen too high for a clock of 2 GHz to show that obtaining a higher ΔV_0 tends to cause a longer overall delay time.

Monte Carlo simulations at an input common-mode voltage of 0.75 V of the proposed comparator circuit have shown, that for a clock of 2 GHz the optimum of improving the influence of mismatch was at *TBIAS* = 0.1 V. This can be seen in

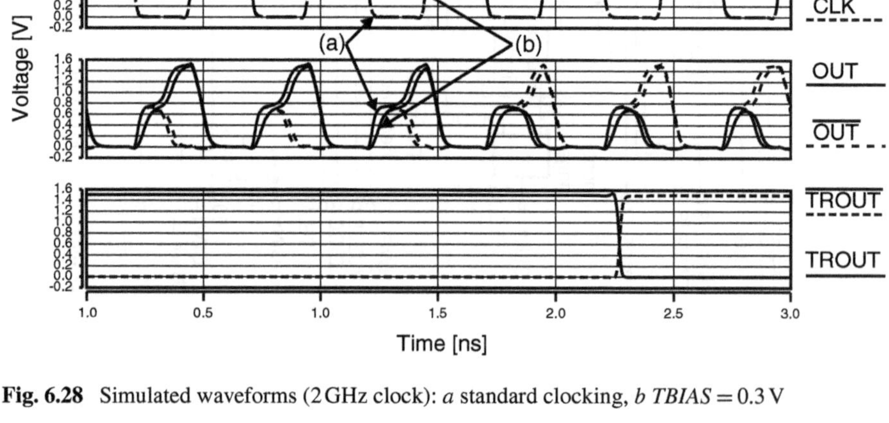

Fig. 6.28 Simulated waveforms (2 GHz clock): *a* standard clocking, *b TBIAS* = 0.3 V

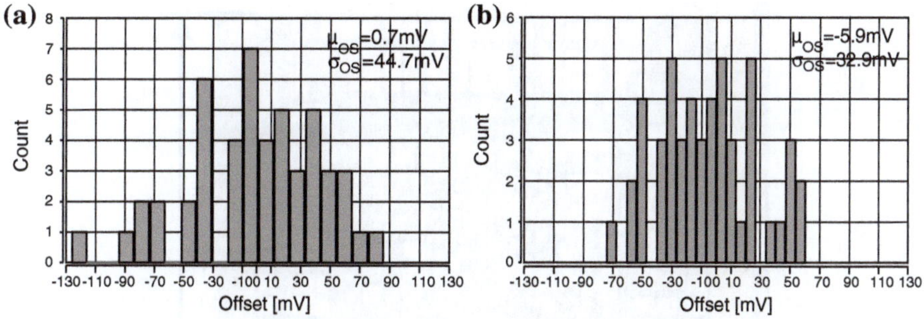

Fig. 6.29 Monte Carlo simulations (50 runs) for an input common mode level of 0.75 V and 2 GHz clock. **a** Standard clocking. **b** $TBIAS = 0.1\,\text{V}$

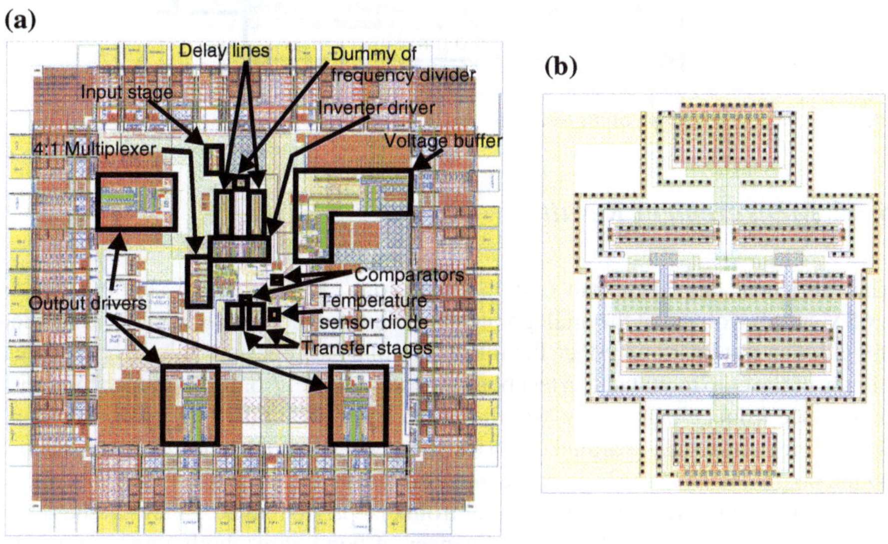

Fig. 6.30 Layout plots. **a** Test chip [6]. **b** Comparator

Fig. 6.29a, b, where the standard deviation of the offset of the comparator is improved from $\sigma_{OS} = 44.7\,\text{mV}$ (standard clocking) to $\sigma_{OS} = 32.9\,\text{mV}$ ($TBIAS = 0.1\,\text{V}$). Other simulations have shown, that the small mean value of $\mu_{OS} = 5.9\,\text{mV}$, which appeared in Fig. 6.29b is not caused by a hysteresis. It might disappear, if the amount of the Monte Carlo runs is enhanced.

In Fig. 6.30a a layout plot of the test chip with the comparator circuits is shown. Instead the die photo does not show any details due to the passivation layer and the planarisation and fill structures of the metal layers in this 0.12 μm CMOS technology, as it can be seen in Fig. 6.31. The area of the whole test chip amounts to $1.38 \times 1.39\,\text{mm}^2$. Thereof $17.5 \times 20.4\,\mu\text{m}^2$ is dedicated to the comparator.

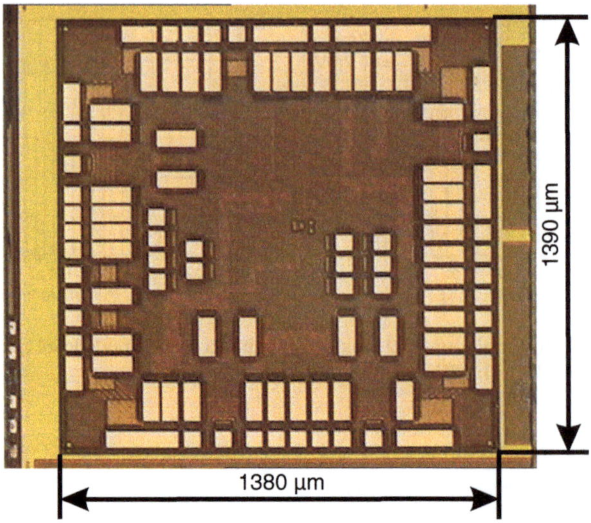

Fig. 6.31 Microphotograph of the test chip

6.3.2 Measurement Results

The fabricated chip (0.12 µm CMOS technology, 1.5 V nominal supply voltage, 1.65 V supply voltage of the analog voltage buffer) was countersunk into and bonded directly to a test board to reduce the influences of a package and its long bond wires. Photos of the chip mounted to a test board are shown in Fig. 6.32. To reduce the series inductance from chip ground V_{SS} to the ground of the test board, more than 20 bond wires were connected in parallel. Also from the power supply pads to the supply lines on the board and from the high-frequency pads to the microstrip lines more than one wire (up to four) were bonded in parallel from one pad to reduce somewhat the series inductances. To further reduce the influences of bond wires, both comparators, the output drivers and different parts of the voltage buffer were supplied with separate pads. On the board 50 Ω microstrip lines were designed for high-frequency input and output signals of the chip (pads *CINP*, *CINN*, *CLKIN*, *COUT*, $\overline{COUT}$, *CREF*, *SIGOUT*). The typical measurement setup is described in Sect. 5.5.

To test the functionality of comparator1 to a first overview a reference voltage of 0.65 V was connected to *CINN* and a saw-tooth with a frequency of 4 MHz with variable offset was applied to *CINP*. The sampling frequency was 2 GHz and the bias voltage at pin *NWELL* was 1.5 V. The oscilloscope picture can be seen in Fig. 6.33a. Due to noise, the comparator switches randomly at the cross-over of *CINN* + offset and *CINP*. Figure 6.33b shows the behavior of the comparator, which was clocked with 4.2 GHz, when a test-wave with a maximum switching rate of a half of the clock frequency was applied (*NWELL* = 1.5 V). The measured on-chip temperature (description in Sect. 5.2) varied between 40 °C (only two digital output drivers active) and 60 °C (all output drivers active).

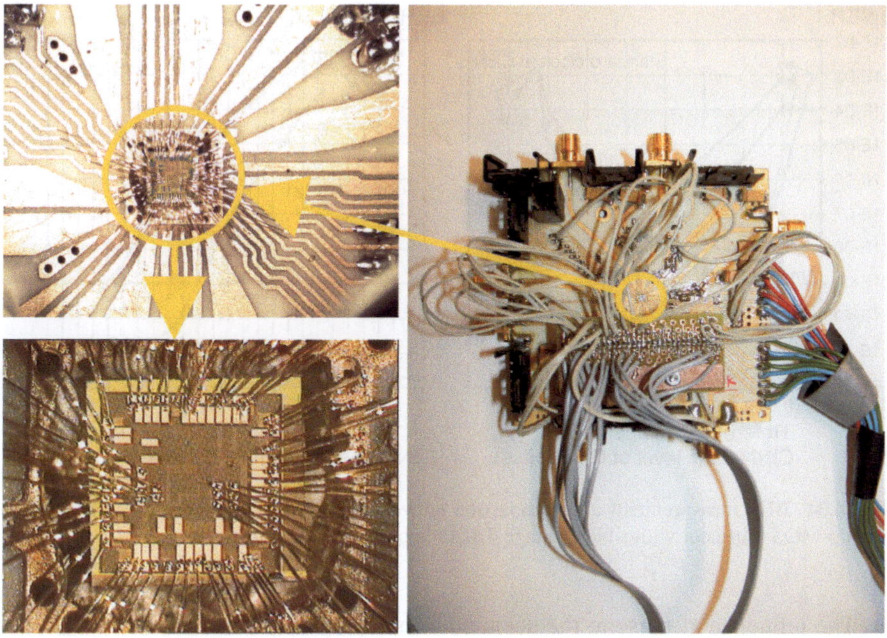

Fig. 6.32 Photos of the test board

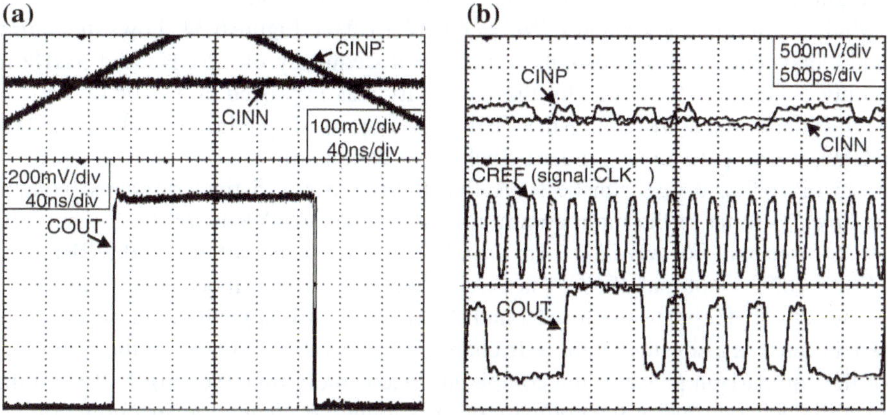

Fig. 6.33 Oscilloscope measurements [6]. **a** Oscilloscope picture of the 2 GHz clocked comparator, where a saw-tooth waveform with 4 MHz is applied to *CINP*, *CINN* = 0.65 V, bias level of *CINP*: variable, ≈0.65V. **b** Oscilloscope picture of the comparator, clocked with 4.2 GHz, when a test wave with max. 2.1 GHz switching rate is applied to *CINP* with amplitude ±130 mV around *CINN* = 0.75 V = Bias level of *CINP*-offset

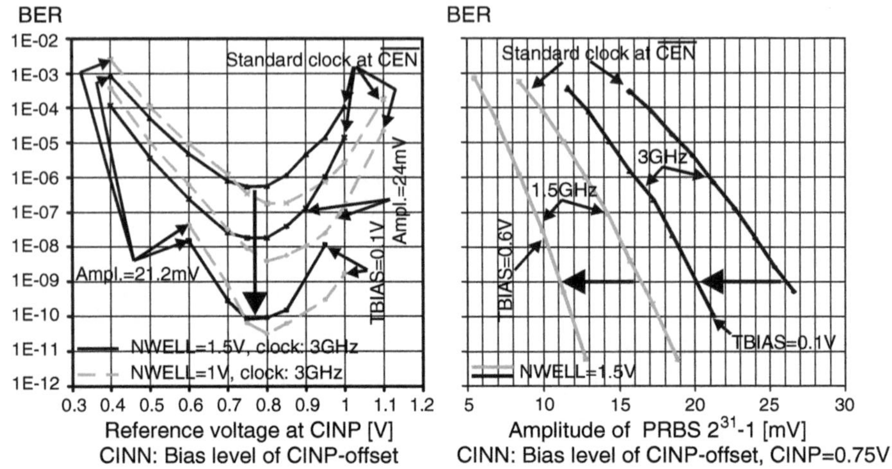

Fig. 6.34 BER measurements and comparison between standard clocking and clocking of $\overline{CEN}$ via a p-MOS transistor with a bias voltage *TBIAS* applied to the gate [6]

The influence of noise to the decision of comparator1 and to determine the sensitivity, BER measurements were done. On the test chip at pad *CINN* a bias voltage and at pad *CINP* a pseudo-random-bit-sequence PRBS $2^{31} - 1$, which was biased with a reference voltage, were applied. *CINN* was always biased so that the offset of the comparator was compensated (best BER occurred). The amplitude of a pseudo-random-bit-sequence PRBS with the length $2^{31} - 1$ means, that the bit-sequence switches ± amplitude around $CINP = CINN + \text{offset}$. Figure 6.34 shows the tuning and enhancement of the resolution at a clock frequency of 1.5 GHz and 3 GHz at a duty-cycle of 50 %. Typically the sensitivity was improved at a BER of 10^{-9} at 1.5 GHz and 3 GHz by about 5 mV (BER was improved by a factor of about 5500 at 3 GHz) in comparison to standard clocking of comparator1.

In Fig. 6.34a it also can be seen, that there exists a distinct reference voltage at *CINN* (similar to the input common-mode voltage (*CINN*+*CINP*)/2), where the BER is optimal. For $NWELL = 1.5$ V and a clock frequency of 3 GHz the BER is at $CINN = 0.75$ V optimal. If *CINN* is raised, the input common-mode level shifts nearer to the cut-off of input transistors *P*1 and *P*2. Here the BER approaches more and more 0.5, because at a constant amplitude charging nodes *OUT* and $\overline{OUT}$ at the beginning of the comparison phase of the comparator needs more time (see Fig. 6.26) and the overall decision time of the comparator lasts longer, even at a higher initial voltage difference ΔV_0 of the latch at the beginning of the time period in Fig. 6.24. This approaches more and more a metastability error, which is additionally influenced by noise (BER→0.5). Because transistor *P*0 is no ideal current source (it works even as switch in the linear region when standard clock is active), its drain current I_0 depends among other things also on the input common-mode voltage $(CINN + CINP)/2$, where as a consequence ΔV_0 depends also on the input common-mode voltage, which is contained in I_0 (see 6.9). So at lower input common-mode voltages I_0 becomes

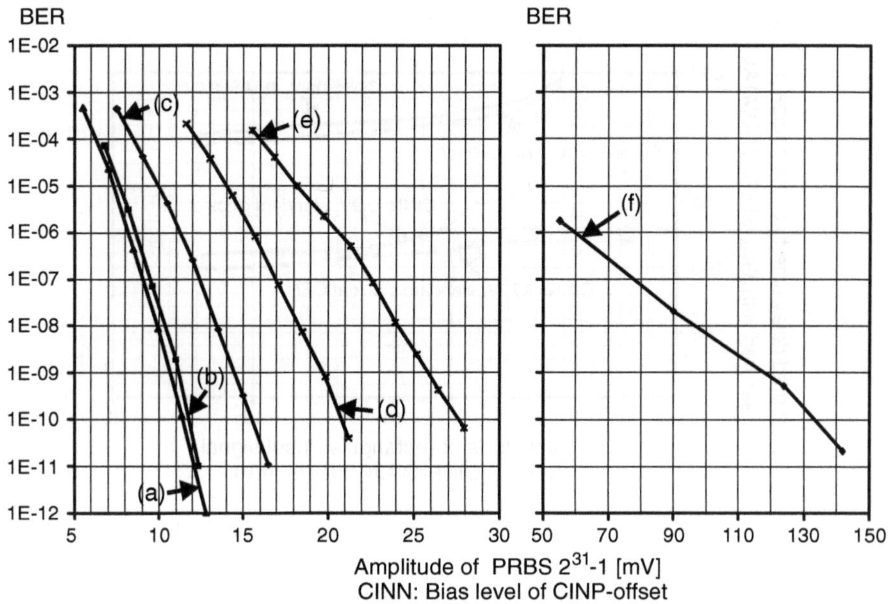

Fig. 6.35 Best achieved BER curves of the comparator [6] Clock frequency: (*a*) 1.5 GHz, (*b*) 2 GHz, (*c*) 2.5 GHz, (*d*) 3 GHz , (*e*) 3.5 GHz, (*f*) 4 GHz, Bias Level of *CINP*: (*a, b, c*) 0.75 V, (*d*) 0.8 V, (*e,f*) 0.75 V, *TBIAS*: (*a*) 0.6 V (*b*) 0.5 V (*c*) 0.35 V (*d*) 0.1 V (*e*) 0.2 V (*f*) 0.21 V, *NWELL*: (*a, b, c, d*) 1 V (*e, f*) 1.5 V / Duty cycle: (*a, c, d, e*) 50% (b) 46% (f) 47%

higher, ΔV_0 degenerates and the BER becomes also worse, because ΔV_0 can be seen as the amplified input voltage difference ΔV_{in}, which is the initial condition, which causes the latch to decide (see also Sects. 2.1 and 2.4.2). The decision of the comparator is more influenced by noise. By lowering the bias voltage at *NWELL* (see Fig. 6.26) the optimal point of BER can be shifted somewhat to higher input common-mode voltages.

In Fig. 6.35 the best achieved curves of BER measurements in dependence on the amplitude of PRBS $2^{31} - 1$ at *CINP* of the comparator are shown. To achieve a BER of 10^{-9}, the presented comparator is able to detect 10.8 mV at 1.5 GHz, 11.2 mV at 2 GHz, 14.5 mV at 2.5 GHz, 20 mV at 3 GHz, 26 mV at 3.5 GHz and 118 mV at 4 GHz. Also the power consumption of one comparator was reduced from 907 μW to 788 μW at a clock of 3.5 GHz in comparison to standard clocking. Due to the fact, that a comparator only needs dynamic current and because the sensitivity tuning technique also reduces the current, the power consumption was 812 μW at a clock frequency of 4 GHz. The measured offset of the comparators was typically in the range of 50 mV, but here no exact statistics of measurement values was done.

The delay time measurements according to Sect. 5.3 of comparator1 can be seen in Fig. 6.36. They were done at a clock frequency of 1.5 GHz and at an supply voltage of $V_{DD} = 1.5$ V. In principle an estimation of the delay time t_d of the comparator of Fig. 6.26 can be done with the delay time of a simple latch (2.3 of Sect. 2.1), where

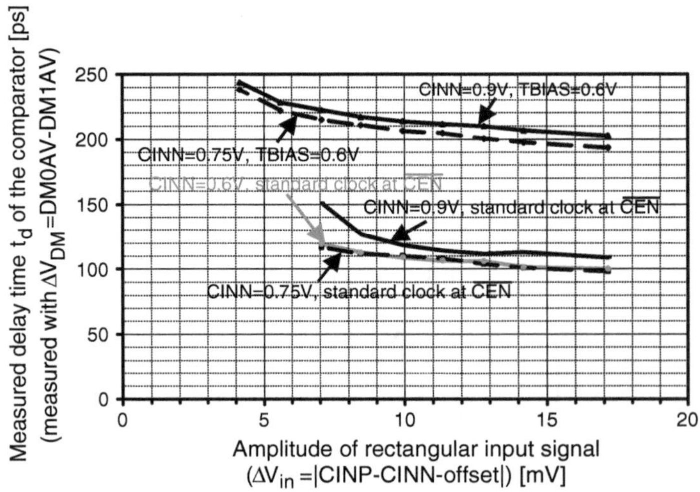

Fig. 6.36 Measured delay time t_d of the comparator versus the amplitude of the input signal

also $t_{Init} \approx 2C_L V_{tn}/(I_0 + I_1)$ has to be added. The result can be seen in 6.11.

$$t_d \approx \frac{2C_L V_{tn}}{I_0 + I_1} + \frac{C_L}{g_m - \frac{1}{r}} ln \left(\frac{\Delta V_{end}}{\Delta V_0} \right) \tag{6.11}$$

In 6.11, g_m is the effective transconductance of the latch (transistors $N0$, $N1$, $P4$, $P5$ in Fig. 6.26), where additionally the positive feedback of transistors $P6$ and $P7$ is also taken into account and r is the effective output resistance. For an first approximation $g_m \approx g_{mn0} + g_{mp4} = g_{mn1} + g_{mp5}$ and $r \approx r_{n0}||r_{p4} = r_{n1}||r_{p5}$ may be assumed, where g_{mn0}, g_{mp4}, g_{mn1} and g_{mp5} are the transconductances and r_{n0}, r_{p4}, r_{n1} and r_{p5} the output resistances of transistors $N0$, $P4$, $N1$ and $P5$ respectively. Additionally $g_{mn0} = g_{mn1}$, $g_{mp4} = g_{mp5}$, $r_{n0} = r_{n1}$ and $r_{p4} = r_{p5}$ is assumed. The final output voltage difference, which is necessary that a following logic gate recognizes the decision of the comparator is denoted by ΔV_{end}. So it can be seen, that if $I_0 + I_1$ becomes smaller, the delay time t_d becomes larger, as it is the case, when sensitivity tuning is done with *TBIAS*. This can be seen also in Fig. 6.36. Furthermore t_d becomes larger, when the amplitude of the rectangular input signal is lowered. This can be also explained with 6.9 and 6.11, where ΔV_0 depends on the input voltage difference $\Delta V_{in} = CINP - CINN + offset$. When ΔV_{in} is lowered, also a lower ΔV_0 is caused and the delay time t_d increases, which can be seen in Fig. 6.36 too. The dependence of t_d on the input common-mode voltage $((CINP + CINN)/2)$ is also predicted by 6.11, because due to the non ideal current source $P0$, which even works as switch in the linear region, if a standard clock is switched on $\overline{CEN}$, I_0 depends also on $(CINP + CINN)/2$ and therefore t_d becomes larger, if the input common-mode level is raised. It also can be said that the input common-mode level approaches to the cut-off point of the input transistors $P2$ and $P3$ of the comparator. Therefore in

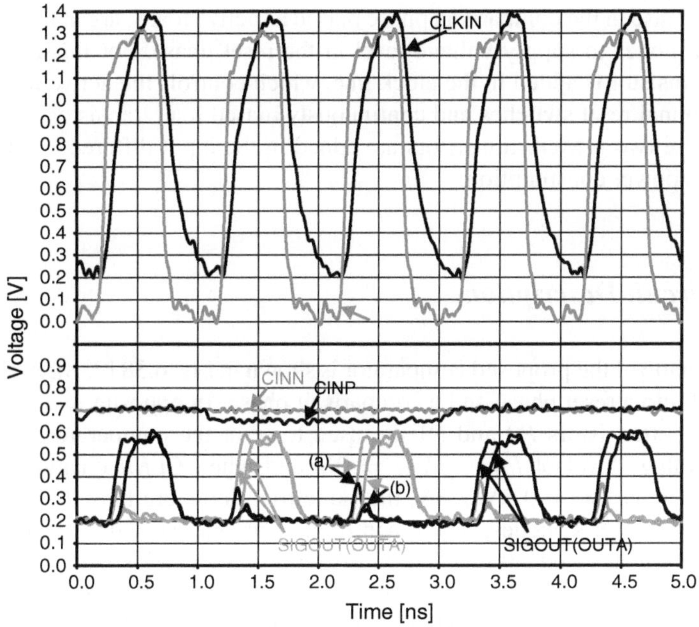

Fig. 6.37 Oscilloscope pictures: *a* Standard clock at *CEN*, *b* *TBIAS* = 0.6 V. The voltage buffer with output *SIGOUT* has an attenuation of about two as described in Sect. 5.1.3

Fig. 6.36 t_d is larger at *CINN* = 0.9 V. Remarkable is also, that the delay time t_d is nearly equal for *CINN* = 0.75 V and *CINN* = 0.6 V (see Fig. 6.36, standard clocking), because t_{Init} and the regeneration time of the latch have an opposite dependence on $(CINP + CINN)/2$, which can be also seen in 6.11 with 6.9 and which is a consequence of I_0 becoming higher (ΔV_0 lower), when the input common-mode voltage is lowered. Similar measurement results within the estimated tolerances of Sect. 5.3 occurred if the delay time was measured at pads *DM2AV* and *DM3AV* instead of *DM0AV* and *DM1AV*.

The measurements with the analog voltage buffer after comparator2 can be seen in Fig. 6.37, which further confirm the effects (higher ΔV_0, higher t_d), when the low voltage level of *CEN* is raised with a bias voltage *TBIAS* at the gate of a p-MOS transistor (see block diagram in Fig. 6.27).

6.4 A Comparator in 120 nm CMOS Requiring 0.5 V and 18 μW at 600 MHz Clock

In this section a comparator with the capability of even working at a supply voltage of 0.5 V is described, where 1.5 V ± 10 % is the nominal supply voltage of the used 120 nm CMOS process. The new circuit extension is, that two p-MOS transistors, which are biase in the comparison phase biased as active loads, are also used to reset the output nodes to supply voltage level in the reset phase. For this an additional n-MOS transistor is added in the clock line, which controls the active loads. Therefore additional reset switches and continuously biased load transistors are avoided. Furthermore two n-MOS transistors are added in the input differential amplifier to reduce the power consumption.

6.4.1 Circuit Description

The schematic of the proposed comparator is shown in Fig. 6.38 [7, 8]. A clock cycle is divided into a reset phase and a comparison phase. In opposite to a conventional comparator, transistors $P0$ and $P1$ are used to reset the comparator when $\overline{CLKR}$ has the voltage level of $low = V_{SS} = 0\,\text{V}$ and when $\overline{CLKR}$ changes to voltage level $high = V_{Comp} = 0.5\,\text{V}$–$1.5\,\text{V}$, the same transistors are biased as active load. So additional reset switches, which contribute parasitic capacitances to the output nodes and reduce speed are avoided. In the reset phase clock signals $\overline{CLKE}$, $\overline{CLKR}$ and $\overline{CLKL}$ (gate of active-load transistors $P0$ and $P1$) have the voltage level of V_{SS}. Transistors $N0$ and $N1$ are turned off and transistors $P0$ and $P1$ are switched on and pull both output nodes OUT and $\overline{OUT}$ towards V_{Comp}.

If $\overline{CLKR}$ and $\overline{CLKE}$ changes to level V_{Comp}, transistors $N0$ and $N1$ are switched on. Because of transistor $N8$ (threshold voltage V_{tn8}), which is biased with an appropriate constant voltage $CBIASN$ at its gate, has been added to the clock line of $\overline{CLKR}$, $\overline{CLKL}$

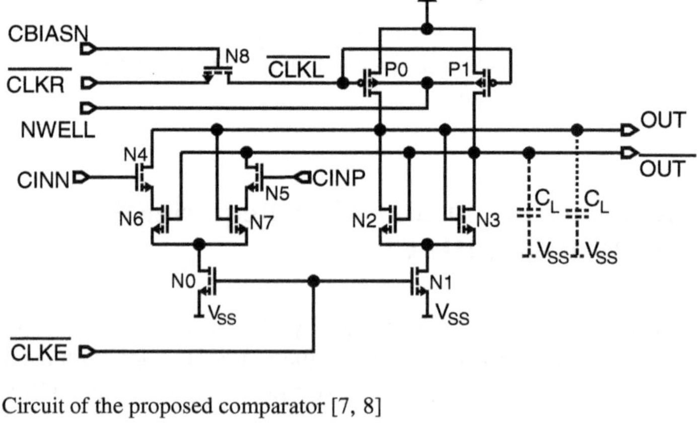

Fig. 6.38 Circuit of the proposed comparator [7, 8]

rises towards $CBIASN - V_{tn8}$, if the case is assumed, that sub-threshold and leakage currents are neglected for simplicity. Then $P0$ and $P1$ are biased to become active loads. At the beginning of the comparison phase the output nodes OUT and $\overline{OUT}$ of the comparator are pre-charged initially to V_{Comp} from the previous reset phase. Transistors $N6$ and $N7$ can be assumed to be on and each of transistors $N2$ and $N3$, which are cross-coupled and builds with the active loads $P0$, $P1$ a latch, has initially a gate-voltage near V_{Comp}. This and the fact that $N2$ to $N8$ are transistors with a lower threshold voltage of about 0.29 V, provided by this CMOS process, it is possible to drive the comparator even down to a supply voltage of 0.5 V. The latch switches due to positive feedback depending on the input-voltage difference at transistors $N4$ and $N5$. To reduce static current flow after the decision of the comparator, transistors $N6$ and $N7$ are added below transistors $N4$ and $N5$. When the voltage level at e.g. OUT is below the threshold voltage of $N7$, the path via $N5$ to $\overline{OUT}$ is cut off. At pin *NWELL* the separated n-well of transistors $P0$ and $P1$ is biased. The simulated operation of the comparator at a supply voltage of $V_{Comp} = 0.5$ V can be seen in Fig. 6.39, where $CBIASN = 0.4$ V and a clock frequency of 500 MHz have been chosen. Two cases of applied input voltage differences at the same input common mode voltage of 0.4 V are depicted in Fig. 6.39. In Fig. 6.39a an amplitude of 25 mV and in Fig. 6.39b an amplitude of 100mV is taken into account. Biasing transistors $P0$ and $P1$ during comparison phase with a voltage level of about $CBIASN - V_{tn8}$ with the help of transistor $N8$ makes it possible to control the current through load transistors $P0$ and $P1$ to be therefore able to adjust an optimal working point with a low power

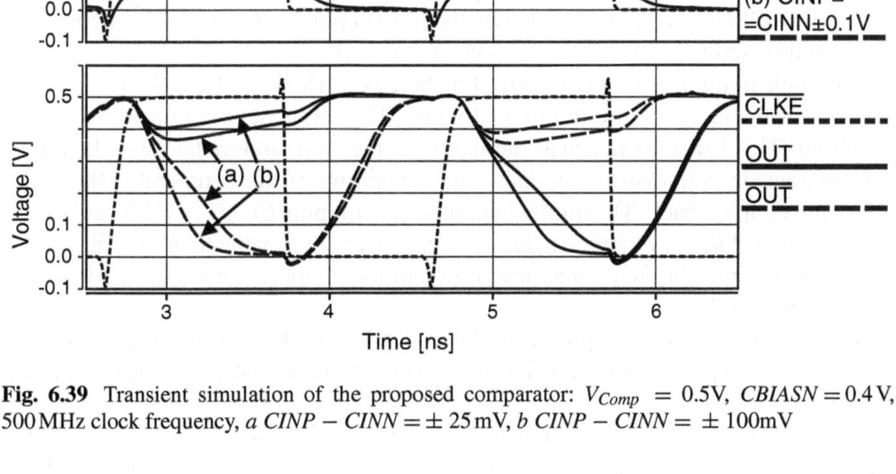

Fig. 6.39 Transient simulation of the proposed comparator: $V_{Comp} = 0.5$V, $CBIASN = 0.4$ V, 500 MHz clock frequency, *a* $CINP - CINN = \pm 25$ mV, *b* $CINP - CINN = \pm 100$mV

consumption. Furthermore current simulation have shown, that adding transistors $N6$ and $N7$ results in power savings of about 12% at a 0.5 V supply and a clock frequency of 500 MHz.

The block diagram of the test chip to investigate the proposed comparator of Fig. 6.38 is shown in Fig. 6.40. The circuit blocks to generate the clock signals *CLKR*, $\overline{CLKR}$, *CLKE* and $\overline{CLKE}$ are already described in Sect. 5.4, where two separate voltage-controlled delay lines are implemented for clock pairs *CLKR*, $\overline{CLKR}$ and *CLKE*, $\overline{CLKE}$, which are controlled with bias voltages at pins *VDELR* and *VDELE* respectively and a bias voltage for the n-wells, which is applied at pin *NWELLDL*. The only difference is, that the last inverter of the line drivers are supplied with a voltage level V_{Comp} so that the delivered clock signals *CLKR*, $\overline{CLKR}$, *CLKE* and $\overline{CLKE}$ switch between $V_{SS} = low = 0\,\text{V}$ and $V_{comp} = high = 0.5\,\text{V}$–$1.5\,\text{V}$. The comparators are placed in a separated area, which is supplied with V_{Comp} to test the functionality of the comparators even at a very low supply voltage. The 4:1 multiplexer and the driver for the reference output *CREF* to monitor the internal clock signals are also described in Sect. 5.4. Like in Sect. 6.3 the test chip consists of two comparators (Comparator1 and Comparator2), where each has the same circuit, which is shown in Fig. 6.38. Comparator1 is connected with its outputs *OUT* and $\overline{OUT}$ via adapters to transfer stages, which are used for BER measurements. The circuit of the transfer stage and the output drivers are already described in Sect. 5.3. The transfer stages are clocked with signals *CLK* and $\overline{CLK}$, which are in principle clock signals *CLKR* and $\overline{CLKR}$ for resetting the comparators, but adapted to the logical voltage levels $V_{SS} = 0\,\text{V}$ and $V_{DD} = 1.5\,\text{V}$ by adapters. The logical decisions of comparator1 at the output nodes *OUT* and $\overline{OUT}$ are also adapted after an inverter, which is supplied with voltage level V_{Comp}, to these normal logical voltage levels to signals *SOUT* and $\overline{SOUT}$. In the adapter an inverter, which is supplied with voltage level V_{Comp}, is placed in front of the real adaptation part to make sure, whether Comparator1 is able to deliver valid logical decisions to a following logic gate. After the adaptation, signals *SOUT* and $\overline{SOUT}$ are applied to the input of the appropriate transfer stage.

The outputs of Comparator2 (*OUTA*, $\overline{OUTA}$) are connected to an analog voltage buffer and to a replica of the buffer to guarantee a symmetrical load (see Sect. 5.1). With pin *DIGBUF* it can be chosen, whether the output *OUTA* or $\overline{OUTA}$ of comparator2 is redirected to output *SIGOUT* to be measured.

The high-frequency inputs *CLKIN* for the clock, *CINP* and *CINN* for the comparators of the test chip are terminated with 50 Ω poly-resistors to achieve a good matching. This low impedance at the inputs of the comparators has also the effect, that no considerable distortions due to the switching of the comparator (kickback noise) are coupled back. The digital output drivers for pins *COUT*, $\overline{COUT}$ and *CREF* and the output *SIGOUT* of the analog voltage buffer are capable to drive a 50 Ω measurement system. The functional description of the temperature sensor diode can be found in Sect. 5.2.

With the help of output pins *CAVP* and *CAVN* the duty cycle of clock signals *CEN* and $\overline{CEN}$ can be measured. The duty cycle can be adjusted by varying the bias voltage at pad *CLKIN*.

In principle it can be said, that the test chip is divided into 2 parts, one with the nominal supply voltage of $V_{DD} = 1.5\,\text{V}$ for optimal functionality of CMOS logic

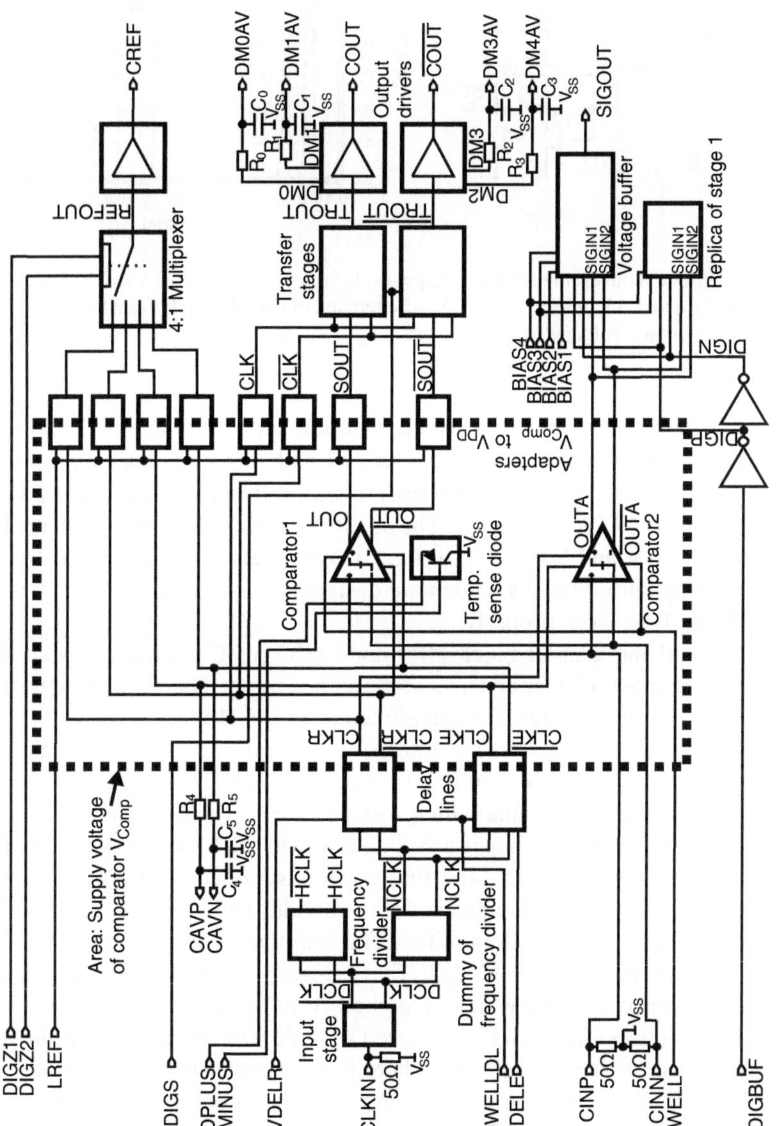

Fig. 6.40 Block diagram of the test chip to investigate the proposed comparator

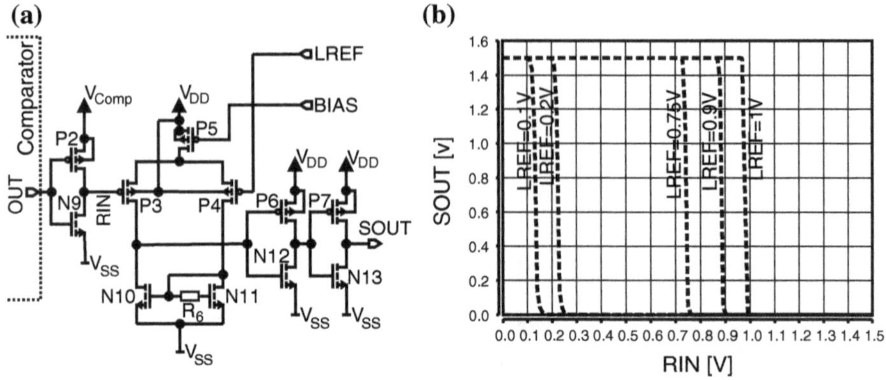

Fig. 6.41 Adapter to convert the logical voltage levels *high* = V_{Comp} = 0.5 V–1.5 V, *low* = V_{SS} = 0 V to *high* = V_{DD} = 1.5 V, *low* = V_{SS} = 0 V. **a** Schematic of the adapter. **b** Transfer characteristics with different reference voltages at *LREF*

needed for measurement purposes, where also the supply voltage of 1.65 V of the analog voltage buffer is added to, and a second with supply voltage V_{Comp}, where both comparators are placed. Comparator1 is clocked with $\overline{CLKR}$ and $\overline{CLKE}$ and Comparator2 with *CLKR* and *CLKE*. The adapters convert the logical voltage levels V_{Comp} = 0.5 V–1.5 V and V_{SS} = 0 V to the voltage levels V_{DD} = 1.5 V and V_{SS} = 0 V so that following circuit blocks, which are designed for the nominal supply voltage V_{DD} = 1.5 V ± 10 %, work properly.

The circuit of the adapter block after node *OUT* of Comparator1 is shown in Fig. 6.41a. Each other adapter consists of the same circuit. The simulation results of the nominal transfer characteristics are plotted in Fig. 6.41b.

After an inverter, which consists of transistors *N*9 and *P*2 and which is supplied by a voltage V_{COMP}, a differential amplifier (transistors *P*3, *P*4, *P*5, *N*10, *N*11 and resistor R_6) compares the logic signal with a voltage *LREF*. The difference is amplified so that the resulting output voltages are compatible to the logic levels of the following inverter with transistors *N*12 and *P*6, which is supplied with V_{DD}. Resistor R_6 is added to the active load to enhance the bandwidth and compensate the transfer characteristic of the differential amplifier. The last inverter consisting of transistors *P*7 and *N*13 is added to drive node *SOUT*, the input of the following transfer stage. *LREF* is a reference voltage, which is applied from outside the chip. The transfer characteristics between node *SOUT* referred to node *RIN* at the input of the differential amplifier is shown in Fig. 6.41, where the parameter is the reference voltage at pad *LREF*. So a digital characterization of Comparator1 with the help of BER measurements can be done at different supply voltages V_{Comp}.

A transient simulation of different waveforms in Fig. 6.40 can be seen in Fig. 6.42, where the bold signals are the case of V_{Comp} = 0.5 V, V_{DD} = 1.5 V, *CBIASN* = 0.4 V and 500 MHz clock frequency. The default printed (normal line width) signals are for the case of V_{Comp} = 1.5 V, V_{DD} = 1.5 V, *CBIASN* = 1 V and 3 GHz clock frequency.

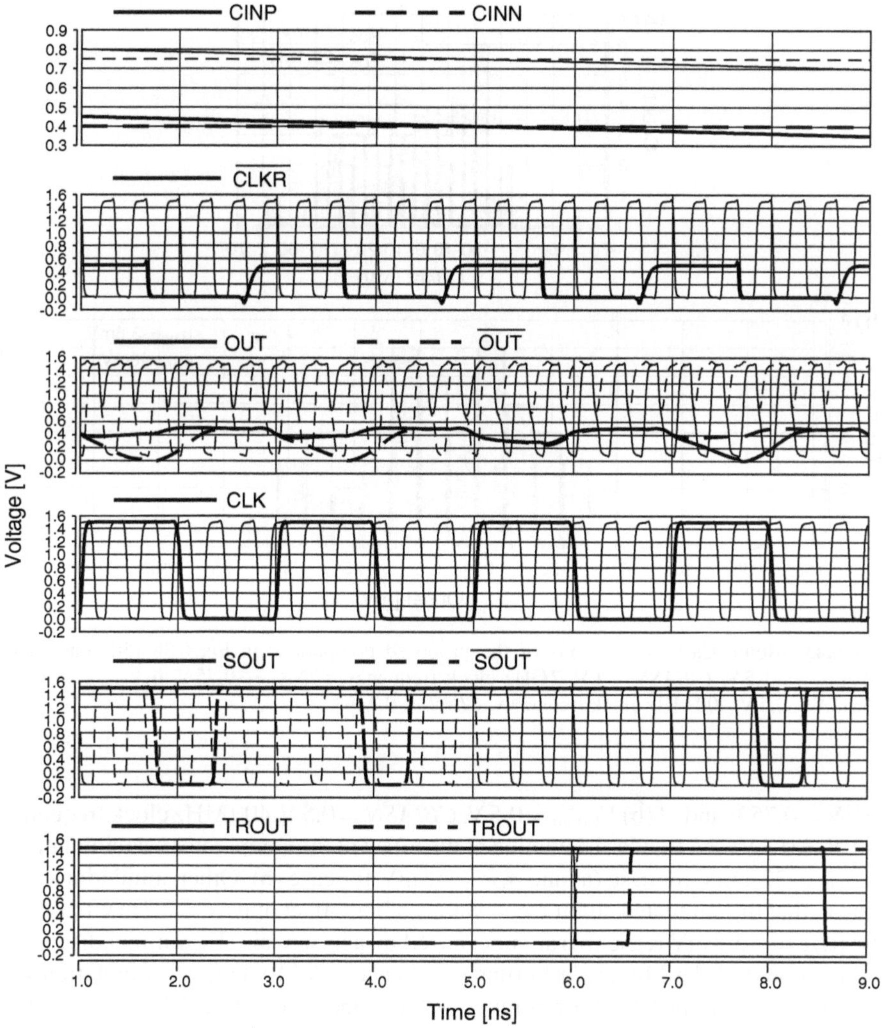

Fig. 6.42 Transient simulation of different signals of the test chip (bold: $V_{Comp} = 0.5$ V, $V_{DD} = 1.5$ V, *CBIASN* = 0.4 V and 500 MHz clock frequency, normal line width: $V_{Comp} = 1.5$ V, $V_{DD} = 1.5$ V, *CBIASN*=1 V and 3 GHz clock frequency)

When the comparator operates at $V_{Comp} = 0.5$ V and 500 MHz clock frequency, it can be seen in Fig. 6.42 that near the crossing point of *CINP* with *CINN*, the input voltage difference is to small for a proper decision, because the crossing point appears at a time, when the comparator has not decided yet (signals *OUT* and $\overline{OUT}$ for $V_{Comp} = 0.5$ V at 5 ns in Fig. 6.42).

Monte Carlo simulations of the comparator structure in Fig. 6.38 are shown in Fig. 6.43 for the cases of (a) $V_{Comp} = 1.5$ V, *CBIASN* = 1 V, 2 GHz clock frequency,

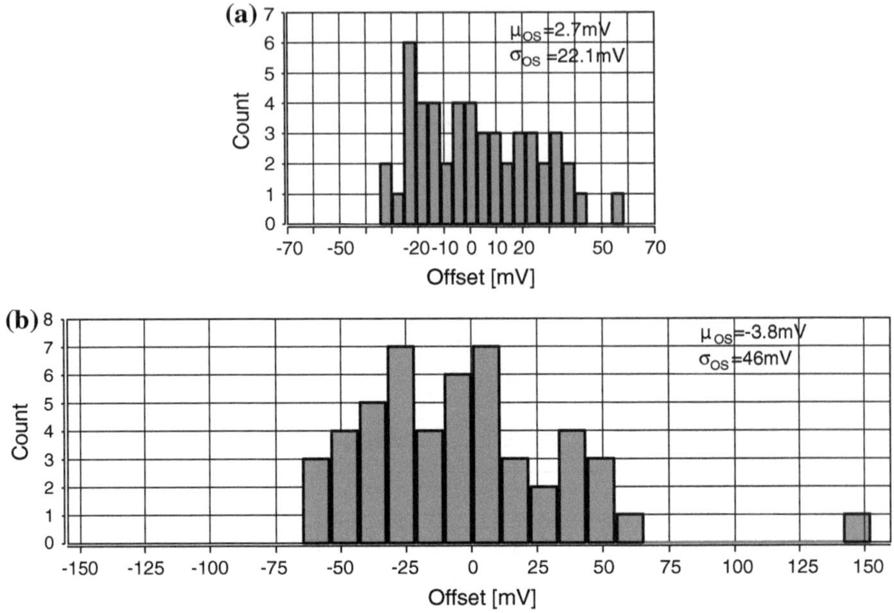

Fig. 6.43 Monte Carlo simulations of the proposed comparator in Fig. 6.38 (50 runs each). **a** V_{Comp} = 1.5 V, *CBIASN*= 1 V, 2 GHz clock frequency, *CINN* = 0.75 V. **b** V_{Comp} = 0.5 V, *CBIASN*= 0.5 V, 400 MHz clock frequency, *CINN* = 0.4 V

CINN = 0.75 V and of (b) V_{Comp} = 0.5 V, *CBIASN* = 0.5 V, 400 MHz clock frequency, *CINN* = 0.4 V. The standard deviation of the offset of the comparator was simulated to be $\sigma_{OS} = 22.1$ mV for case (a) and $\sigma_{OS} = 46$ mV for case (b). Other simulations have shown, that the simulated slight mean values of the offset of $\mu_{OS} = 2.7$ mV (case (a)) and $\mu_{OS} = -3.8$ mV (case (b)) are not caused by a hysteresis. They might disappear, if the amount of the Monte Carlo runs is enhanced. A Monte Carlo simulation was done by applying at *CINN* a reference voltage and at *CINP* a rising ramp with a distinct small slope, where e.g. 1 mV/ns for a clock of 3 GHz was chosen. Then this slope and the duration of the period of the clock frequency determines the sensitivity. To exclude the influence of a possible hysteresis, also the falling ramp was considered.

In Fig. 6.44a a layout plot of the test chip consisting of the comparator circuits is shown. In Fig. 6.44b the layout plot of a comparator is shown. Instead the die photo does not show any details due to the passivation layer and the planarisation and fill structures of the metal layers in this 0.12 μm CMOS technology, as it can be seen in Fig. 6.45. The area of the whole test chip amounts to $1.38 \times 1.39\,\text{mm}^2$. Thereof $22 \times 21\,\mu\text{m}^2$ is dedicated to a comparator.

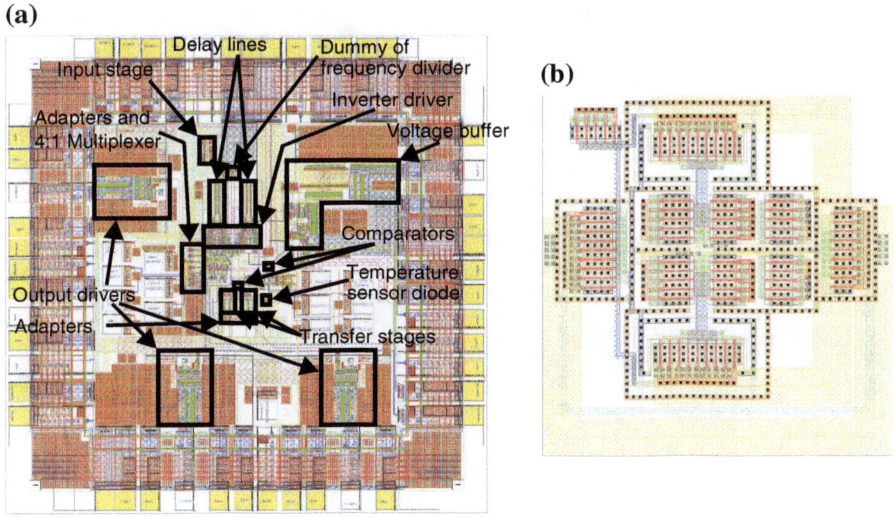

Fig. 6.44 Layout plots. **a** Test chip. **b** Comparator

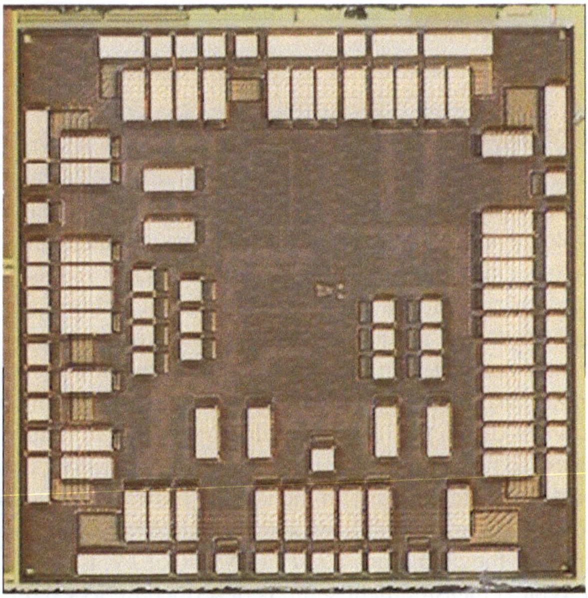

Fig. 6.45 Microphotograph of the test chip

Fig. 6.46 Photos of the test board

6.4.2 Measurement Results

The fabricated chip (0.12 μm CMOS technology, 1.5 V nominal supply voltage, 1.65V supply voltage of the analog voltage buffer) was countersunk into and bonded directly to a test board to reduce the influences of a package and its long bond wires. Photos of the chip mounted to a test board are shown in Fig. 6.46. To reduce the series inductance from chip ground V_{SS} to the ground of the test board, more than 20 bond wires were connected in parallel. Also from the power supply pads to the supply lines on the board and from the high-frequency pads to the microchip lines more than one wires (up to four) were bonded in parallel to reduce somewhat the series inductances. To further reduce the influences of bond wires, both comparators, the output drivers and different parts of the voltage buffer were supplied by separate pads. On the board 50 Ω microstrip lines were designed for the high-frequency input and output signals of the chip (pads *CINP*, *CINN*, *CLKIN*, *COUT*, $\overline{COUT}$, *CREF*, *SIGOUT*). The typical measurement setup is described in Sect. 5.5.

To test the functionality of comparator1 to a first overview a oscilloscope measurements were done, which can be seen in Fig. 6.47. It shows the behavior of the comparator, which was clocked with 6 GHz and supplied with $V_{Comp} = 1.5$ V, when a test-wave with a maximum switching rate of a half of the clock frequency was applied to *CINN* (*NWELL* = 1.5 V). The test signal applied to *CINN* was biased witch *CINP*-offset and. At *CINP* a reference voltage of 0.8 V was applied.

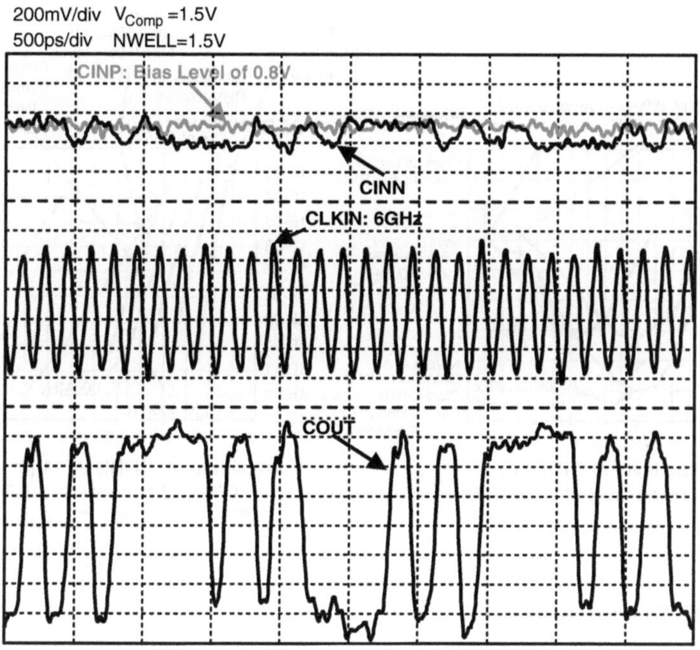

Fig. 6.47 Oscilloscope measurements [7]

The measured on-chip temperature (description in Sect. 5.2) varied between 40 °C (only two digital output drivers active) and 60 °C (all output drivers active).

To determine the influence of noise to the decision of comparator1 and to measure the sensitivity, BER measurements were done and depicted in Fig. 6.48. On the test chip at pad *CINP* a reference voltage and at pad *CINN* a pseudo-random-bit-sequence PRBS $2^{31}-1$, which was biased with a bias voltage, were applied. *CINP* was always biased so that the offset of the comparator was compensated (best BER occurred). The amplitude of a pseudo-random-bit-sequence PRBS with the length $2^{31}-1$ means, that the bit-sequence switches $\pm$ amplitude around $CINN = CINP$-offset. The measured comparator worked down to a supply voltage of $V_{Comp} = 0.5$ V with a maximal clock frequency of 600 MHz consuming 18 μW. For a BER better than 10^{-9} a minimal amplitude of 60.5 mV at 600 MHz, 25.8 mV at 500 MHz and 21.2 mV at 400 MHz has to be applied to the comparator. At a supply voltage of $V_{Comp} = 1.5$ V these values are 38 mV at 5.5 GHz, 29.4 mV at 5 GHz, 16.5 mV at 4 GHz and 11.2 mV at 3 GHz. At 6 GHz a BER of 10^{-6} is achieved at an amplitude of 150mV. The comparator has a power consumption of 2.65 mW at $V_{Comp} = 1.5$ V with an operating clock of 6 GHz and 2.17 mW at $V_{Comp} = 1.5$ V with an operating clock of 4 GHz, which was measured at a separate pad to supply both comparators of the test chip. The output voltage swing is sufficient to drive a following CMOS logic gate. In opposite to the proposed comparator of Sect. 6.3, during the decision phase always a current flows from V_{Comp} to V_{SS} in the comparator described here

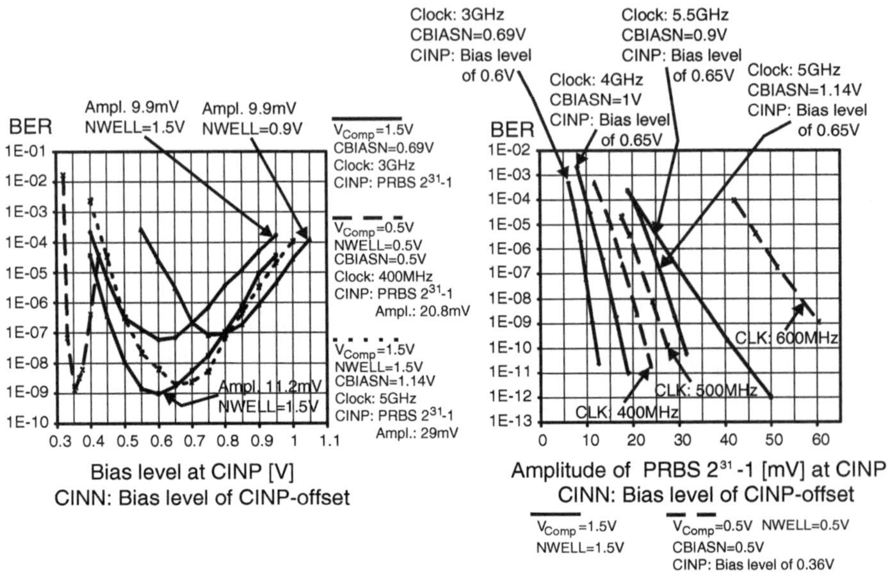

Fig. 6.48 BER measurements [7]

(see Fig. 6.38), because of the biased load transistors $P0$ and $P1$ and the turned on tail transistors $N0$ and $N1$. So the latch (transistors $N2$, $N3$ with load transistors $P0$ and $P1$) is already at the beginning of the comparison phase completely active. To a first approximation, where transistors $N6$ and $N7$ are assumed to be turned on and therefore are neglected for the beginning of comparison, 2.8 of Sect. 2.2 can be used to describe the output voltage difference $\Delta V_{OUT}(t) = OUT(t) - \overline{OUT}(t)$ with a small-signal expression. Adapted to the proposed comparator in Fig. 6.38, 6.12 can be written, where $\Delta V_{in} = CINP - CINN$ is the input voltage difference, which is assumed to be constant and $g_m \approx g_{mn2} = g_{mn3}$ and $r \approx r_{p0}||r_{n2} = r_{p1}||r_{n3}$ are the effective transconductance and the output resistance of the latch respectively.

$$\Delta V_{OUT}(t) = g_{mn4} \Delta V_{in} \left(\frac{1}{\frac{1}{r} - g_m} + \frac{1}{g_m - \frac{1}{r}} e^{\frac{g_m - \frac{1}{r}}{C_L} t} \right) \tag{6.12}$$

Transconductance g_m may also include the additional positive feedback of transistors $N6$ and $N7$. The transconductances $g_{mn4} = g_{mn5}$ are assumed to be equal at the beginning of the comparison phase, when both $N6$ and $N7$ are turned on (because during reset the both output nodes have been pulled to voltage level V_{Comp}). With 6.12 the BER measurement results of Fig. 6.48 can be explained. So a higher ΔV_{in} causes at a distinct time a higher output voltage difference thus reducing the probability of a metastability error. Also BER is then reduced, because if a metastability error occurs, in reality the decision of the comparator is additionally influenced by noise if *CINN* is near *CINP*-offset (see Sect. 2.4.2). The dependence of the BER on the input common mode voltage (similar to the bias level of *CINP*) can be explained with the dependence of $g_{mn4} = g_{mn5}$ on the input common mode voltage. So there exists an optimal working point at *CINP* = 0.6 V for the case of V_{Comp} = 1.5 V, *CBIASN* = 0.69 V, *NWELL* = 1.5 V and a clock frequency of 3 GHz, where BER is optimal, because here the input transistors *N*4 and *N*5 have best transconductances, because of e.g. the degeneration caused by transistors *N*6 and *N*7 at the beginning of comparison or e.g. the dependence of the tail current through switch transistor *N*0 on the input common mode voltage. Equation 6.12 explains also the right-shift of the optimal working point to *CINP* = 0.65 V at a clock frequency of 5 GHz (dotted line in Fig. 6.48), because at a higher frequency the probability of a metastability error increases at lower *CINP* due to the shorter time duration for a comparison phase. A tendency of a shift of the optimal operating point to higher values of *CINP* and also a tendency of a shift of the whole BER curve towards a worse BER were measured, when the bias voltage at *NWELL* was lowered towards 0.9 V. A lower voltage at *NWELL* causes in common also a lower amount of the threshold voltage of transistors *P*0 and *P*1, which increases their drain currents, if a constant bias voltage at the gates is considered. This reduces *r* of 6.12 and BER becomes worse because of switching of the latch tends to last longer. If *P*0 and *P*1 deliver more current, then also more current has to be taken away from the output nodes by transistors *N*4 and *N*5 which may explain the shift of the optimal working point. At V_{Comp} = 0.5 V the comparator shows a similar dependence of BER on the bias voltage level at *CINP* as for V_{Comp} = 1.5 V, but where the whole BER curve is compressed into the range $V_{tn4} < CINP \leq 0.5$ V.

The measurements with the analog voltage buffer after comparator2 can be seen in Fig. 6.49 (see also Fig. 6.40). With a distance between *low* and *high* of about 260 mV

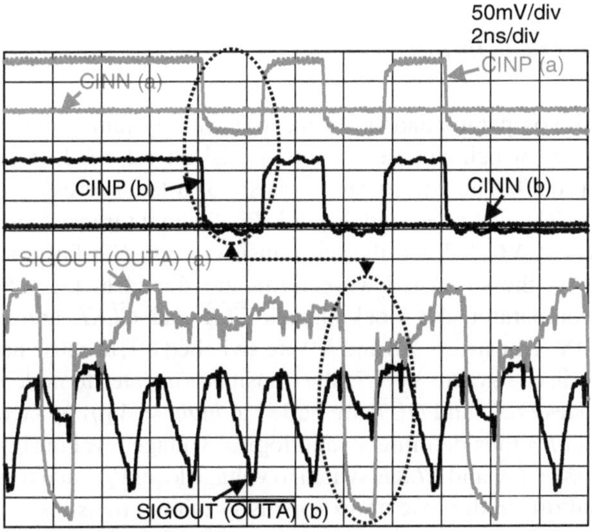

Fig. 6.49 Oscilloscope pictures [8]: *a* V_{Comp} = 1 V *b* V_{Comp} = 0.5 V. The voltage buffer with output *SIGOUT* has an attenuation of about two as described in Sect. 5.1.3. The output signals are delayed by about 6 ns from *CINP*

(2×130 mV due to attenuation of the voltage buffer, see signal $SIGOUT(\overline{OUT})$ in Fig. 6.49b) at $V_{Comp} = 0.5$ V a proper operation is guaranteed. The short peaks of signals $SIGOUT(OUT)$ and $SIGOUT(\overline{OUT})$ are caused by the clock, which is coupled via parasitic capacitances.

The comparator worked at maximal clock frequency of 600 MHz at $V_{Comp} = 0.5$ V consuming 18 μW, where $CBIASN = 0.5$ V was applied. At 1.5 V and $CBIASN = 0.9$ V a maximal clock of 6 GHz was achieved with a power consumption of 2.65 mW.

6.5 A Comparator in 120 nm CMOS with Advanced Sensitivity Tuning

In flash ADCs or other applications, where many comparators are implemented, optimizations of existing comparator circuits are important to save chip area and power consumption. This section investigates an extended tuning schemata of that presented in Sect. 6.3. Here in contrast to Sect. 6.3 the currents through the input amplifier path and the latch can be controlled separately and a complementary comparator circuit is used to take advantage of a typically higher transconductance of a n-MOS transistor. As a consequence of measurement results and experiences of the test chip, a short extension of the comparator is finally proposed, where only a resistor is added to the circuit to reduce the influence of mismatch and to enhance the sensitivity in a simple way.

6.5.1 Circuit Description

The circuit of the comparator and the extended sensitivity-tuning schemata (separated with a dotted line), which consist of the transmission gates $N4$, $P4$ and $N5$, $P5$ are depicted in Fig. 6.50. The comparator consists of a latch (cross-coupled inverters $N0$, $P0$ and $N1$, $P1$), two reset switches $P2$, $P3$, the input transistors $N2$, $N3$, cross-coupled transistors $N6$, $N7$ and the tail transistors $N8$, $N9$. For an explanation of sensitivity tuning, the simulated wave forms are shown in Fig. 6.51. In the reset phase of the comparator, the clock signals $\overline{CLK}$ and $\overline{CLKE}$ are at voltage level $low = V_{SS} = 0$ V. Transistors $N8$ and $N9$ are switched off, because nodes $\overline{CENA}$ and $\overline{CENL}$ are also discharged to V_{SS} via the transmission gates to node $\overline{CLKE} = V_{SS}$. Both output nodes OUT and $\overline{OUT}$ are pulled to $high = V_{DD}$ by switching on reset switches $P2$ and $P3$ thus defining a valid logical voltage level during reset. For the comparison phase $\overline{CLK}$ and $\overline{CLKE}$ switch to voltage level V_{DD} and transistors $P2$ and $P3$ are switched off. In the case of no sensitivity tuning, transmission gates $N4$, $P4$ and $N5$, $P5$ are turned completely on ($NAMP = NLAT = V_{DD}$, $PAMP = PLAT = V_{SS}$), nodes $\overline{CENA}$ and $\overline{CENL}$ are pulled to V_{DD} too and transistors $N8$ and $N9$ are

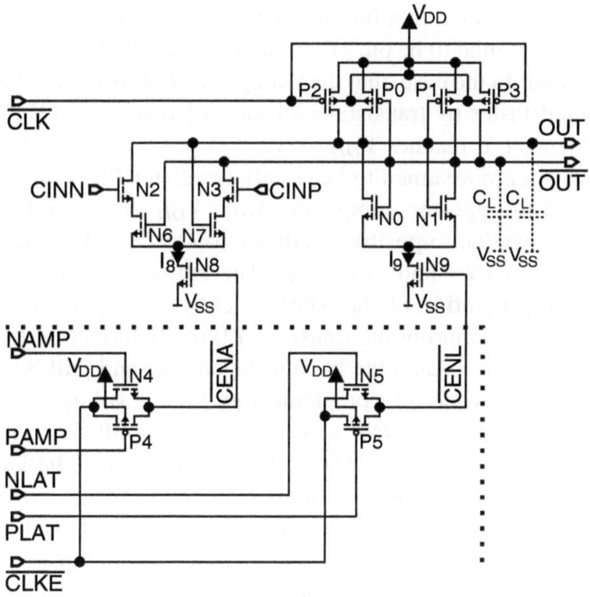

Fig. 6.50 Circuit of the proposed comparator [9]

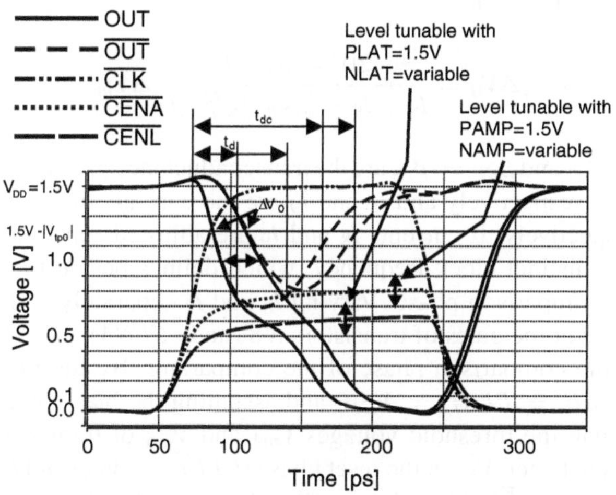

Fig. 6.51 Transient simulations [9]: ΔV_0 can be tuned by adjusting tail currents I_8 of transistor $N8$ and I_9 of transistor $N9$ separately. To do that, a voltage $PLAT = PAMP = 1.5\,\text{V}$ is applied to turn off transistors $P4$ and $P5$ and $NAMP$ and $NLAT$ are biased with a variable reference voltage individually

switched on. Due to the fact, that initially both output nodes are at V_{DD}, transistors $N6$ and $N7$ can be assumed to be on. Transistors $N6$ and $N7$ are added to avoid static current consumption. In the case, that the voltage at $CINP$ is higher than $CINN$, node $\overline{OUT}$ is discharged faster by transistors $N1$ and $N3$ than node OUT by transistors $N0$ and $N2$. When $\overline{OUT}$ reaches $V_{DD} - |V_{tp0}|$ ($V_{tp0} = V_{tp1}$ = threshold voltage of $P0$ and $P1$, which are assumed to be equal), transistor $P0$ is turned on, a current flow is established through the comparator from V_{DD} to V_{SS} and the whole latch switches (regenerates) due to positive feedback to $OUT = V_{DD}$ and $\overline{OUT} = V_{SS}$ in the case $CINP > CINN$ and vice versa. The output voltage difference ΔV_0 at the beginning of regeneration of the whole latch (one output node has reached a voltage level so that the appropriate transistor $P0$ or $P1$ turns on) defines the initial voltage difference, which causes the latch to regenerate and pull both output nodes of the comparator to appropriate, complementary logical voltage levels. Depending on whether $CINP > CINN$ or $CINP < CINN$, OUT is pulled to V_{DD} and $\overline{OUT}$ to V_{SS} or OUT is pulled to V_{SS} and $\overline{OUT}$ to V_{DD} respectively. The larger the output voltage difference ΔV_0 a short time before regeneration of the latch starts, the larger is the robustness of the comparator against noise and mismatch [3–5] and thus the sensitivity [6] of the comparator becomes better. In praxis the sensitivity is defined by all kinds of noise, which influences the decision of the comparator. Even at an input voltage difference, where a metastability error would occur in an ideal simulation, noise causes the comparator to make a false or right decision. Adapted to the circuit of Fig. 6.50 with neglecting transistors $N6$ and $N7$, ΔV_0 can be estimated with 6.13 (see also 6.9 in Sect. 6.3.1).

$$\Delta V_0 = \frac{2|V_{tp0}|\sqrt{\beta_{n2} I_8}}{I_8 + I_9 - 2|V_{tp0}|\sqrt{\beta_{n0} I_9}} \Delta V_{in} \tag{6.13}$$

In 6.13, $\beta_{n0} = \beta_{n1}$ and $\beta_{n2} = \beta_{n3}$ are the transconductance parameters of transistors $N0$, $N1$, $N2$, $N3$ respectively and I_8 and I_9 are the currents through tail transistors $N8$ and $N9$ respectively. If currents I_8 and I_9 are adjusted separately, ΔV_0 can be optimized for a distinct clock frequency, which defines beside the duty cycle the duration of the comparison phase. Varying I_8 and I_9 separately is done by varying the voltage levels at the gates of transistors $N8$ (node $\overline{CENA}$) and $N9$ (node $\overline{CENL}$) separately for the comparison phase of the comparator. By turning off transistors $P4$ and $P5$ ($PAMP = PLAT = V_{DD}$) and assuming the bias voltages $NAMP$ and $NLAT$ higher than the threshold voltages V_{tn4} and V_{tn5} of transistors $N4$ and $N5$, $\overline{CENA}$ switches between V_{SS} in the reset phase ($\overline{CLKE} = V_{SS}$) and $NAMP - V_{tn4}$ in the comparison phase ($\overline{CLKE} = V_{DD}$) in the ideal case, if leakage and sub-threshold currents are neglected for simplicity. $\overline{CENL}$ switches between V_{SS} and $NLAT - V_{tn5}$ respectively. In Fig. 6.50 transistors $P4$ and $P5$ of the transmission gates are only added to have the possibility to compare the measurements with the case of clocking transistors $N8$ and $N9$ directly with signal $\overline{CLKE}$ by turning on the transmission gates (standard clocking). As a trade-off the delay time t_{dc} ($0.5V_{DD}$ at $\overline{CLK} = \overline{CLKE}$ to $0.1V_{DD}$) or depending on definition t_d ($0.5V_{DD}$ at $\overline{CLK} = \overline{CLKE}$ to $0.5V_{DD}$ at

OUT if $CINP < CINN$ or at $\overline{OUT}$ if $CINP > CINN$) of the comparator tends to become larger when adjusting ΔV_0 to larger values. Thus the sensitivity is restricted by the duration of the comparison time and therefore also by the clock frequency. In Fig. 6.50 all transistors except tail transistors *N*8 and *N*9 and reset transistors *P*2 and *P*3, which are regular threshold transistors, are transistors with a lower threshold voltage of about 0.29 V, which are also provided by this 0.12 μm CMOS process with a nominal supply voltage of 1.5 V ± 10 %.

In Fig. 6.52 the block diagram of the test chip to investigate and characterize the proposed comparator of Fig. 6.50 is shown. The circuit blocks to generate the clock signals *CLKR*, $\overline{CLKR}$, *CLKE* and $\overline{CLKE}$ are already described in Sect. 5.4, where two separated voltage-controlled delay lines are implemented for clock pairs *CLKR*, $\overline{CLKR}$ and *CLKE*, $\overline{CLKE}$, which are controlled with bias voltages at pins *VDELR*, *VDELE* respectively and a bias voltage for the n-wells, which is applied at pin *NWELLDL*. The 4:1 multiplexer and the driver for the reference output *CREF* to monitor the internal clock signals are also described in Sect. 5.4.

The test chip consists of two comparators (Comparator1 and Comparator2), where each has the same circuit, which is shown in Fig. 6.50. Comparator1 is connected with its outputs *OUT* and $\overline{OUT}$ to transfer stages, which are used for BER measurements. The circuit of the transfer stage and the output drivers with the additional outputs *DM*0*AV* *DM*1*AV* *DM*2*AV* and *DM*3*AV* for delay time measurement are described in Sect. 5.3. The outputs of Comparator2 (*OUTA*, $\overline{OUTA}$) are connected to an analog voltage buffer and to a replica of the buffer to guarantee a symmetrical load (see Sect. 5.1). With pin *DIGBUF* it can be chosen, whether the output *OUTA* or $\overline{OUTA}$ of Comparator2 is redirected to output *SIGOUT* to be measured.

The functional description of the temperature sensor diode can be found in Sect. 5.2.

Comparator1 is clocked with $\overline{CLK}$, $\overline{CENA}$ and $\overline{CENL}$ and comparator2 with *CLK*, *CENA* and *CENL*, which are in principle the clock signals $\overline{CLKR}$, $\overline{CLKE}$ and *CLKR* and *CLKE* respectively, but where each of it is leaded through a transmission gate. The transmission gates between $\overline{CLKR}$ and $\overline{CLK}$ and between *CLKR* and *CLK* are always on and are only added to compensate the delay time differences between the clock signals. Two additional open transmission gates are added additionally to nodes *CLKR* and $\overline{CLKR}$ to have nearly the same load as at nodes *CLKE* and $\overline{CLKE}$. $\overline{CLK}$ and *CLK* are used to reset Comparator1 and Comparator2 respectively and are the clock signals, which are also connected to the transmission gates. Clock signals $\overline{CENA}$ and $\overline{CENL}$ to enable transistor *N*8 (amplifier part of the comparator) and *N*9 (latch part of the comparator) of compartor1 respectively are in principle clock signal $\overline{CLKE}$, but where an individual transmission gate (transistors *N*4, *P*4 for $\overline{CENA}$ and transistors *N*5, *P*5 for $\overline{CENL}$) is switched between. Bias voltages *NAMP* and *PAMP* are applied from outside the test chip and are the gate voltages of transistors *N*4 and *P*4 respectively. In a similar way bias voltages *NLAT* and *PLAT* are the gate voltages of transistors *N*5 and *P*5 respectively. Comparator2 is clocked in the same way as Comparator1 but where instead clock signal *CLK* is used for reset and clock signals *CENA* and *CENL*, which are in principle clock signal *CLKE* but leaded via two separate transmission gates, are used for enable. The gates of the transistors of

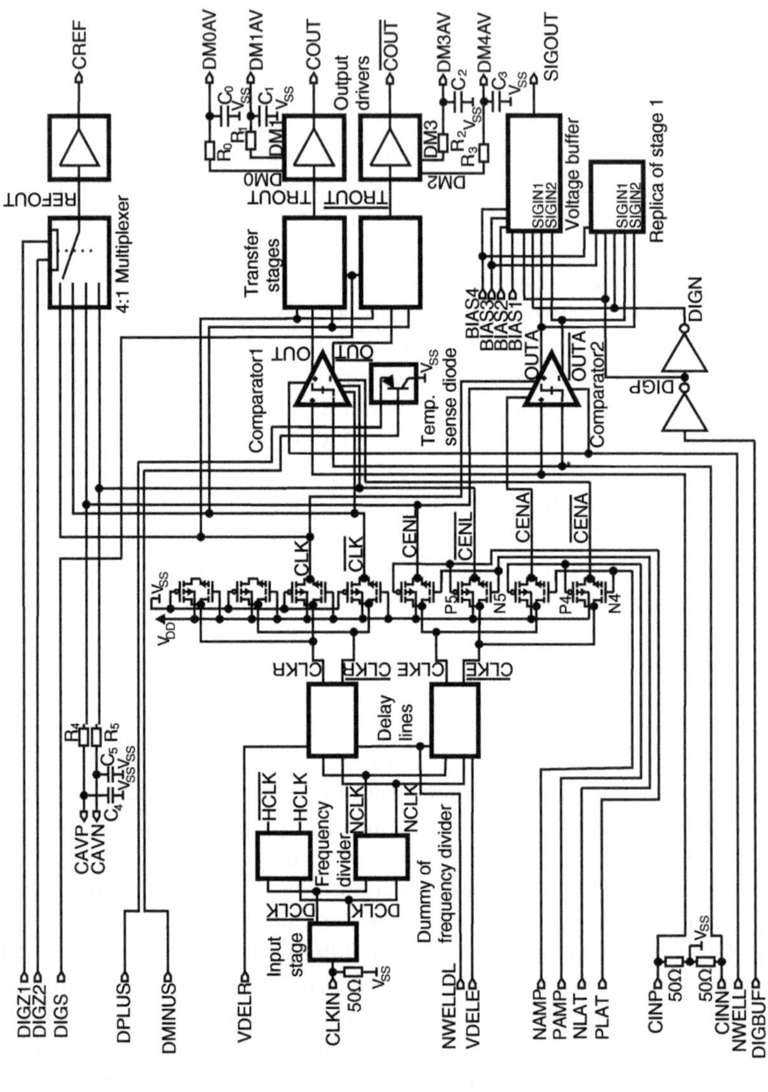

Fig. 6.52 Block diagram of the test chip to measure the proposed comparator circuit of Fig. 6.50

these transmission gates are also biased with bias voltages *NAMP*, *PAMP*, *NLAT* and *PLAT* like the transmission gates for Comparator1. So depending on these bias voltages it can be chosen between standard clocking of comparator1 and comparator2 (*PAMP* = *PLAT* = $V_{SS} = 0\,\text{V}$ and *NAMP* = *NLAT* = $V_{DD} = 1.5\,\text{V}$), where all transmission gates are open or tuning sensitivity, where the currents through tail transistors *N*8 and *N*9 (see Fig. 6.50) are controlled individually by varying their voltage level at their gates during the comparison phase (see Fig. 6.51) by switching of the p-MOS transistors (*PAMP* = *PLAT* = $V_{DD} = 1.5\,\text{V}$) and applying a bias voltage at the gates of the n-MOS transistors (*NAMP*, *NLAT* variable bias voltage) of the appropriate transmission gates (see Fig. 6.52).

With the help of output pins *CAVP* and *CAVN* the duty cycle of clock signals *CENL* and $\overline{CENL}$ can be measured. The duty cycle can be adjusted by varying the bias voltage at pad *CLKIN*.

The high-frequency inputs *CLKIN* for the clock, *CINP* and *CINN* for the comparators of the test chip are terminated with $50\,\Omega$ polysilicon resistors to achieve a good matching. This low impedance at the inputs of the comparators has also the effect, that no considerable distortions due to the switching of the comparator (kickback noise) are coupled back. The digital output drivers for pins *COUT*, $\overline{COUT}$ and *CREF* and the output *SIGOUT* of the analog voltage buffer are capable to drive a $50\,\Omega$ measurement system.

A simulation of different signals is shown in Fig. 6.53 for Comparator1. Similar results would be achieved for comparator2. In Fig. 6.53a the case that $\overline{CENA}$ and $\overline{CENL}$ switch in the same way as $\overline{CLK}$ between $V_{SS} = 0\,\text{V}$ (reset phase) and $V_{DD} = 1.5\,\text{V}$ (comparison phase) is shown (standard clocking). In Fig. 6.53b the case with *PAMP* = *PLAT* = 1.5 V *NAMP* = 1.5 V and *NLAT* = 0.9 V can be seen, where $\overline{CENA}$ switches between $V_{SS} = 0\,\text{V}$ (reset phase) and about 1.1 V (comparison phase), $\overline{CENL}$ switches between $V_{SS} = 0\,\text{V}$ (reset phase) and about 0.6 V (comparison phase) and $\overline{CLK}$ switches as usual between $V_{DD} = 1.5\,\text{V}$ (comparison phase) and $V_{SS} = 0\,\text{V}$ (reset phase). For the demonstration simulation in Fig. 6.53c, extreme values of *NAMP* and *NLAT* were chosen for a clock of 2 GHz to show that obtaining a higher ΔV_0 tends to cause a longer overall delay time. Monte Carlo simulations of the comparator structure in Fig. 6.50 are shown in Fig. 6.54 for the cases of (a) standard clocking with 2 GHz clock frequency (*PAMP*= *PLAT* = 0 V, *NAMP* = *NLAT* = 1.5 V), *CINN* = 0.75 V and of (b) 2 GHz clock frequency (*PAMP* = PLAT = 1.5 V, *NAMP* = 1.5 V, *NLAT* = 1 V), *CINN* = 0.75 V. The standard deviation of the offset of the comparator was simulated to be $\sigma_{OS} = 25.8\,\text{mV}$ for case (a) and $\sigma_{OS} = 14.5\,\text{mV}$ for case (b), which shows that with sensitivity tuning the influence of mismatch can be reduced. Other simulations have shown, that the simulated slight mean values of the offset of $\mu_{OS} = 4.5\,\text{mV}$ (case (a)) and $\mu_{OS} = 1.3\,\text{mV}$ (case (b)) are not caused by a hysteresis. They might disappear, if the amount of the Monte Carlo runs is enhanced. A Monte Carlo simulation was done by applying at *CINN* a reference voltage and at *CINP* a rising ramp with a distinct small slope, where e.g. 1 mV/ns for a clock of 2 GHz was chosen.
Then this slope and the duration of the period of the clock frequency determines the sensitivity. To exclude the influence of a possible hysteresis, also the falling ramp

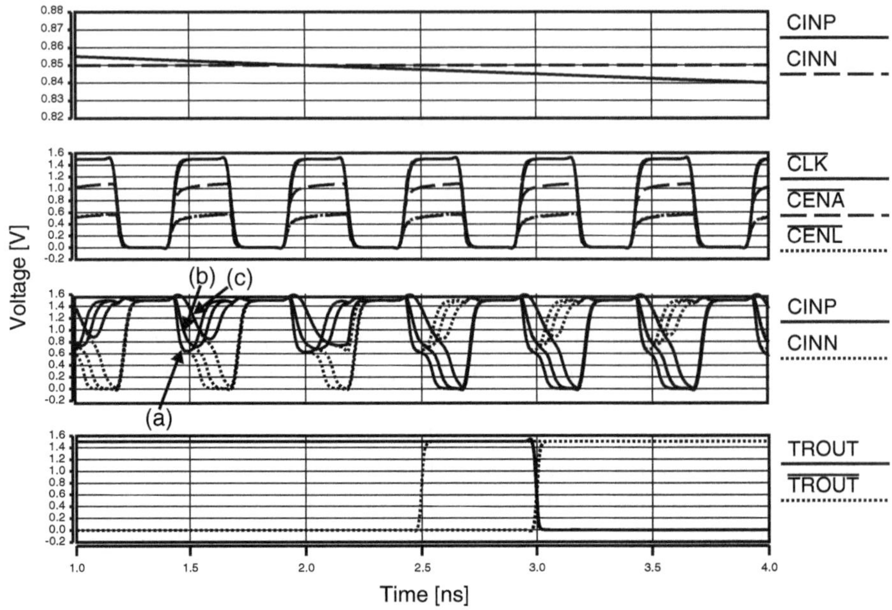

Fig. 6.53 Transient simulation (2 GHz clock, *DIGS* = 0 V): *a* standard clocking (*NAMP* = *NLAT* = 1.5 V, *PAMP* = *PLAT* = 0 V), *b* *PAMP* = *PLAT* = 1.5 V *NAMP* = 1.5 V, *NLAT* = 0.9 V is applied and *c* *PAMP* = *PLAT* = 1.5 V *NAMP* = *NLAT* = 0.9 V is applied

was considered. Another Monte Carlo simulation (50 runs) for *CINN* = 0.85 V and a clock frequency of 3 GHz shows also an improvement of the standard deviation of the offset from $\sigma_{OS} = 29$ mV at standard clocking to $\sigma_{OS} = 16.1$ mV for *PAMP* = *PLAT* = 1.5 V, *NAMP* = 1.5 V and *NLAT* = 1 V. Figure 6.55 shows ΔV_0 and the delay time t_{dc} of the comparator in dependence on the bias voltages at *NAMP* and *NLAT*, extracted from a post-layout simulation of the whole test chip (*PAMP* = *PLAT* = 1.5 V). It can be seen, that adjusting a larger ΔV_0 at the given comparator structure of Fig. 6.50 needs typically also a larger delay time t_{dc} (or t_d). So at a distinct clock period the duration of the comparison phase defines the adjustment range of ΔV_0. For example if at a clock frequency of 3 GHz a duration of the comparison phase of about 167 ps (50 % duty cycle) is assumed the range of bias voltages *NAMP* of about 1.5 V–0.9 V or *NLAT* of about 1.5 V–0.9 V for adjusting ΔV_0 can be read out of Fig. 6.55. If *NAMP* or *NLAT* are lowered towards the lower limit the decision of the comparator changes more and more towards a metastability error and the sensitivity and the BER becomes worse due to the higher influence of noise (random switching of the comparator due to noise). Otherwise it also has to be considered that the delay time t_{dc} is defined from 50 % voltage level of the clock till one output node of the comparator has reached 10 % of V_{DD}. So for using this time in Fig. 6.55, t_{dc} may be assumed larger than the duration of the comparison phase, because a following inverter may even recognize logical voltage levels higher than 10 % of V_{DD} (depends on the switching threshold V_{STH} of the inverter as described in 1.1 in Sect. 1.2.1). So

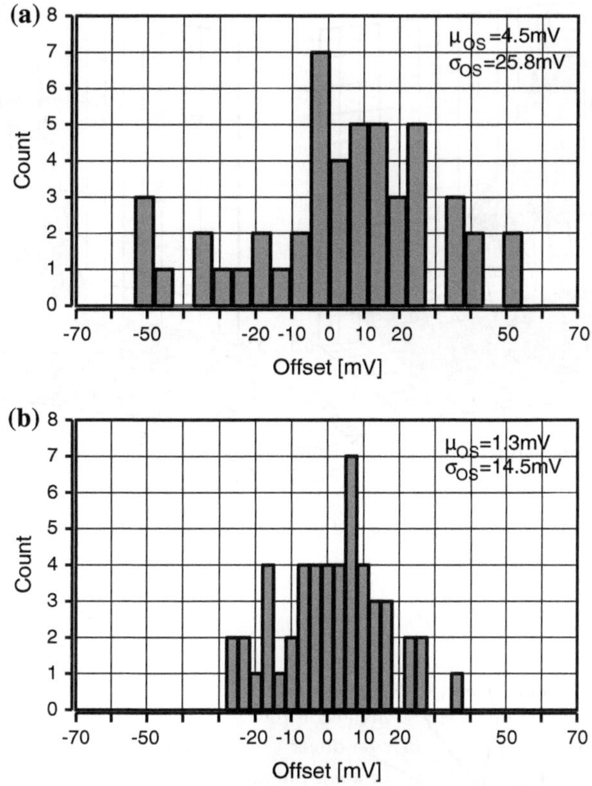

Fig. 6.54 Monte Carlo simulations of the proposed comparator in Fig. 6.50 (50 runs each), where the clock frequency was 2 GHz. **a** Standard clocking, *PAMP* = *PLAT* = 0 V, *NAMP* = *NLAT* = 1.5 V, *CINN* = 0.75 V. **b** *PAMP* = *PLAT* = 1.5 V, *NAMP* = 1.5 V, *NLAT* = 1 V, *CINN* = 0.75 V

e.g. t_d (50 % voltage level of clock to 50 % voltage level of one output node) may only be chosen to be smaller than the duration of the comparison phase, when the switching threshold of a following inverter is at about 50 % of V_{DD} or higher and as a consequence t_{dc} may than result in a larger amount than the duration of the comparison phase.

In Fig. 6.56a a layout plot of the test chip with the comparator circuits. Instead the die photo does not show any details due to the passivation layer and the planarisation and fill structures of the metal layers in this 0.12 μm CMOS technology, as it can be seen in Fig. 6.57. The area of the whole test chip amounts to $1.38 \times 1.39\,\text{mm}^2$. Thereof $18 \times 19\,\mu\text{m}^2$ is dedicated to the comparator.

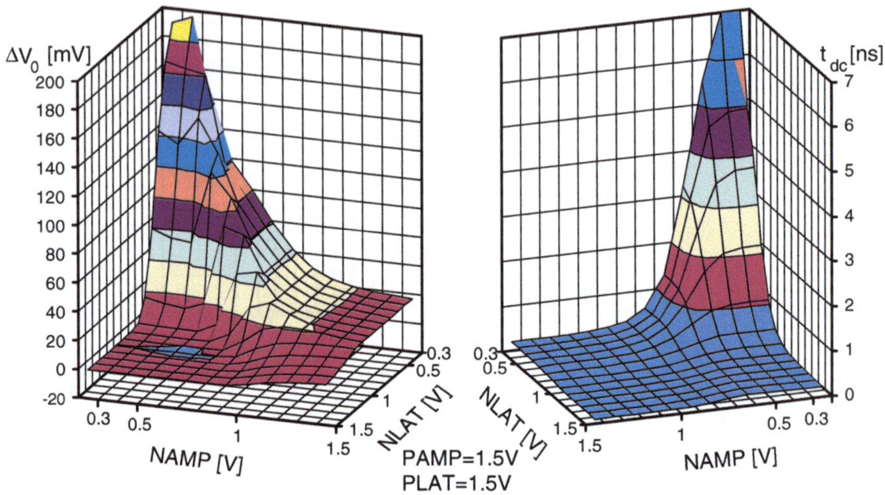

Fig. 6.55 Post layout simulation of ΔV_0 and t_{dc} in dependence on bias voltages *NLAT* and *NAMP* (*PAMP* = *PLAT* = 1.5 V) [9]

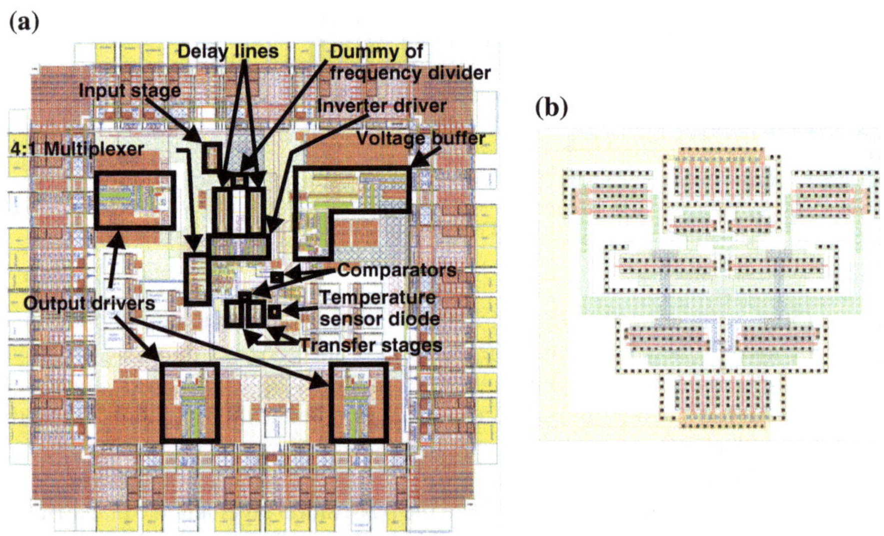

Fig. 6.56 Layout plots. **a** Test chip. **b** Comparator

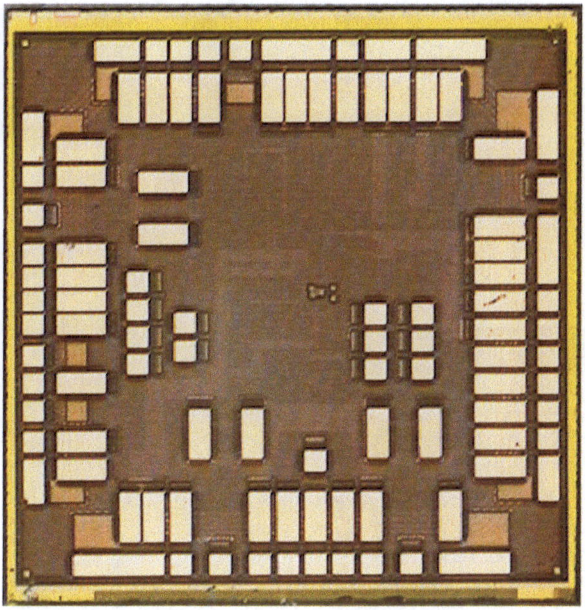

Fig. 6.57 Microphotograph of the test chip

6.5.2 Measurement Results

The fabricated chip (0.12 μm CMOS technology, 1.5 V nominal supply voltage, 1.65 V supply voltage of the analog voltage buffer) was countersunk into and bonded directly to a test board to reduce the influences of a package and its long bond wires. Photos of the chip mounted to a test board are shown in Fig. 6.58.

To reduce a the series inductance from chip ground V_{SS} to the ground of the test board, more than 20 bond wires were connected in parallel. Also from the power supply pads to the supply lines on the board and from the high-frequency pads to the microchip lines more than one wires (up to four) were bonded in parallel to reduce somewhat the series inductances. To further reduce the influences of bond wires, both comparators, the output drivers and different parts of the voltage buffer were supplied by separate pads. On the board 50 Ω microstrip lines were designed for high-frequency input and output signals of the chip (pads *CINP*, *CINN*, *CLKIN*, *COUT*, $\overline{COUT}$, *CREF*, *SIGOUT*). The typical measurement setup is described in Sect. 5.5. The duty cycle was determined with the DC voltages measured at output pads *CAVP* and *CAVN* and always adjusted to typically 50 % with changing the bias voltage at input pad *CLKIN*.

To test the functionality of comparator1 to a first overview a reference voltage of about 0.8 V was connected to *CINN* and a saw-tooth with a frequency of about 1 MHz with variable offset was applied to *CINP*. The sampling frequency was 3 GHz. The oscilloscope picture can be seen in Fig. 6.59a. Due to noise the comparator switches

Fig. 6.58 Photos of the test board

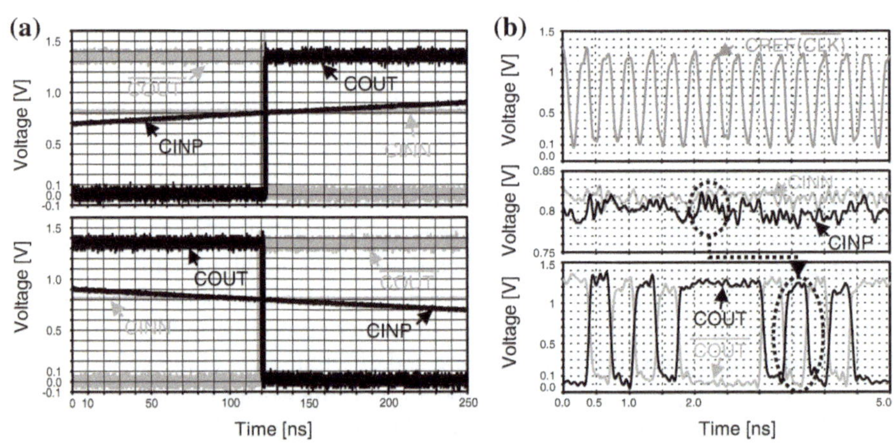

Fig. 6.59 Oscilloscope measurements. **a** Oscilloscope picture of the 3 GHz clocked comparator, where a saw-tooth waveform with about 1 MHz is applied to *CINP*, *CINN* = 0.8 V, bias level of *CINP*: variable, ≈0.8 V, *NAMP* = *PAMP* = *PLAT* = 1.5 V, *NLAT* = 1.07 V. **b** Oscilloscope picture of the comparator, clocked with 3 GHz, when a test wave with max. 1.5 GHz switching rate is applied to *CINP* with amplitude ±8mV around *CINN* = 0.82 V= Bias level of *CINP*-offset, *NAMP* = *PAMP* = *PLAT* = 1.5 V, *NLAT* = 1.07 V

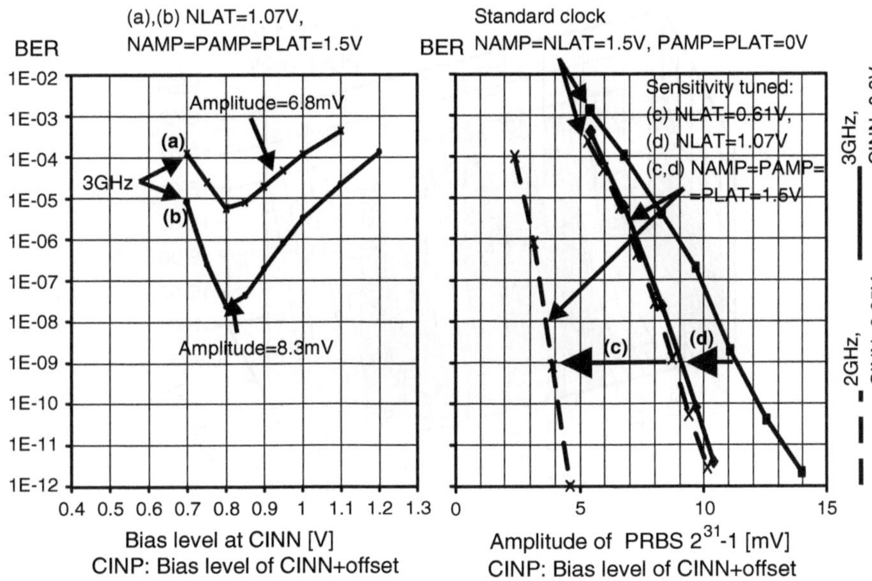

Fig. 6.60 BER measurements [9]: The effects of sensitivity tuning can be seen in *(c)* and *(d)*

randomly at the cross-over of *CINN* + offset and *CINP*. Figure 6.59b shows the behavior of the comparator, which was clocked with 3 GHz, when a test-wave with a maximum switching rate of a half of the clock frequency was applied.

The measured on-chip temperature (description in Sect. 5.2) varied between 40 °C (only two digital output drivers active) and 60 °C (all output drivers active).

The influence of noise to the decision of comparator1 and to determine the sensitivity, BER measurements were done. On the test chip at pad *CINN* a reference voltage and at pad *CINP* a pseudo-random-bit-sequence PRBS $2^{31} - 1$, which was biased with a DC voltage, were applied. *CINP* was always biased so that the offset of the comparator was compensated (best BER occurred). The amplitude of a pseudo-random-bit-sequence PRBS with the length $2^{31} - 1$ means, that the bit-sequence switches ± amplitude around the bias voltage $CINP = CINN +$ offset.

Figure 6.60 shows BER measurement results, where in Fig. 6.60a, b the BER versus the bias voltage of *CINN* is shown and where in Fig. 6.60a the amplitude of the superimposed PRBS $2^{31} - 1$ amounts to 6.8 mV and in Fig. 6.60b 8.3 mV. The clock frequency was 3 GHz, $NAMP = PAMP = PLAT = 1.5$ V and $NLAT = 1.07$ V. Of why an optimal working point with optimal BER appears is already explained in Sect. 6.3.2 for the complementary circuit structure. The improvement of sensitivity can be seen in Fig. 6.60c, d for the clock frequencies of (c) 2 GHz and (d) 3 GHz when $NAMP = PAMP = PLAT = 1.5$ V and $NLAT = 1.07$ V are applied compared with standard clocking ($PAMP = PLAT = 0$ V and $NAMP = NLAT = 1.5$ V).

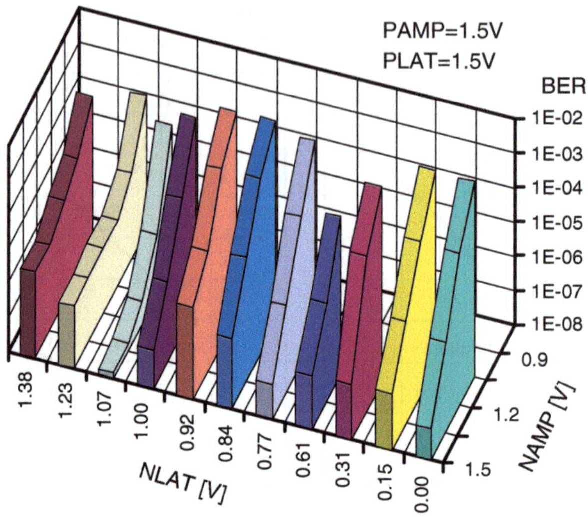

Fig. 6.61 BER in dependence on *NLAT* and *NAMP* at a clock frequency of 3 GHz, an amplitude of 8.3 mV with PRBS $2^{31} - 1$ at *CINP*, which was biased *CINN* +offset and $CINN = 0.85$ V [9]

For a BER of 10^{-9} a minimal amplitude at the input of 3.9 mV (2 GHz) and 9.2 mV (3 GHz) is necessary. The power consumption of a comparator was measured to be 422 μW at a clock frequency of 2 GHz and 584 μW at a clock frequency of 3 GHz.

Figure 6.61 shows the behavior of the BER when varying the bias voltages at *NLAT* and *NAMP* when *PLAT* and *PAMP* are set to 1.5 V (p-MOS transistors of the transmission gates are turned off), at 3 GHz clock and an input amplitude of PRBS $2^{31} - 1$ of 8.3 mV. *CINN* was set to 0.85 V and the bias level of *CINP*, where the PRBS was superimposed, was *CINN*+ offset. The optimal BER was achieved at $NLAT = 1.07$ V and $NAMP = 1.5$ V. When *NLAT* was raised higher than about 0.9 V and *NAMP* was lowered below about 0.9 V, ΔV_0 got smaller and BER got worse. At lower voltage levels than $NAMP \approx 0.9$ V and at *NLAT* smaller than about 0.9 V, the BER increased and became bad, because the comparator's decisions approach to metastability errors, where also noise may influence the decision (BER→0.5, see Sect. 2.4.2). At lower voltage values than $NLAT \approx 0.31$V it can be said that only the input amplifier ($N2$, $N3$, $N6$, $N7$ and $N8$) with its cross-coupled transistors $N6$, $N7$ and $P0$, $P1$ did the comparison work. Other simulations showed that for $NLAT < 0.31$ V a bias voltage was established at node $\overline{CENL}$ (*CENL* respectively) due to leakage currents so that transistors $N0$, $N1$ and $N2$ of the comparator (see Fig. 6.50) also contributed somewhat to the decision. So the measured results for $NLAT < 0.31$V should not be trusted.

The delay time measurements according to Sect. 5.3 of comparator1 can be seen in Fig. 6.62 and which were done at a clock frequency of 1.5 GHz at an supply voltage of $V_{DD} = 1.54$ V. Here t_d is defined the time duration from when the reset is released

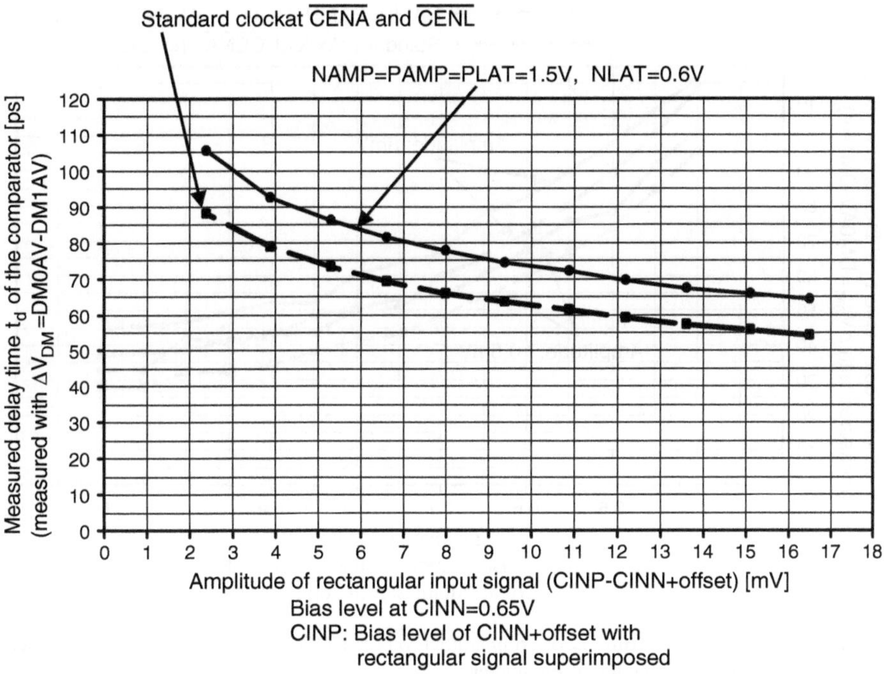

Fig. 6.62 Measured delay time t_d of comparator1 versus the amplitude of the input signal

and the ramp has reached 50 % of V_{DD} till one output node of the comparator has reached a voltage level, where the following inverter is able to switch from the logical reset state of comparator1 (*low* $=$ V_{SS} at the output of the inverter) to *high* $=$ V_{DD} at the output of the inverter. For this one output node has to reach a voltage level, which is typical a little bit lower than 50 % of V_{DD}. In Fig. 6.62 it can be seen that the delay time of the comparator lasts longer, when the amplitude of the rectangular input signal is lowered. Furthermore it should be stated that as expected the delay time is larger in the case of sensitivity tuning than in the case of a standard clock is used. The explanation of why the delay time has such a behavior can be found in Sect. 6.3.2 in the case of the complementary comparator circuit and in Sect. 6.5.1. The measurement of the delay time of comparator1 versus *CINN* (similar to the input common-mode voltage $(CINP + CINN)/2$) can be seen in Fig. 6.63. According to the explanations of Sect. 6.3.2 the delay time of comparator1 raises if the voltage level of *CINN* is lowered towards the cut-off of the input transistors. Furthermore as described in Sect. 6.5.1 the delay time is larger in the case of sensitivity tuning than in the case of a standard clock is used.

The measurements with the analog voltage buffer after comparator2 can be seen in Fig. 6.64 (see also the block diagram in Fig. 6.52). Unfortunately the effect of a larger delay time in the case of sensitivity tuning could not be observed here, because firstly

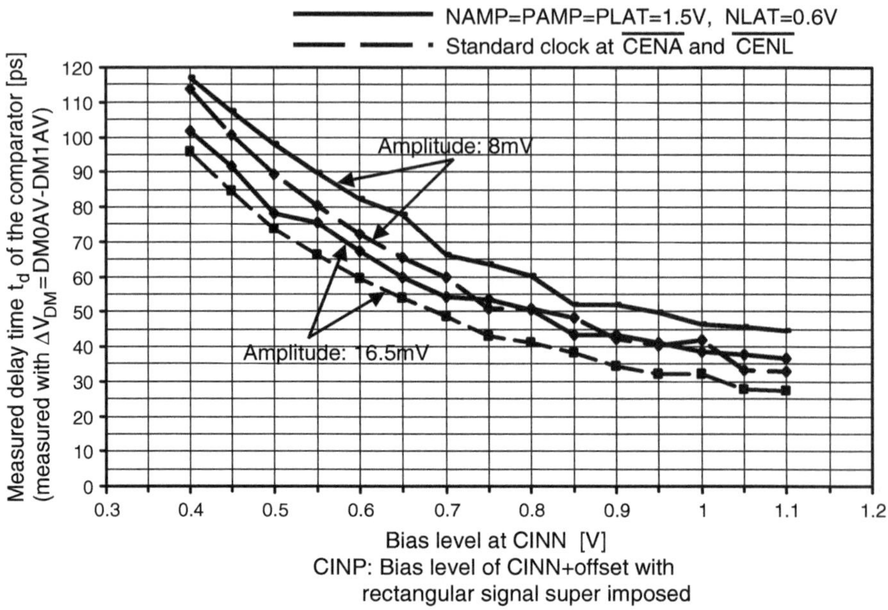

Fig. 6.63 Measured delay time t_d of comparator1 versus *CINN* (*CINP* $=$ *CINN* $+$ offset) with the amplitude of the input signal as parameter

due to the fast n-MOS transistors at the inputs of the comparator which causes a lower delay time difference between standard clocking and sensitivity tuning (bandwidth of the voltage buffer is too low) and secondly the tendency of the voltage buffer to oscillations at higher input common-mode levels.

6.5.3 *Short Extension to Reduce the Influence of Noise and Mismatch*

Another possibility to achieve a better robustness against noise and mismatch is shown in Fig. 6.65 as a consequence of measurements and 6.13 in Sects. 6.5.1 and 6.5.2. Here with an additional poly-resistor $R_6 = 900\,\Omega$ a delay between *CLKG* and *CLK* is introduced with the gate capacitance of transistor *N*9 due to a finite ramp of the clock edges of *CLK*. The other function of the comparator is the same as the comparator in Sect. 6.5.1 during standard clocking. With this delay in 6.13 it can be assumed, that initially a short time duration $I_9 \approx 0$ (or discharging of nodes *OUT* and $\overline{OUT}$ by cross-coupled transistors *N*0, *N*1 is delayed), thus enhancing ΔV_0 at the beginning of the regeneration of the whole latch and reducing the influence of mismatch and noise with the trade-off of a slight larger delay time of the decision of the comparator. A transient simulation of the comparator without and with resistor

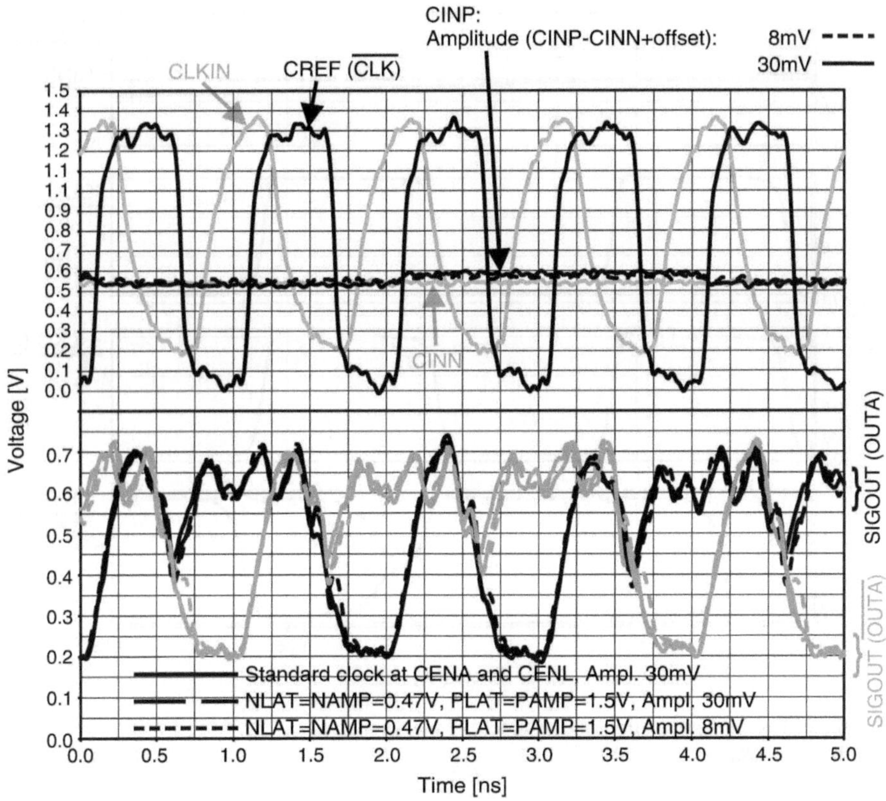

Fig. 6.64 Oscilloscope pictures (the voltage buffer with output *SIGOUT* has an attenuation of about two as described in Sect. 5.1.3)

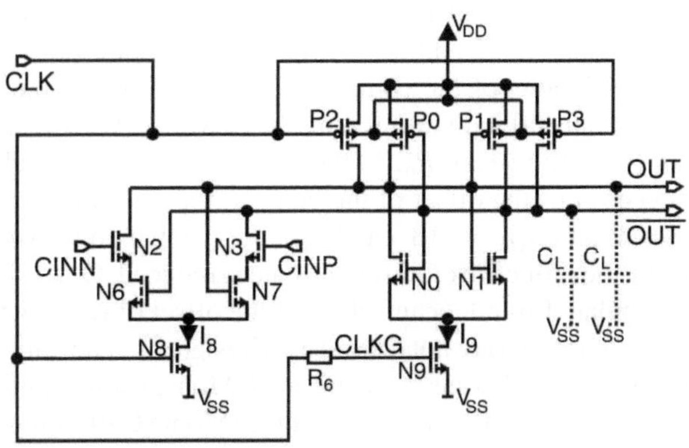

Fig. 6.65 Simple circuit extension (resistor R_6 to reduce the influence of mismatch and noise) [9]

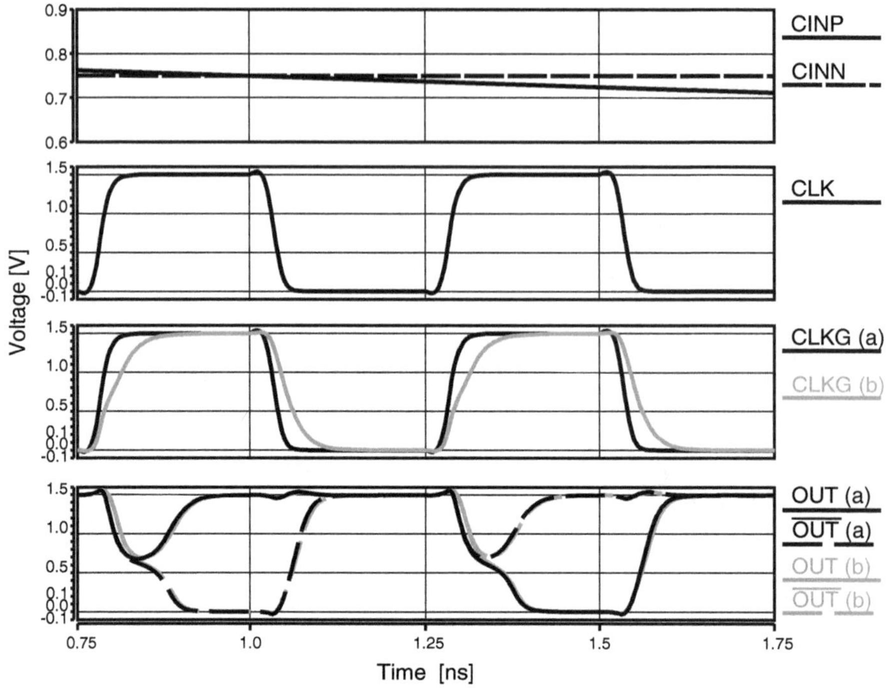

Fig. 6.66 Transient simulation (2 GHz clock) of the comparator in Fig. 6.65: *a* without resistor R_6 *b* with resistor R_6

R_6 is shown in Fig. 6.66. There the delay of signal *CLKG* referred to *CLK* can be seen in the case of an additional resistor R_6. This delay causes as explained a larger ΔV_0 to reduce the influence of mismatch and noise to the cost of a slight larger delay time of the output signals *OUT* and $\overline{OUT}$ of the comparator.

Monte Carlo simulations of the comparator structure in Fig. 6.65 are shown in Fig. 6.67 for the cases of (a) without resistor R_6, $CINN = 0.75$ V and 2 GHz clock frequency and of (b) with resistor R_6, $CINN = 0.75$ V and 2 GHz clock frequency. The standard deviation of the offset of the comparator was simulated to be $\sigma_{OS} = 25.8$ mV for case (a) and $\sigma_{OS} = 18.7$ mV for case (b), which shows that with only adding a resistor R_6 the influence of mismatch can be reduced. Other simulations have shown, that the simulated small mean values of the offset of $\mu = 4.5$ mV (case (a)) and $\mu = -1.9$ mV (case (b)) are not caused by a hysteresis. They might disappear, if the amount of the Monte Carlo runs is enhanced. A Monte Carlo simulation was done in the same way as explained in Sect. 6.5.1. Another Monte Carlo simulation (50 runs) for $CINN = 0.85$ V and a clock frequency of 3 GHz shows also an improvement of the standard deviation of the offset from $\sigma_{OS} = 29$ mV at standard clocking to $\sigma_{OS} = 18.3$ mV with additional resistor R_6.

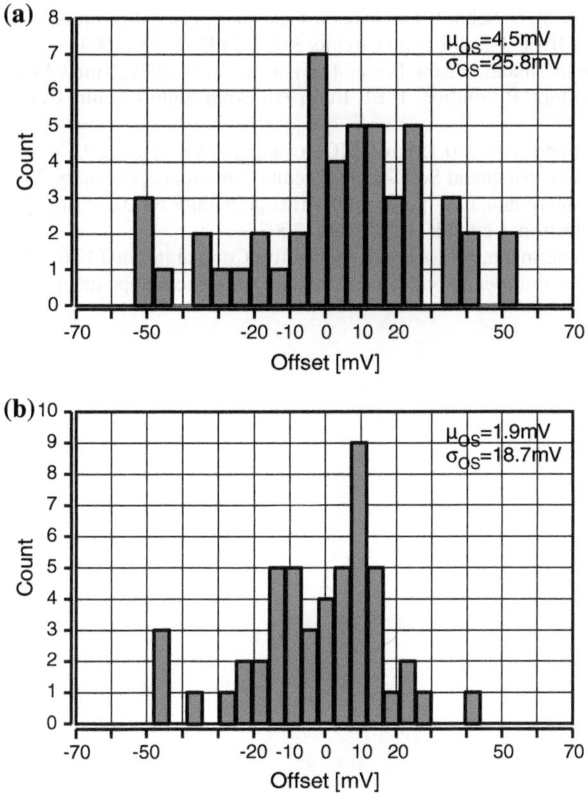

Fig. 6.67 Monte Carlo simulations of the comparator in Fig. 6.65 (50 runs each), where the clock frequency was 2 GHz and $CINN = 0.75$ V. **a** Without resistor R_6. **b** With resistor R_6

The tuning scheme described above does not simply introduce a delayed regeneration of the latch as can be seen in Sect. 6.3, where in the comparator circuit of Fig. 6.26 the gates of the tail transistors were connected together.

References

1. B. Goll, H. Zimmermann, A Low Power 1.4 GSample/s Comparator for Flash-ADCs in 120 nm CMOS Technology, Austrochip 2004, October 2005, pp. 39–42
2. B. Goll, H. Zimmermann, A Low-Power 2-GSample/s Comparator in 120 nm CMOS Technology, IEEE European Solid-State Circuits Conference, September 2005, pp. 507–510
3. B. Wicht, T. Nirschl, D. Schmitt-Landsiedel, A Yield-Optimized Latch-Type SRAM Sense Amplifier, IEEE European Solid-State Circuits Conference, September 2003, pp. 409–412
4. B. Wicht, *Current Sense Amplifiers for Embedded SRAM in High-Performance System-on-a-Chip Designs* (Springer, Berlin, 2003)

5. B. Wicht, T. Nirschl, D. Schmitt-Landsiedel, Yield and speed optimization of a latch-type voltage sense amplifier. IEEE J. Solid-State Circuits **39**(7), 1148–1158 (2004)
6. B. Goll, H. Zimmermann, A Low-Power 4 GHz Comparator in 120 nm CMOS Technology with a Technique to tune Resolution, IEEE European Solid-State Circuits Conference, September 2006, pp. 320–323
7. B. Goll, H. Zimmermann, A 0.12 μm CMOS Comparator Requiring 0.5 V at 600 MHz and 1.5 V at 6 GHz, IEEE International Solid-State Circuits Conference, February 2007, pp. 316–317
8. B. Goll, H. Zimmermann, Low-power 600 MHz comparator for 0.5 V supply voltage in 0.12 μm CMOS. IET Electron. Lett. **43**(7), 388–390 (2007)
9. B. Goll, H. Zimmermann, A Clocked, Regenerative Comparator in 0.12 μm CMOS with Tunable Sensitivity, IEEE European Solid-State Circuits Conference, September 2007, pp. 408–411

Chapter 7
Comparators in 65 nm CMOS

Two comparators in 65 nm low-power CMOS are introduced in this chapter. First a high-speed comparator for a sample rate of 7 GHz inclusive measured results is presented. The second comparator is optimized with respect to a low offset voltage, but nevertheless achieves a maximum clock frequency of 5 GHz.

7.1 A 7 GHz Comparator

The comparator presented here is designed to work at a high clock frequency of 7 GHz at a nominal supply voltage of 1.2 V. Even when the supply voltage is reduced to 0.6 V, the circuit is able to manage a clock frequency of 700 MHz. This has been achieved with a modified latch, where in contrast to the widely used circuit structure in Fig. 2.7a, which needs a sufficiently large supply voltage for proper operation, the stack of transistors between V_{DD} and V_{SS} is avoided.

7.1.1 Circuit Description

The schematic of the presented comparator with the modified latch is depicted in Fig. 7.1 [1]. The circuit has been designed in 1.2 V 65 nm Low-Power CMOS, where the threshold voltage of a p-MOS transistor, as well as of a n-MOS transistor amounts to $\approx$0.4 V. In opposite to the circuit in Fig. 2.7a, which consists of a latch with cross-coupled inverters placed above two input transistors, the proposed latch in the presented comparator is expanded into two paths between the supply rails with an input stage in parallel. The transient simulations of the presented comparator with the proposed latch are shown in Fig. 7.2.

In the upper left of Fig. 7.1 transistors in transmission gates for tuning the sensitivity are shown. Nodes *CLK* and $\overline{CLK}$ delivers a non-inverted and an inverted rectangular clock with logical voltage levels of V_{SS} and V_{Co}. When connecting node *PM* to V_{Co} (transistors $P12$ and $P13$ are turned off) and nodes *NL* and *NA* each to a

B. Goll and H. Zimmermann, *Comparators in Nanometer CMOS Technology*,
Springer Series in Advanced Microelectronics 50, DOI 10.1007/978-3-662-44482-5_7

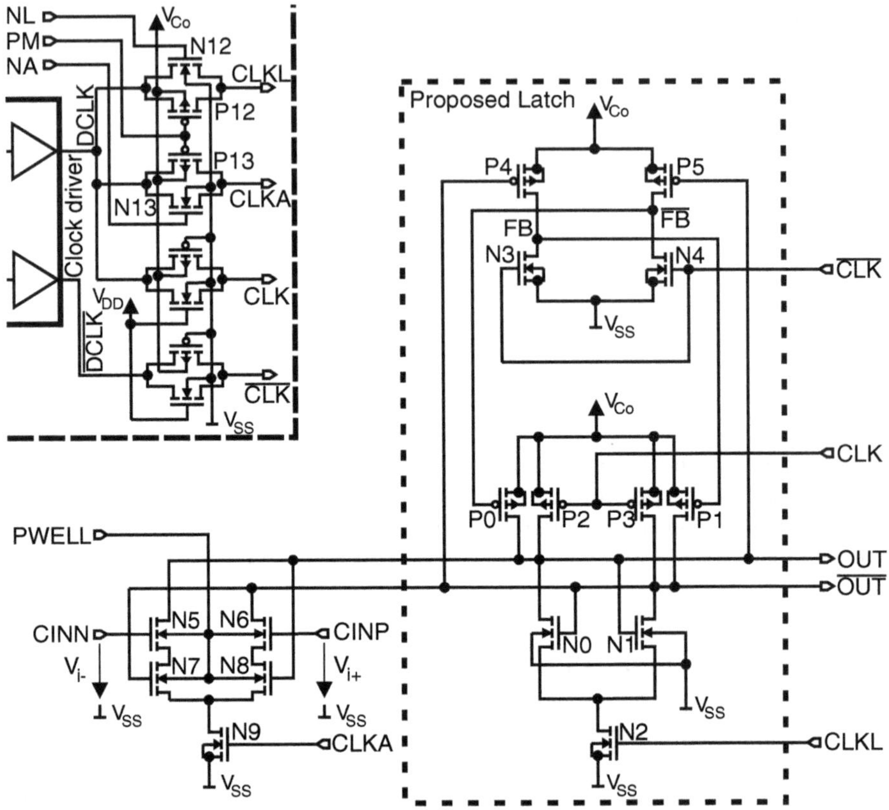

Fig. 7.1 Schematic of the comparator with the proposed latch [1]

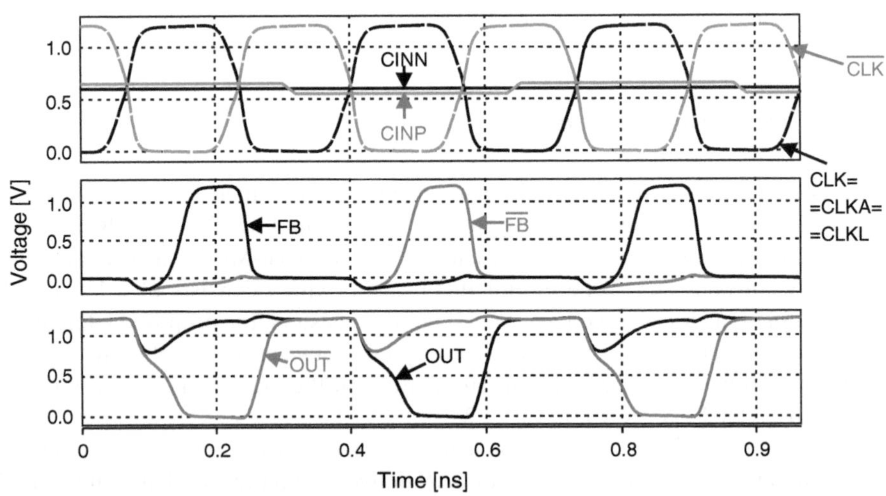

Fig. 7.2 Transient simulation of the comparator for functional illustration without tuning the sensitivity ($NL = NA = V_{Co} = 1.2\,V$, $PM = V_{SS} = 0\,V$)

reference voltage, the upper logic voltage levels of the clock signals *CLKA* and *CLKL* can be adjusted to a value below V_{Co}. Thus the sensitivity of the comparator can be optimized for a distinct clock frequency as already described in Sect. 6.5. In general and without tuning the sensitivity, a clock period is subdivided into a reset phase and a comparison phase. In the reset phase ($CLK = CLKA = CLKL = V_{SS}$) the comparator is forced into a metastable point with turned on transistors $P0$, $P1$, $N3$ and $N4$. Transistors $N2$ and $N9$ are switched off and reset transistors $P2$ and $P3$ pull both output nodes OUT and $\overline{OUT}$ to V_{Co}, which as a consequence turn $P4$ and $P5$ off. Nodes FB and $\overline{FB}$ are discharged to V_{SS} with transistors $N3$ and $N4$. The gates of transistors $N7$ and $N8$ are at voltage level V_{Co} as well, so that $N7$ and $N8$ are initially on, when the comparison phase starts. A start condition of $OUT = \overline{OUT} = V_{Co}$ is build up at the output nodes and the internal nodes are set to $FB = \overline{FB} = V_{SS}$. In the following comparison phase ($CLK = CLKA = CLKL = V_{Co}$) the voltages V_{i+} at *CINP* and V_{i-} at *CINN* are compared. For comparison of the input voltages, transistors $N2$ and $N9$ are turned on and $P2$, $P3$, $N3$ and $N4$ are switched off. In the case of $V_{i+} > V_{i-}$ $\overline{OUT}$ is discharged with transistor $N6$ more than OUT with $N5$ and positive feedback is started. Hence output node $\overline{OUT}$ is pulled towards V_{SS} by $N6$ and $N1$ more than OUT by $N5$ and $N0$. As a consequence $P4$ is turned on before $P5$ and FB is pulled towards V_{Co} while $\overline{OUT}$ remains near V_{SS} when $|V_{i+} - V_{i-}|$ is sufficiently large so that no metastability error occurs. Finally transistors $P1$ and $N0$ are switched off, $P0$ and $N1$ are turned on and OUT is pulled to V_{Co}. Node $\overline{OUT}$ switches to V_{SS} and FB is at voltage level V_{Co}. Because transistors $N7$ and $N8$ avoid static current flow through the input part via $N9$ due to the fact that $\overline{OUT} = V_{SS}$ and $N7$ is turned off, no static current can flow and the decision of the comparator is done. In the opposite case of $V_{i+} < V_{i-}$, the decision works in a similar way with the result that OUT is pulled to V_{SS} and $\overline{OUT}$ to V_{Co}. There exists a separated p-well of n-MOS transistors, because the comparator is designed in a triple-well process. So the body effect can be used to reduce the threshold voltages of transistors $N5$ to $N8$ by applying maximal 0.7 V to *PWELL*.

Due to the fact that transistors $P0$ and $P1$ are turned on and build a lower resistive load at the beginning of comparison, on the one hand a fast amplification and on the other hand a proper operation to supply voltages below of the sum of the amount of the threshold voltages of p-MOS and n-MOS transistors ($V_{tn} + |V_{tp}| \approx 0.8$V in the technology used) are possible. The standard latch in Fig. 2.1 gets significantly slower when the supply voltage is lower than $V_{tn} + |V_{tp}|$. Another effect occurs during comparison: the amplification caused by e.g. $P5$ and $P4$ in the p-MOS part is multiplied with the amplification of $P0$ and $P1$, respectively. The comparison of the proposed latch with a standard latch, which is implemented in typical comparators, is shown in Fig. 7.4. For simulation the comparator of Fig. 7.1 without sensitivity tuning section is compared with a standard comparator in Fig. 7.3. There the same input stages (transistors $N5$ to $N9$) are implemented. The diagram in Fig. 7.4 shows the simulated delay time. The delay time is defined here as the time between 50 % of V_{Co} has been reached in the rising edge of *CLK* to the time point when $|OUT - \overline{OUT}|$ has reached 50 % of V_{Co}). Both latches were designed similar so that they have the same delay time at $V_{Co} = 1.2$ V. In Fig. 7.4 it can be seen that if the supply voltage

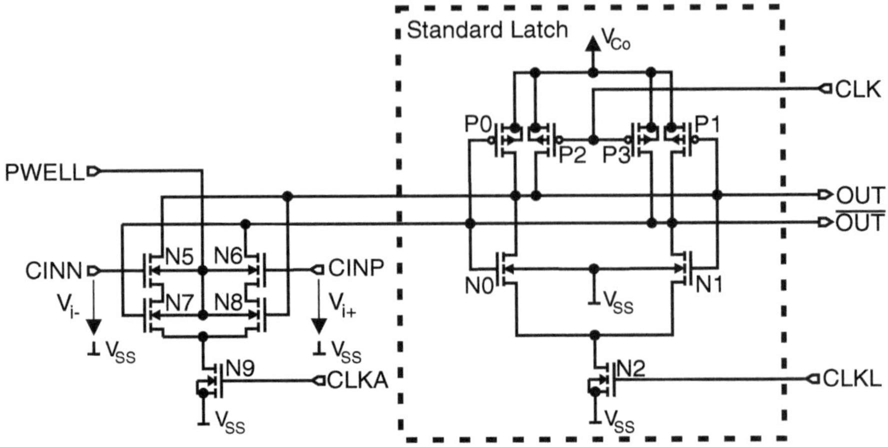

Fig. 7.3 Comparator with a standard latch

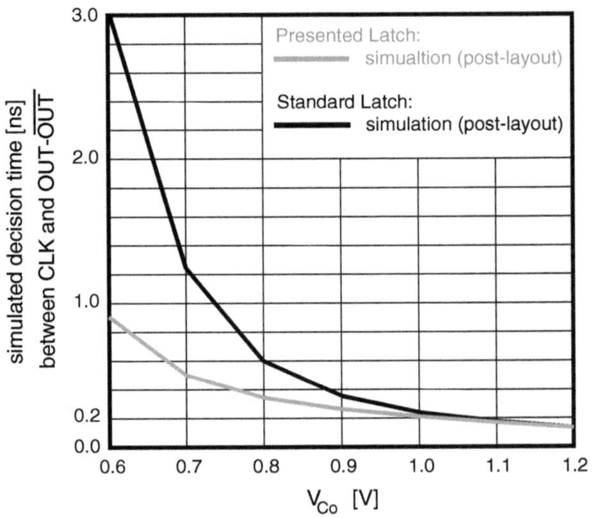

Fig. 7.4 Comparison of the standard latch with the presented latch

V_{Co} is reduced to below 0.8 V, then the delay time of the standard latch is much longer than the delay time of the proposed comparator. The simulation resulted in a difference of 3 ns compared to 0.9 ns at $V_{Co} = 0.6\,\text{V}$. The block diagram of the whole test chip ist depicted in Fig.7.5. The proposed comparator of Fig. 7.1 with the transmission gates (T-Gates) for tuning of the sensitivity is placed in an area with a separate supply voltage V_{Co}. This is done to investigate the behavior of the comparator for lower supply voltages. With adapters the logic voltage levels of the clock signals as well the logic voltage levels of the comparators output are converted in that way that they switch between V_{SS} and the actual supply voltage. A detailed

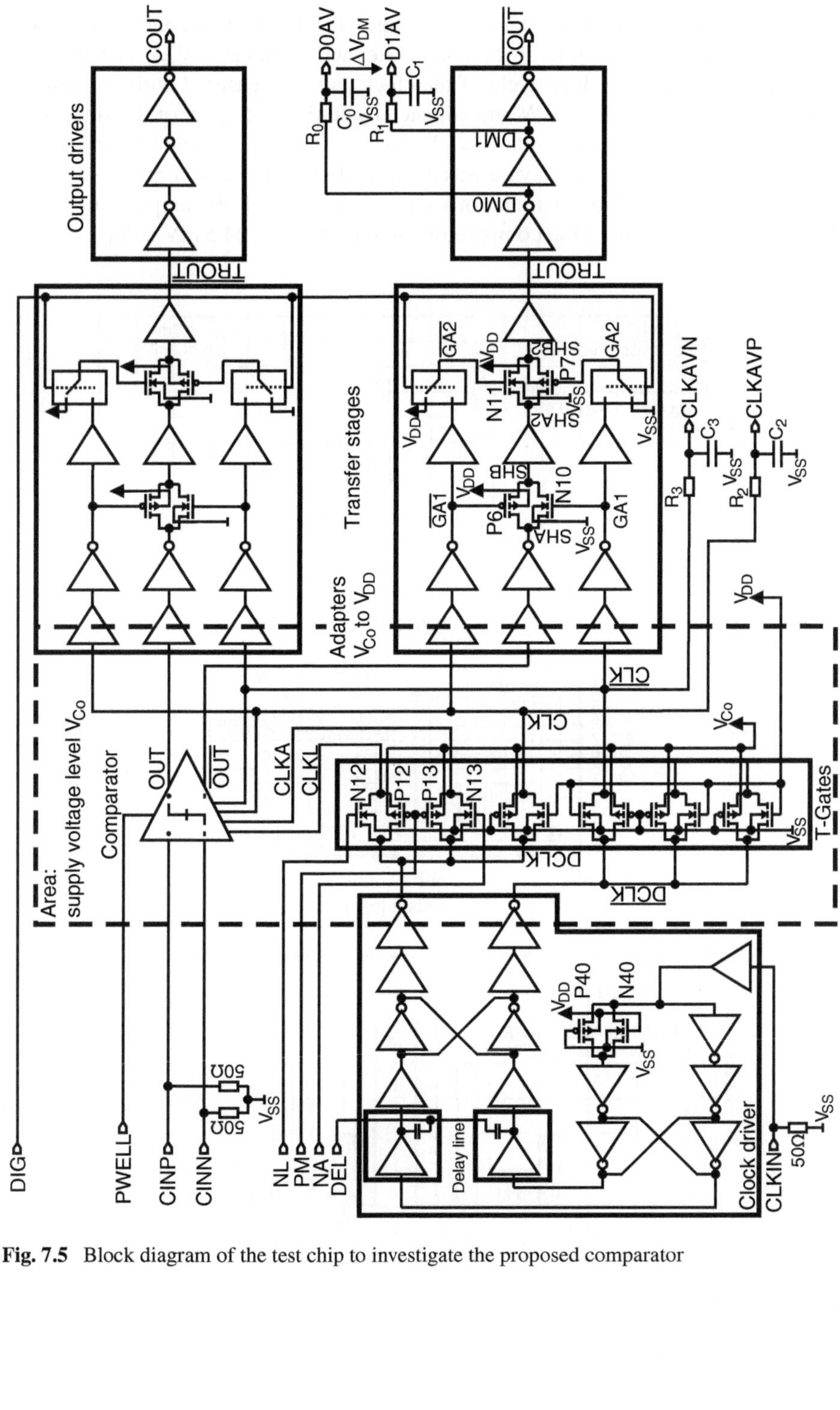

Fig. 7.5 Block diagram of the test chip to investigate the proposed comparator

description of the clock driver, the transfer stages, the output driver and the technique for measuring the delay time of the comparator with voltage ΔV_{DM} can be found in Chap. 5. The internal duty cycle of the clock can be measured with the voltages at nodes *CLKAVP* and *CLKAVN* and adjusted by setting the offset of the sine signal, which is applied off-chip to node *CLKIN* to create on-chip complementary square-wave clock signals. The duty cycle was always adjusted to be near 50 %.

In Fig. 7.6 Monte Carlo simulations with each a run of 50 samples are shown. The standard deviation of the comparators offset is $\sigma_{OS} = 24.5\,\text{mV}$ at $V_{Co} = 1.2\,\text{V}$

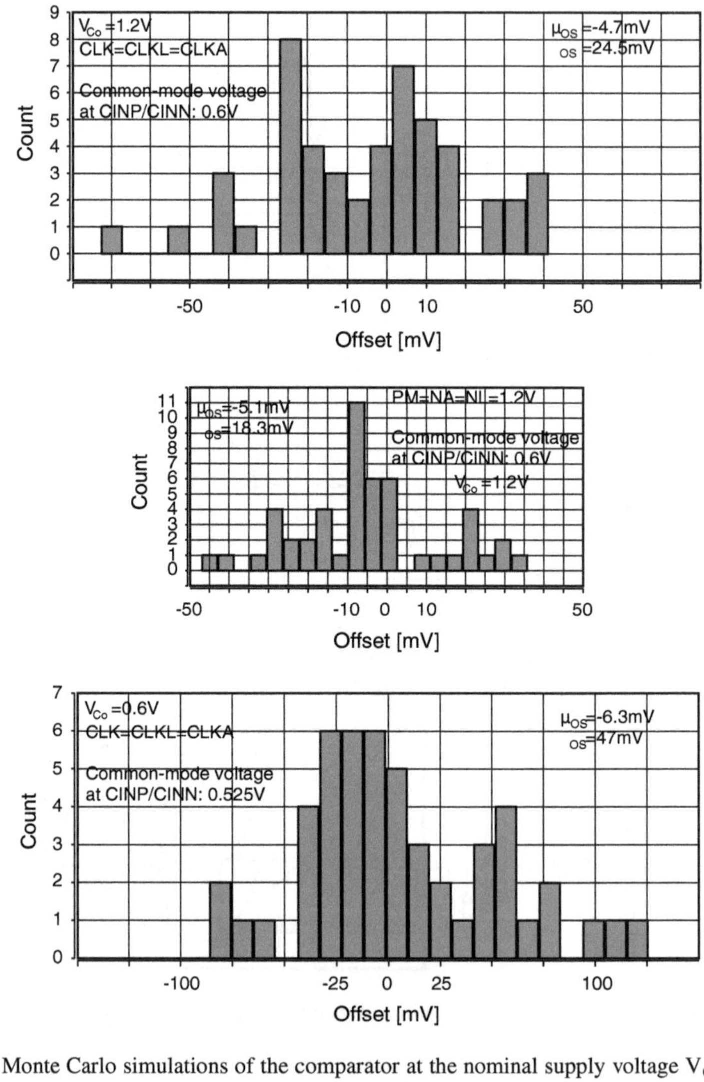

Fig. 7.6 Monte Carlo simulations of the comparator at the nominal supply voltage $V_{Co} = 1.2\,\text{V}$ (*top*), when improving the sensitivity with the help of the T-Gates (*middle*) and at the low supply voltage of $V_{Co} = 0.6\,\text{V}$ (*bottom*)

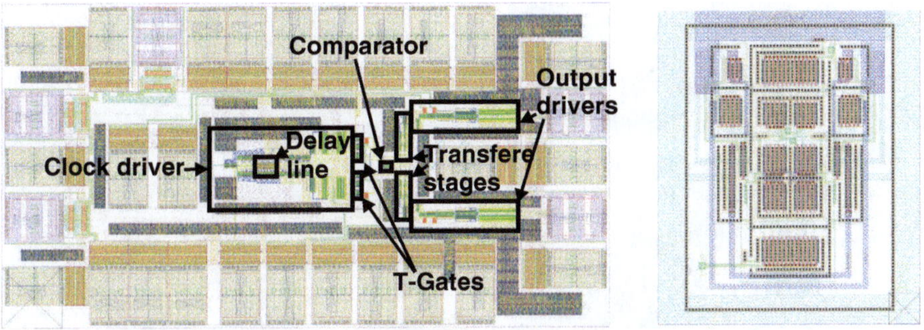

Fig. 7.7 Layout plots of the test chip (*left side*) and the comparator (*right side*)

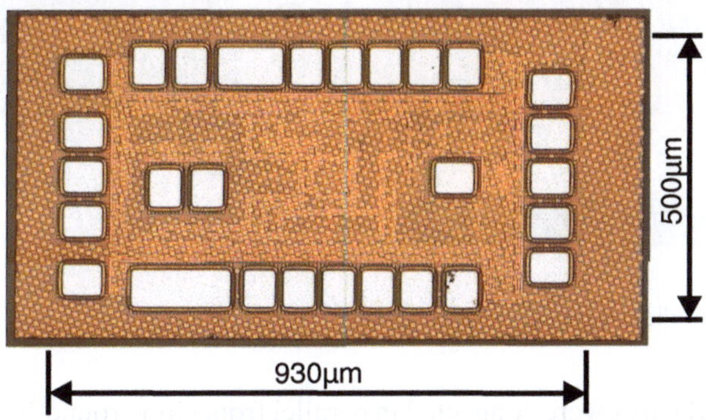

Fig. 7.8 Microphotograph of the test chip

and $\sigma_{OS} = 47$ mV at $V_{Co} = 0.6$ V without tuning the sensitivity. A simulation at $V_{Co} = 1.2$ V when sensitivity tuning is used is shown in the middle of Fig. 7.6. Here the standard deviation of the offset amounts to $\sigma_{OS} = 18.3$ mV which is smaller compared to the previous $\sigma_{OS} = 24.5$ mV.

The layout plot of the test chip is depicted in Fig. 7.7 on the left side and the layout plot of the comparator without the T-Gates for sensitivity tuning is shown on the right side. The size of the test chip amounts to 930×500 µm and the comparator occupies a chip area of 19.6×16.3 µm. A microphotograph of the test chip can be seen in Fig. 7.8. The test chip is fabricated in a 1.2 V 65 nm low-power CMOS process with a threshold voltage of n-MOS and p-MOS transistors of $V_{tn} \approx |V_{tp}| \approx 0.4$ V.

7.1.2 Measurement Results

The fabricated chip (65 nm CMOS technology with 1.2 V nominal supply voltage) was bonded directly to a test board to reduce the influences of a package and its long bond wires. Photos of the chip mounted to a test board are shown in Fig. 7.9.

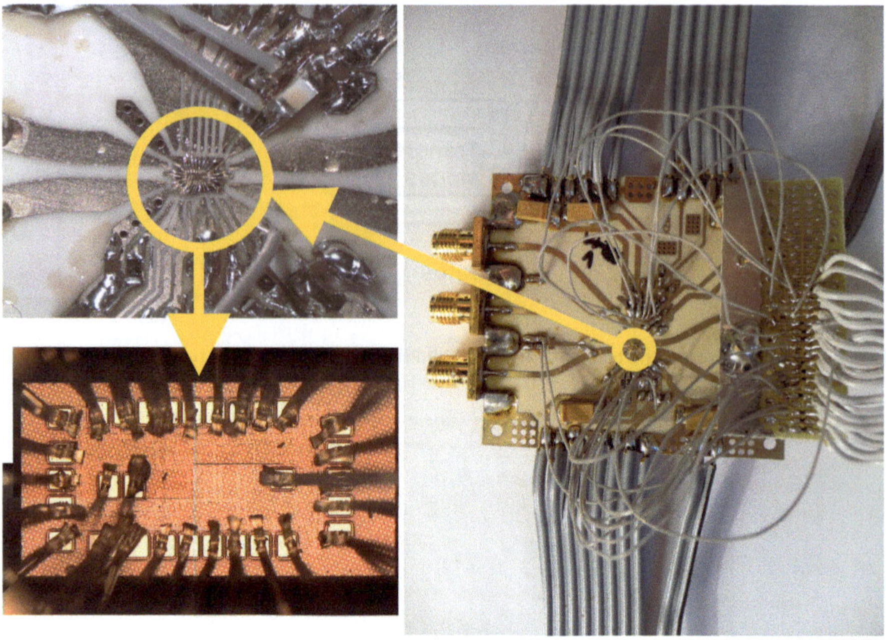

Fig. 7.9 Photos of the test board

Several bond wires were connected in parallel from chip ground V_{SS} to the ground of the test board to reduce the series inductance and hence the ground bounce. Additionally from the power supply pads on the chip to the supply lines on the board and from the high-frequency pads on the chip to the microstrip lines two or more wires were bonded in parallel to reduce somewhat the series inductances. There were different supply voltages for the output drivers, for the comparator and for the logic circuitry to furthermore reduce disturbances. On the board 50 Ω microstrip lines were designed for the high-frequency input and output signals of the chip (pads *CINP*, *CINN*, *CLKIN*, *COUT*, $\overline{COUT}$). A typical measurement setup is described in Sect. 5.5. The duty cycle was determined with the DC voltages measured at output pads *CLKAVP* and *CLKAVN* and always adjusted to typically 50 % with changing the bias voltage at input pad *CLKIN*. Figure 7.10 shows screen shots of the test chip with the comparator at the maximum clock frequency of 7 GHz. At *CLKIN* a 7 GHz sine signal, to input pad *CINP* a test signal overlayed on 0.65 V DC voltage and to pad *CINN* a reference voltage of 0.65 V minus the offset of the comparator were applied. This proved the functionality of the test chip.

To determine the sensitivity as a consequence of the influence of noise to the decision of a comparator BER measurements were done. At pad *CINP* a pseudo-random-bit-sequence $2^{31} - 1$, which is biased with a DC voltage and at pad *CINN* a reference voltage were applied. Here the amplitude of a pseudo-random-bit-sequence is defined that the bit-sequence switches +/− amplitude around the bias voltage *CINP*, which equals the sum of the reference voltage at *CINN* and the offset of the

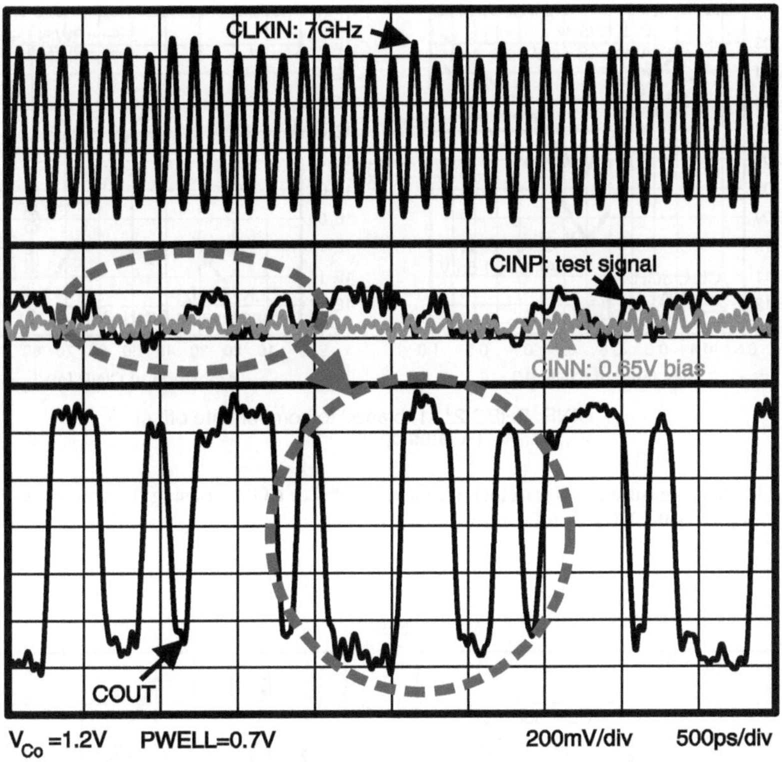

Fig. 7.10 Oscilloscope screen shots at 7 GHz clock frequency

comparator. So the bias voltage at *CINP* was varied until the best BER occurred. The measurement results for applying a standard rectangular digital clock with logic voltage levels $V_{Co} = 1.2\,\text{V}$ and $V_{SS} = 0\,\text{V}$, 50 % duty cycle at 5 and 6 GHz can be seen in Fig. 7.11. The black lines (solid and dotted) are for a voltage $PWELL = 0\,\text{V}$ and the grey ones are for the case when *PWELL* is raised to 0.7 V to use the back-gate effect of n-MOS transistors of the input stage in their separated p-well. On the left side of Fig. 7.11 it can be seen that for an amplitude of 26.5 mV (5 GHz clock frequency), the point with the optimal BER is shifted to lower input common-mode voltages and has a ten times better BER due to reduction of the threshold voltage of the n-MOS transistors. On the right side of Fig. 7.11 the BER versus the amplitude of the input data signal at *CINP* is shown for the $CINN = 0.6\,\text{V}$, where best BER occurred when setting $PWELL = 0.7\,\text{V}$ for 5 GHz clock frequency (grey line). To reach a BER of 10^{-9} an amplitude of 27.2 mV is needed. In the same figure the measurement result for a clock frequency of 6 GHz ($PWELL = 0\,\text{V}$) is depicted. There the data signal at the input needs an amplitude of 63 mV to get a BER of 10^{-9}.

The BER measurements of the comparator when the supply voltage V_{Co} is lowered to 0.6 V are shown in Fig. 7.12. The p-wells of transistors $N5$ to $N7$ of the input stage

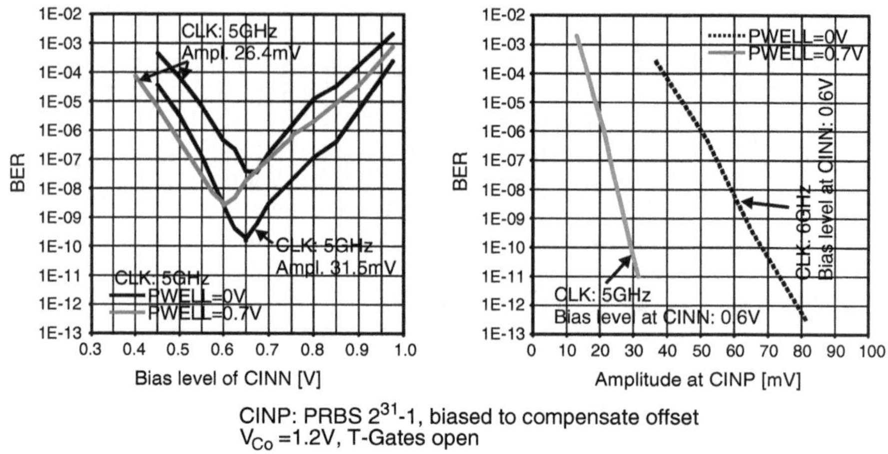

Fig. 7.11 BER measurements for clock frequencies of 5 or 6 GHz: raising the voltage of *PWELL* to 0.7 V improves the BER by a factor of 10

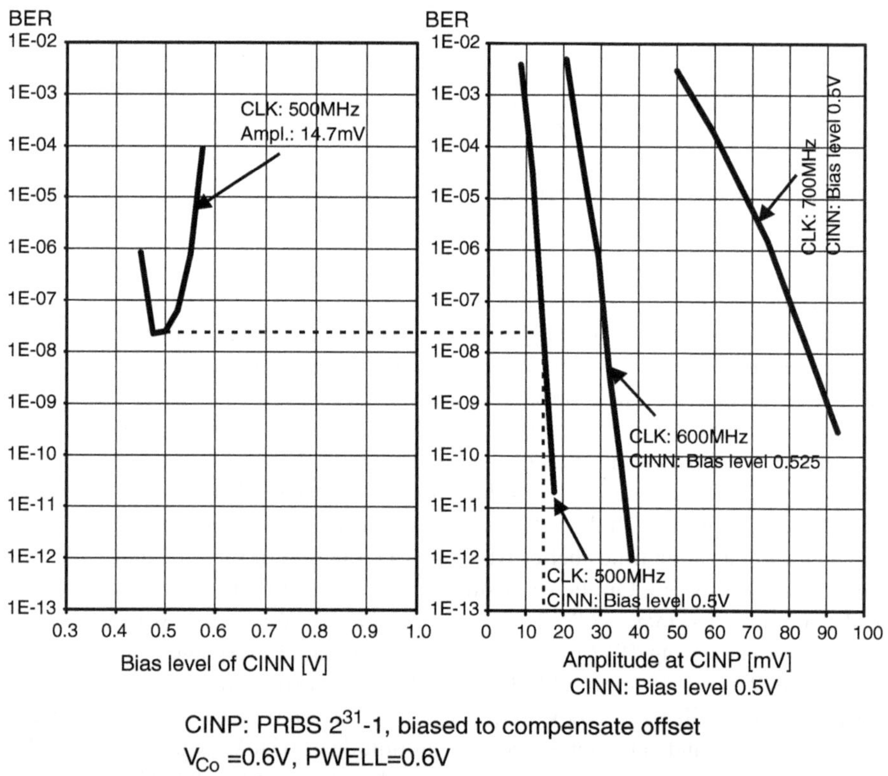

Fig. 7.12 BER measurements for a supply voltage of $V_{Co} = 0.6\,V$ of the comparator

of the comparator were set to $PWELL = 0.6\,\text{V}$, but in the proposed latch no back-gate effect has been used. The measurements were done for clock frequencies of 500, 600 and 700 MHz. The left diagram in Fig. 7.12 reveals, that the best BER is achieved at a reference voltage around 0.5 V at *CINN*. The right side shows the dependencies of the BER versus the amplitude of the data signal at *CINP*. Without tuning the sensitivity (T-Gates are completely open) the sensitivity (BER $= 10^{-9}$) is 16 mV at 500 MHz, 34 mV at 600 MHz and 90.2 mV at 700 MHz.

For lower clock frequencies the sensitivity can be tuned to better values (see Sect. 6.5). In the case of Fig. 7.13 the gates of the transistors in the transmission gates are set to $PM = NA = NL = 1.2\,\text{V}$ for a clock frequency of 3 GHz. Hence the clock signal *CLKA* as well as *CLKL* switches between V_{SS} and $\approx V_{Co} - V_{tn}$ (subthreshold effects and leakage currents are neglected). The reduced speed of discharging the output nodes OUT and $\overline{OUT}$ especially of the input stage (transistors $N5$ to $N9$) results into a higher sensitivity. This can be seen on the left side of Fig. 7.13. There the effect of raising *PWELL* from 0 V to an extreme value of 0.745 V on the one hand, which is on the boarder when a p-n junction starts to conduct considering the triple well process and the drain-source voltage drop of transistor $N9$ (see Fig. 7.1) and setting the gate voltages of transistors $N12$, $P12$, $N13$ and $P13$ to $PM = NA = NL = 1.2\,\text{V}$ is shown. For comparison also the cases $PWELL = 0\,\text{V}$ and fully open T-gates are depicted. Raising the voltage of PWELL to a maximum

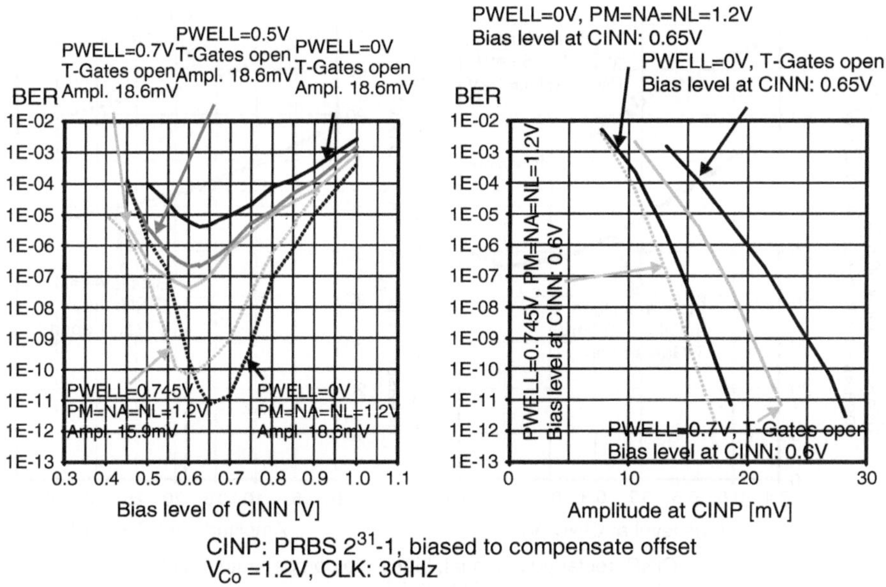

Fig. 7.13 BER measurements for a clock frequency of 3 GHz: raising the voltage of PWELL to maximum value of 0.745 V, which includes the voltage drop of transistor $N9$ and additionally reducing the comparator speed with setting $PM = NA = NL = 1.2\,\text{V}$) shifts the sensitivity by 10 mV to lower input voltage amplitudes. The lower the clock frequency of the comparator is reduced below the maximum frequency, the more the sensitivity can be enhanced

value of 0.745 V and setting $PM = NA = NL = 1.2$ V instead of $PM = 0$ V and $NA = NL = 1.2$ V for open transmission gates (T-gates) shifts the sensitivity 10 mV to lower amplitudes at 3 GHz, which is shown on the right side of Fig. 7.13 when considering the input common-mode voltages with the optimal BER. For a clock frequency of 3 and 4 GHz a sensitivity ($BER = 10^{-9}$) of 15 and 20 mV, respectively have been achieved with the help of sensitivity tuning. At a clock frequency of 7 GHz a sensitivity of 281 mV has been achieved.

The measured delay time at a clock frequency of 3 GHz and $PWELL = 0.7$ V of the comparator can be seen in Fig. 7.14. The black lines represent the case when no sensitivity tuning is applied. The grey lines depict results of post layout simulations to verify the measurement. It can be seen that the measured result is similar to the simulated one. The delay time of the comparator is e.g. 64 ps at $CINN = 0.6$ V and 18.6 mV amplitude. The dotted line shows the case of setting $PM = NA = NL = 1.2$ V to get a better sensitivity but with the trade-off that the delay time of the comparator is around 20 ps longer. At lower input amplitudes, the measurements are influenced by more random switching of the comparator.

The measured power consumptions amounted for a supply voltage $V_{Co} = 1.2$ V of the comparator to 1.3 mW at 7 GHz and to 1.19 mW at 6 GHz. For a supply voltage $V_{Co} = 0.6$ V the comparator consumed 41 μW at 500 MHz and 47 μW at 700 MHz.

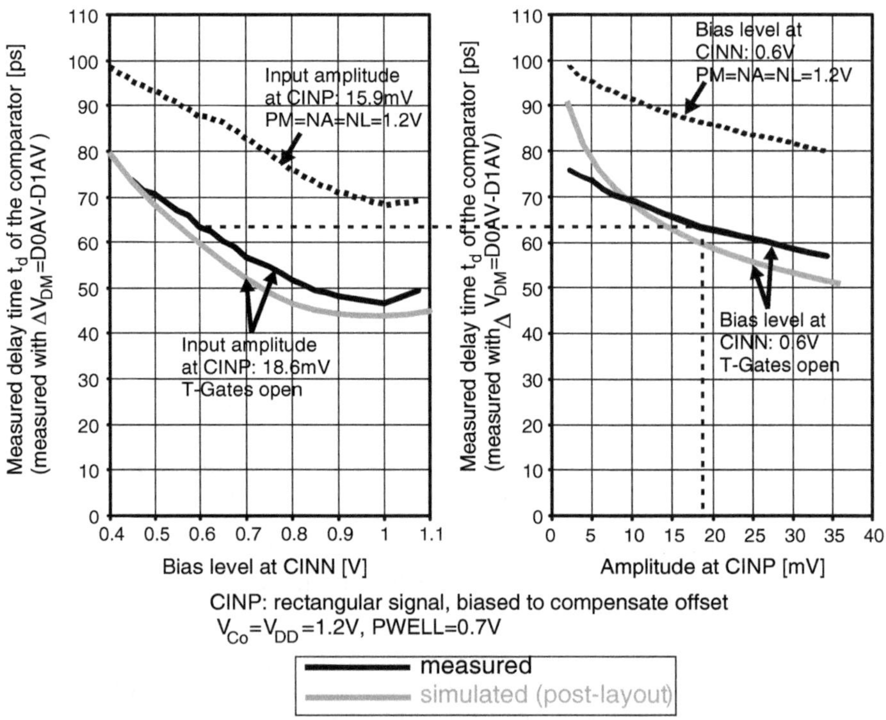

Fig. 7.14 Measured delay time t_d of the comparator

7.2 A Low-Offset Comparator

The comparator circuit shown in Fig. 2.7a is a widely used standard circuit with rail-to-rail output swing, with high-impedance input and no static power consumption. Furthermore this circuit has the advantage of a good robustness against noise and mismatch, because among other reasons it can be also designed with large input transistors $N2$, $N3$ e.g. to minimize the offset, where their larger parasitic capacitances do not directly affect switching speed, which depends primarily on the load capacitances at the output nodes OUT and $\overline{OUT}$. Disadvantageous is the fact that due to the many stacked transistors a sufficiently high supply voltage is needed for a proper delay time [2]. In low-voltage UDSM-CMOS technologies this may cause problems, where even a stand-alone latch (e.g. used in [2]) as shown in Fig. 2.1 with its two cross-coupled inverters, suffers from a longer delay time, if the supply voltage is reduced below two times the threshold voltage and if a low-power process with higher transistor threshold voltages is used. For the standard comparator in Fig. 2.7a, after the reset phase the initial condition of the comparison phase is $OUT = \overline{OUT} = V_{DD}$. So at the beginning of the decision only transistors $N0$ and $N1$ of the latch contribute to positive feedback until the voltage level of one output node has dropped enough to turn on transistor $P0$ or $P1$ to start complete regeneration.

7.2.1 Circuit Description

To achieve low-voltage operation a proposed comparator design with the modified latch of Sect. 7.1 is presented, where the latch of the conventional circuit in Fig. 2.7a has been replaced by a new latch for low supply-voltage operation, where the advantages of the resulting comparator circuit are again a high-impedance input, a rail-to-rail output swing, no static power consumption and non-direct influence of parasitic capacitances of the input transistors on the output nodes and therefore to switching speed have been kept. The circuit of a comparator with the modified new latch of section 7.1 is shown in Fig. 7.15, where in opposite to the conventional latch used in Fig. 2.7a, where only $N0$ and $N1$ are initially on, the latch of the proposed comparator is expanded into two paths between the supply rails (transistors $N0$, $N1$, $P0$, $P1$, $P4$, $P5$) [3]. So at the beginning of the comparison phase, where both output nodes have the initial condition $OUT = \overline{OUT} = V_{Co}$, transistors $N0$, $N1$ are turned on. But also transistors $P0$, $P1$ are turned on and build together with the input transistors $N2$ and $N3$ an amplifier with a distinct operating point. Finally the complete positive feedback is done with $N0$, $N1$, $P0$, $P1$, where $P4$ and $P5$ help with additional amplification. The advantages of a high-impedance input, a rail-to-rail output swing, no static power consumption and no direct influence of parasitic capacitances of $N2$, $N3$ to the output nodes are kept. The transient simulations of the comparator are shown in Fig. 7.16. A clock period is divided into a reset phase and a comparison phase, where the voltage at *CINP* is compared with that at *CINN*. In the reset phase ($CLK = V_{SS}$) an initial condition $OUT = \overline{OUT} = V_{Co}$ for the following

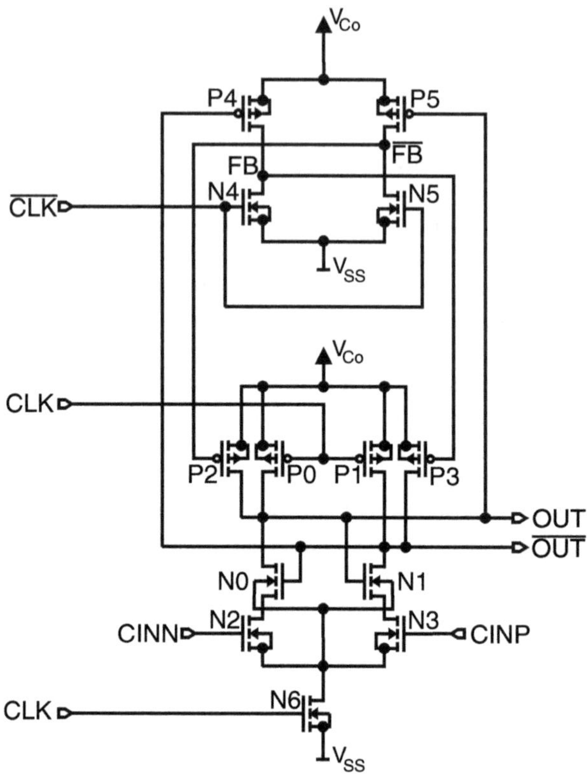

Fig. 7.15 Schematic of the comparator with the proposed latch [3]

comparison phase ($CLK = V_{Co}$, V_{Co} is the positive supply voltage of the comparator) is established. During reset, transistor $N6$ is switched off and transistors $P0$, $P1$, $N4$ and $N5$ are on. Consequently the output nodes OUT and $\overline{OUT}$ are pulled towards V_{Co} by reset transistors $P0$ and $P1$. This causes transistors $P4$ and $P5$ to be switched off. Reset transistors $N4$ and $N5$ are switched on and pull both internal nodes FB and $\overline{FB}$ to V_{SS}. As a consequence transistors $P2$ and $P3$ are turned on and help pulling OUT and $\overline{OUT}$ to the final voltage level V_{Co}. Comparison of the input voltages *CINP* with *CINN* is started, when *CLK* switches to voltage level V_{Co} (comparison phase) thus transistor $N6$ is turned on and $P0$, $P1$, $N4$ and $N5$ are turned off. At the very beginning, transistors $P4$, $P5$ are switched off, transistors $P2$ and $P3$ work in linear region and act as load for an amplifier with $N2$, $N3$. Transistors $N0$ and $N1$ are initially on (initially $OUT = \overline{OUT} = V_{Co}$ and $FB = \overline{FB} = V_{SS}$). Assuming that the voltage at *CINP* is larger than the voltage at *CINN*, then transistor $N3$ pulls the voltage level at output node $\overline{OUT}$ down faster than $N2$ does it at output node OUT. As a consequence transistor $P4$ begins to conduct when the potential at $\overline{OUT}$ becomes smaller than $V_{Co} - |V_{tP4}|$, where V_{tP4} is the threshold voltage of $P4$. In this initial time period also a small amount of positive feedback is contributed by transistors $N0$ and $N1$. When $P4$ begins to conduct, node *FB* is charged towards V_{Co} ($N4$, $N5$

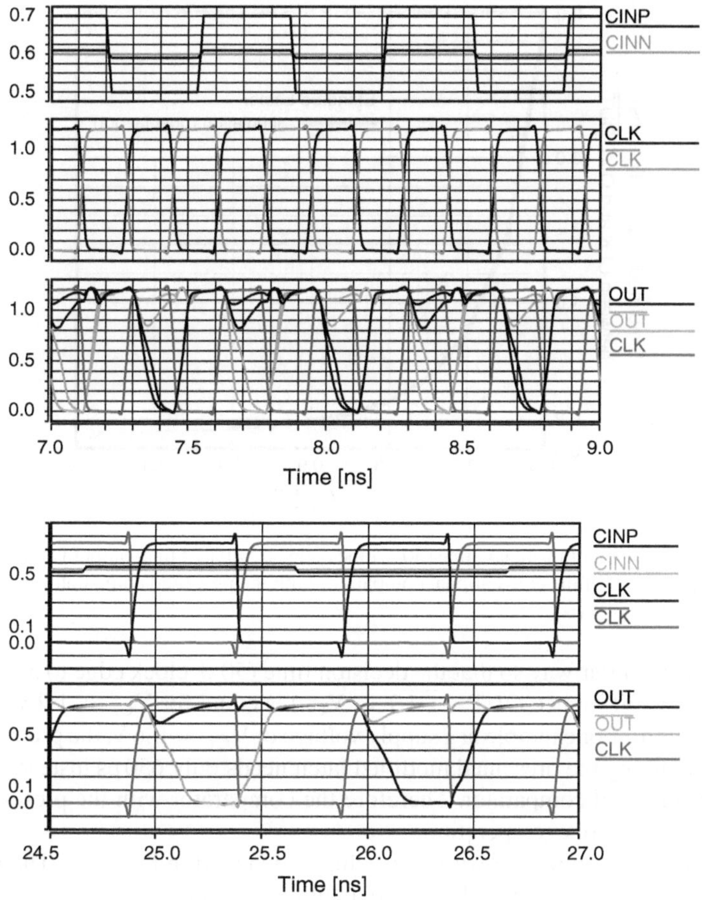

Fig. 7.16 Transient simulation of the comparator for the illustration of its function at a supply voltage $V_{Co} = 1.2V$ when two amplitudes of the data (10 mV and 0.1 V) are applied at the input nodes (*top*) and at a supply voltage $V_{Co} = 0.75$ V for an amplitude of 25 mV at 1 GHz clock frequency (*bottom*)

are off) and finally complete positive feedback is started. Transistor $P3$ is turned off and $P2$ keeps conducting, because of node OUT is pulled to V_{Co} thus $P5$ keeps off and $\overline{FB}$ remains near V_{SS} (The input voltage difference $CINP - CINN$ is assumed to be sufficiently high so that no metastability error occurs). When the comparison has finished, transistors $N1$, $P4$ and $P2$ are switched on and $N0$, $P3$ and $P5$ are turned off. Output node OUT is at voltage level V_{Co} and $\overline{OUT}$ at V_{SS}. No static current can flow after decision. When considering the other case of $CINP > CINN$, OUT is pulled to V_{SS} and $\overline{OUT}$ to V_{Co}, respectively.

For comparison of the decision times at different supply voltages of the proposed comparator containing the presented latch of Fig. 7.15 with the comparator of Fig. 2.7a have been simulated and are shown in Fig. 7.17. Both comparators were

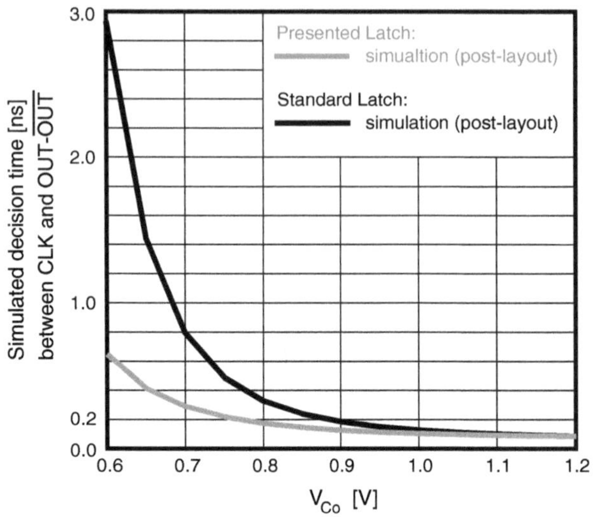

Fig. 7.17 Comparison of the delay time (50 % clock edge to 50 % of final output voltage difference $OUT - \overline{OUT} = V_{Co}$) of the conventional comparator with the proposed comparator in dependence on the supply voltage V_{Co}

designed in a similar way so that the decision time (50 % clock edge to 50 % of final output voltage difference $OUT - \overline{OUT} = V_{Co}$) is equal for $V_{Co} = 1.2$ V and same input conditions. For example at a supply voltage of $V_{Co} = 0.6$ V, the proposed comparator containing the presented modified latch needs only 650 ps instead of 2.95 ns of the conventional comparator. Therefore the comparator with the presented latch is capable of low-voltage operation.

The block diagram of the whole test chip is depicted in Fig. 7.18. The proposed comparator of Fig. 7.1 is placed in an area with a separate supply voltage V_{Co}. This is done to investigate the behavior of the comparator for lower supply voltages. With adapters the logic voltage levels of the clock signals as well the logic voltage levels of the comparators output are converted in that way that they switch between V_{SS} and the actual supply voltage. All transmission gates (T-Gates) are always turned on in this test chip. The T-Gates are included because the test environment of the previous test chip in Sect. 7.1 is used here as well to save additional time for drawing a new clock driver for this chip. So it has to be considered that the clock signals *CLKA* and *CLKL* equal *CLK*. A detailed description of the clock driver, the transfer stages, the output driver and the technique for measuring the delay time of the comparator with voltage ΔV_{DM} can be found in Chap. 5. The internal duty cycle of the clock can be measured with the voltages at nodes *CLKAVP* and *CLKAVN* and adjusted by setting the offset of the sine signal, which is applied off-chip to node *CLKIN* to create on-chip complementary square-wave clock signals. The duty cycle was always adjusted to be near 50 %.

In Fig. 7.19 Monte Carlo simulations with each a run of 50 samples are shown for the supply voltage of $V_{Co} = 1.2$ V and $V_{Co} = 0.65$ V of the comparator are depicted. In the comparator design the input transistors $N2$ and $N3$ (see Fig. 7.15)

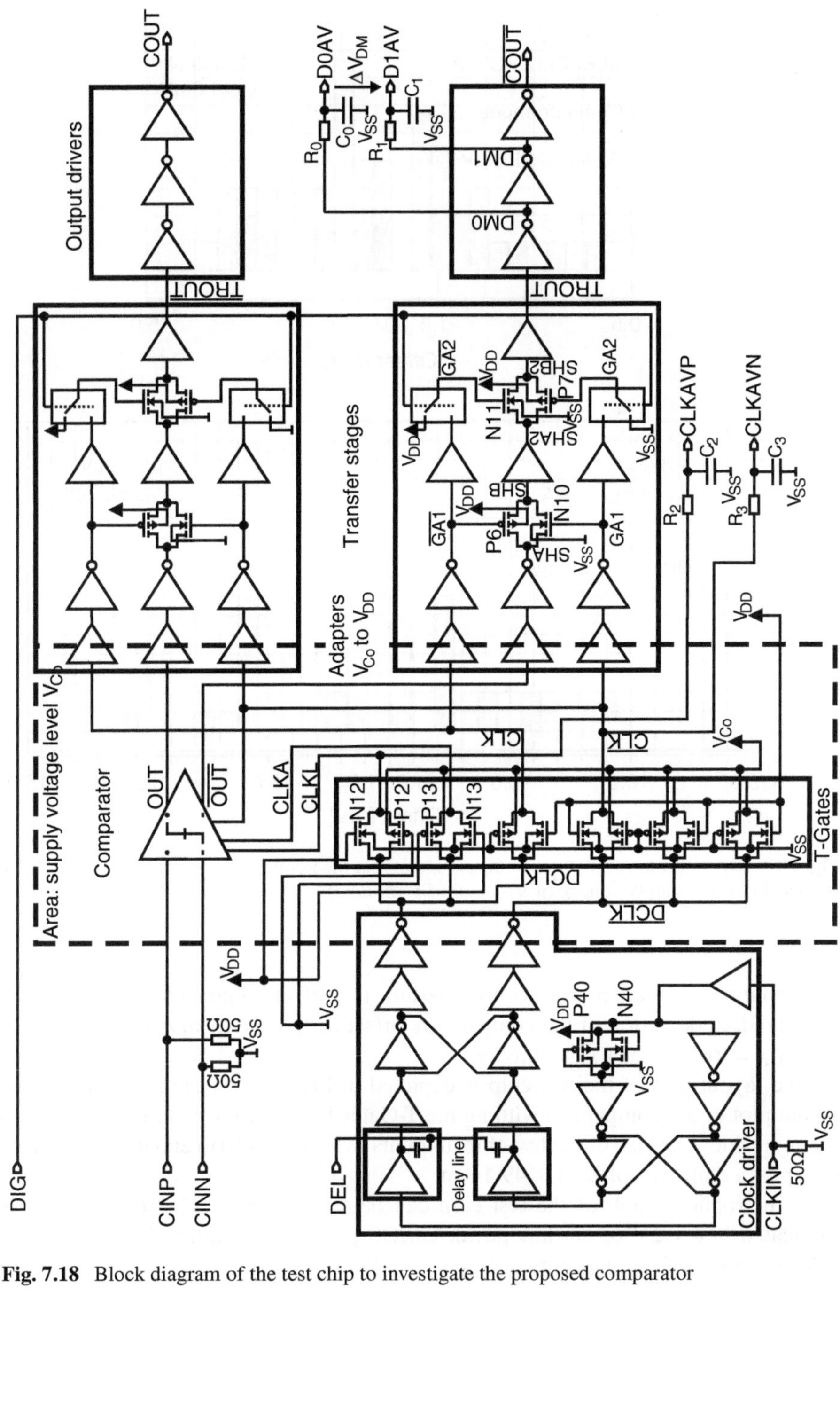

Fig. 7.18 Block diagram of the test chip to investigate the proposed comparator

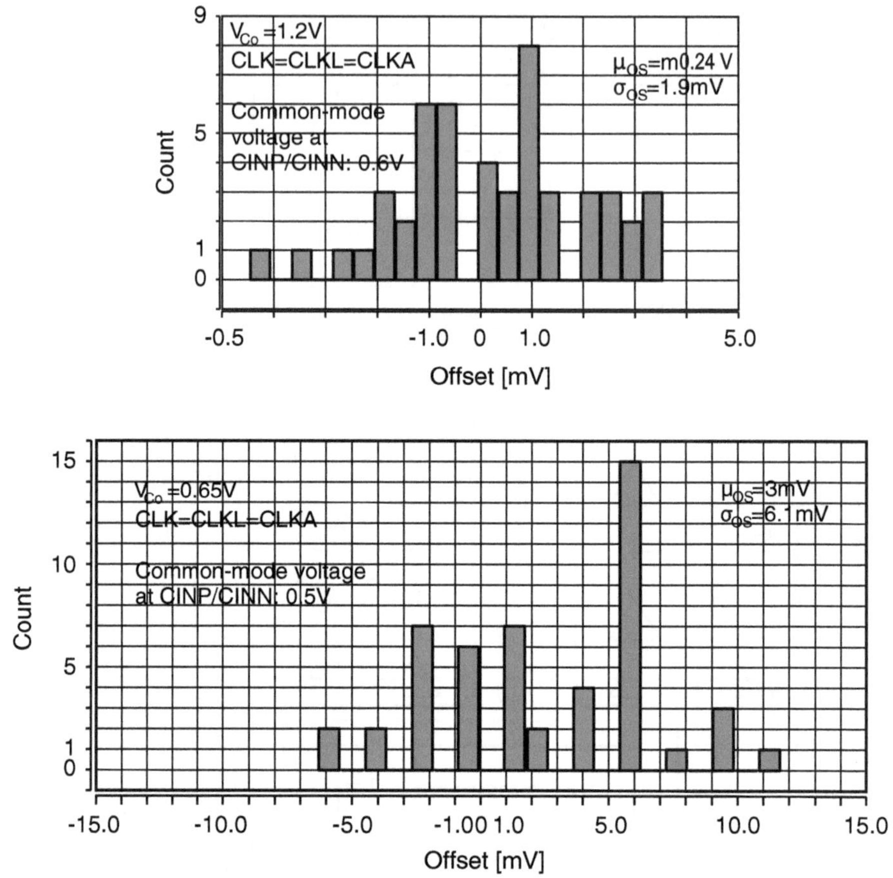

Fig. 7.19 Monte Carlo simulations of the comparator at the nominal supply voltage $V_{Co} = 1.2\,V$ (*top*) and at a low supply voltage of $V_{Co} = 0.65\,V$ (*bottom*)

are designed with larger gate width to reduce the offset voltage of the comparator. The standard deviation of the comparators offset is $\sigma_{OS} = 1.9\,mV$ at $V_{Co} = 1.2\,V$ and $\sigma_{OS} = 6.1\,mV$ at $V_{Co} = 0.65\,V$.

The layout plot of the test chip is depicted in Fig. 7.20 on the left side and the layout plot of the comparator without the T-Gates for sensitivity tuning is shown on the right side. The size of the test chip amounts to $930 \times 500\,\mu m$ and the comparator occupies a chip area of $28.4 \times 49.1\,\mu m$.

A microphotograph of the test chip can be seen in Fig. 7.21. The test chip is fabricated in a 1.2 V 65 nm low-power CMOS process with a threshold voltage of n-MOS and p-MOS transistors of $V_{tn} \approx |V_{tp}| \approx 0.4\,V$.

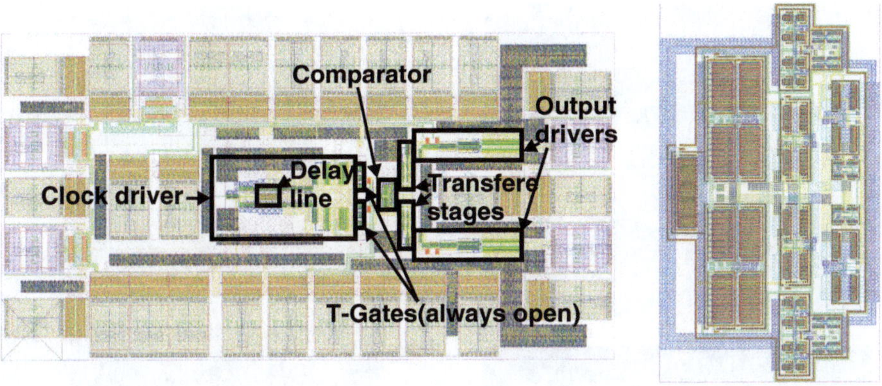

Fig. 7.20 Layout plots of the test chip (*left side*) and the comparator (*right side*)

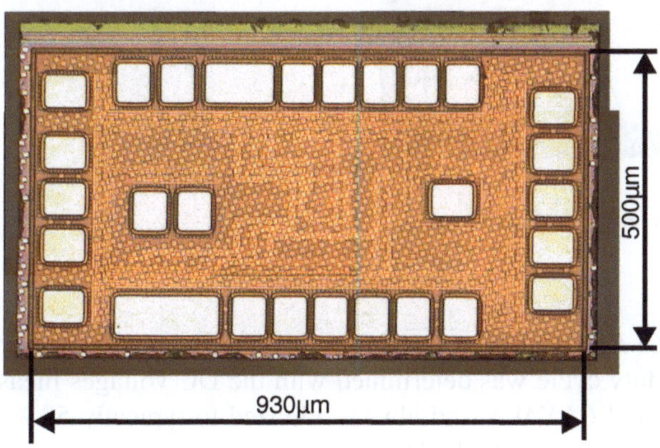

Fig. 7.21 Microphotograph of the test chip

7.2.2 Measurement Results

The fabricated chip (65 nm CMOS technology with 1.2 V nominal supply voltage) was bonded directly to a test board to reduce the influences of a package and its long bond wires. Photos of the chip mounted to a test board are shown in Fig. 7.22.

More than one bond wires were connected in parallel from chip ground V_{SS} to the ground of the test board to reduce a the series inductance and hence the ground bounce. Additionally from the power supply pads on the chip to the supply lines on the board and from the high-frequency pads on the chip to the microstrip lines two or more wires were bonded in parallel to reduce somewhat the series inductances. There were different supply voltages for the output drivers, for the comparator and for the logic circuitry to furthermore reduce disturbances. On the test board 50 Ω microstrip lines were designed for the high-frequency input and output signals of the chip (pads

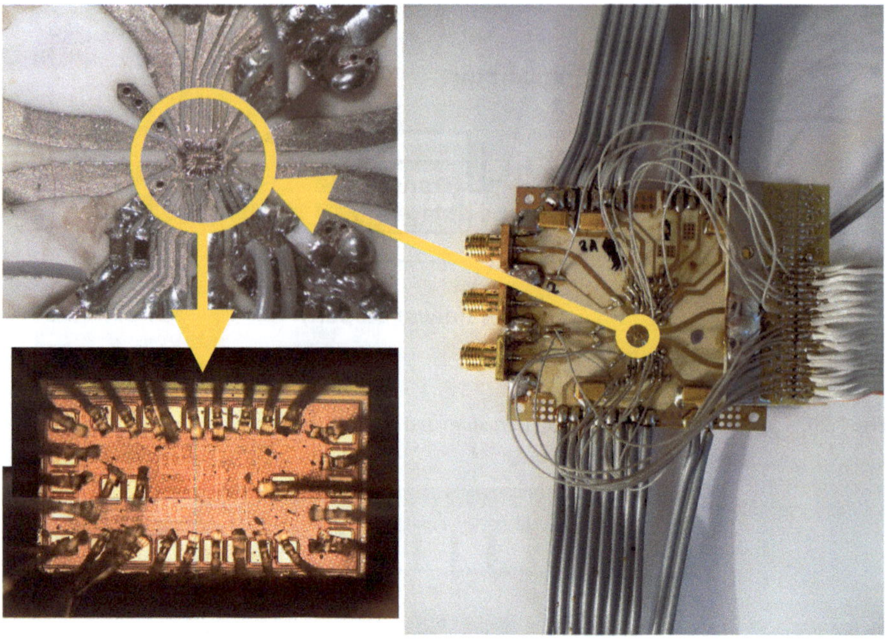

Fig. 7.22 Photos of the test board

CINP, *CINN*, *CLKIN*, *COUT*, $\overline{COUT}$). A typical measurement setup is described in Sect. 5.5. The duty cycle was determined with the DC voltages measured at output pads *CLKAVP* and *CLKAVN* and always adjusted to typically 50 % with changing the bias voltage at input pad *CLKIN*.

To test the functionality of the test chip, at *CLKIN* a sine signal, to input pad *CINP* a test signal overlayed on a DC voltage to compensate the offset of the comparator and to pad *CINN* a reference voltage of 0.65 V of the comparator were applied. Figure 7.23 shows screen shots of the test chip with the comparator at the maximum clock frequency of 5 GHz.

The influence of noise on the decision of the comparator and thus the sensitivity can be characterized with the help of statistical measurements by measuring the BER with an appropriate pattern generator and receiver. At *CINN* a reference voltage and at *CINP* a PRBS $2^{31} - 1$ were applied, which was superimposed to a bias voltage of *CINN* + offset to compensate the offset of the comparator. The amplitude of a bit-sequence is defined here, that the bit-sequence switches at *CINP* ± amplitude around voltage level *CINN* + offset, while at CINN the reference voltage is applied. Measured BER results of a typical test chip for lower supply voltages of $V_{Co} = 0.75\,V$ and $V_{Co} = 0.65\,V$ than the nominal supply voltage of 1.2 V are shown in Fig. 7.24. On the left side in Fig. 7.24 the BER versus the bias voltage at *CINN*, which is similar to the common-mode input voltage, is plotted. The points of optimal BER are located at *CINN* = 0.5 V for 0.5 GHz clock frequency and $V_{Co} = 0.65\,V$

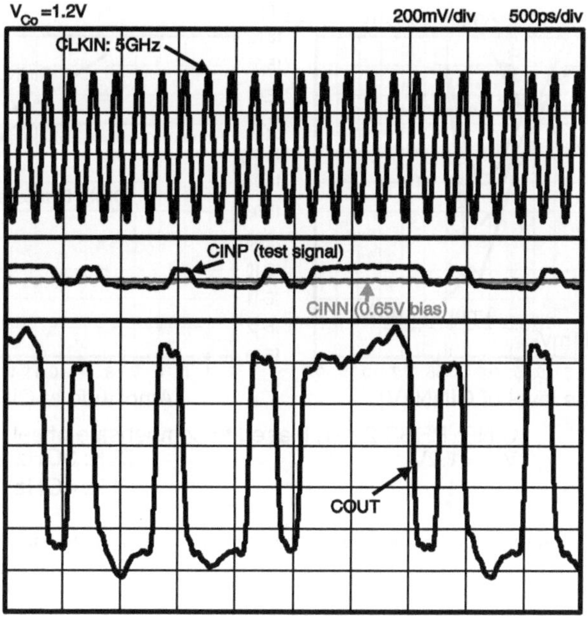

Fig. 7.23 Oscilloscope screen shots at 5 GHz clock frequency

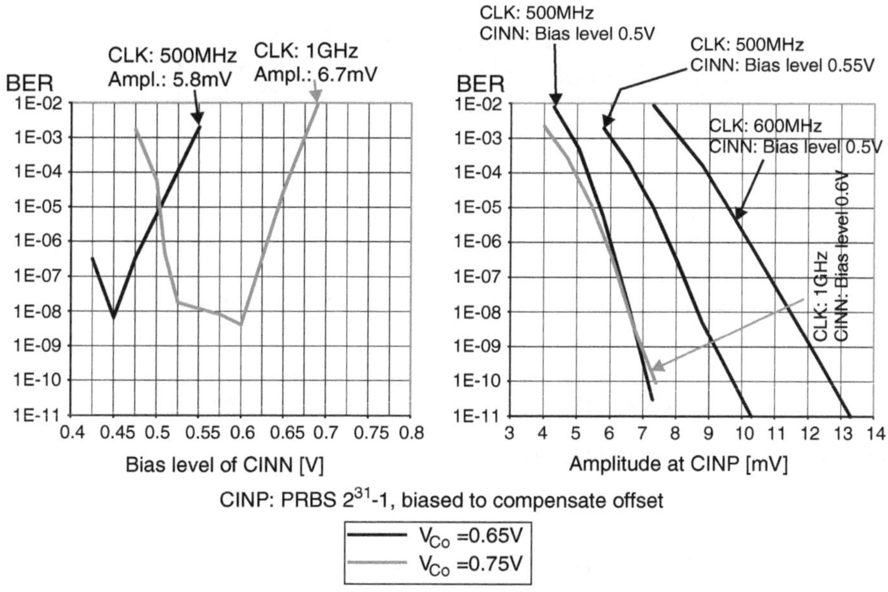

Fig. 7.24 BER measurements for supply voltages of $V_{Co} = 0.65\,V$ and $V_{Co} = 0.75\,V$ of the comparator

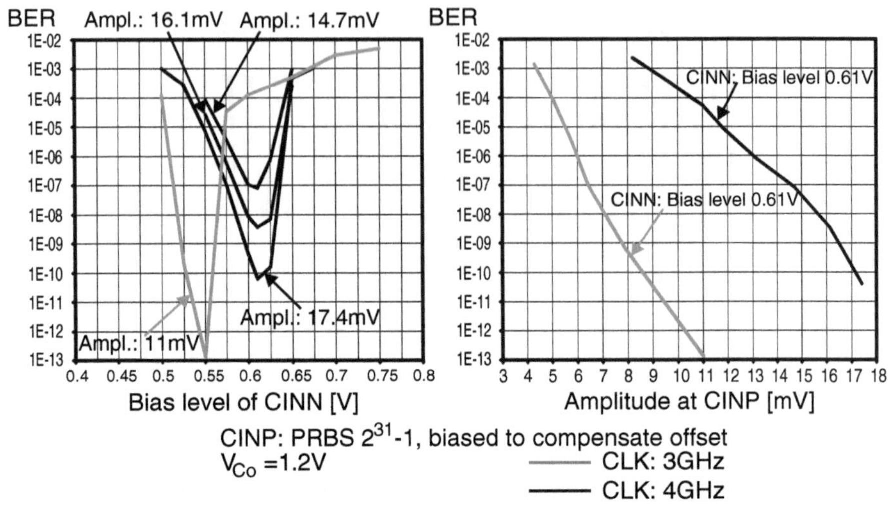

Fig. 7.25 BER measurements for clock frequencies of 3 and 4 GHz at the nominal supply voltage of $V_{Co} = 1.2V$

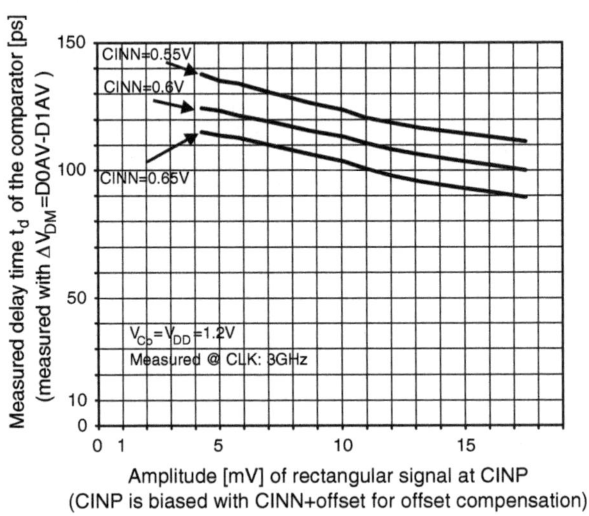

Fig. 7.26 Measured delay time t_d of the comparator

and *CINN* = 0.6 V for 1 GHz clock frequency and $V_{Co} = 0.75$ V. On the right side in Fig. 7.24 several curves of BER versus the amplitude of the data signal applied at *CINP* at a distinct *CINN* are depicted for different supply voltages V_{Co} of the comparator and clock frequencies. The measured sensitivities (BER $= 10^{-9}$) are 7 mV at 1 GHz/0.75 V, 6.9 mV at 0.5 GHz/0.65 V and 12.1 mV at 0.6 GHz/0.65 V.

The measurement results for the nominal supply voltage $V_{Co} = 1.2\,V$ and clock frequencies of 3 and 4 GHz are shown in Fig. 7.24. To achieve a BER $= 10^{-9}$ at 1.2 V supply, an amplitude of 7.8 mV at 3 GHz and 16.5 mV at 4 GHz are needed (Fig. 7.25). For a clock frequency of 5 GHz and $V_{Co} = 1.2\,V$, an amplitude of 145 mV has to be applied.

The results of the mean delay time of the comparator, which has been determined by a measurement of 10 chip samples are shown in Fig. 7.26. For an amplitude of e.g. 15 mV at the input of the comparator the mean delay times ($\sigma \approx 7\,ps$) were 93 ps at *CINN* = 0.65 V, 104 ps at *CINN* = 0.6 V and 115 ps at *CINN* = 0.55 V at $V_{Co} = 1.2\,V$.

The measured power consumption of the comparator is 2.88 mW at 5 GHz (1.2 V supply voltage), 295 μW at 1 GHz (0.75 V supply voltage), 128 μW at 0.6 GHz (0.65 V supply voltage).

References

1. B. Goll, H. Zimmermann, A 65nm CMOS comparator with modified latch to achieve 7 GHz/1.3 mW at 1.2 V and 700 MHz/47 μW at 0.6 V, in IEEE International Solid-State Circuits Conference, pp. 328–329 (2009)
2. D. Schinkel, E. Mensink, E. Klumperink, E. van Tuijl, B. Nauta, A double-tail latch-type voltage sense amplifier with 18 ps setup-hold time, in IEEE International Solid-State Circuits Conference, pp. 314–315 (2007)
3. B. Goll, H. Zimmermann, A comparator with reduced delay time in 65-nm CMOS for Supply voltages down to 0.65 V. IEEE Trans. Circuits Syst-II: Express Briefs **56**(11), 810–814 (2009)

Chapter 8
Conclusion and Comparison

A key element in an ADC is the comparator. It compares an input signal with a reference voltage and has as a result a logic stage, which indicates whether the signal is lower or higher. In most cases in fast ADCs, e.g. a flash ADC or a folding converter system [1, 2], many parallel clocked regenerative comparators are implemented, which work in parallel. In ADCs typically additional circuitry to the latch [3, 4] and pre-amplifiers, which are mostly offset compensated [5], are added, which enhance the resolution and reduce in some cases the overall offset. Also techniques to cancel the offset of a comparator may be inserted [6, 7]. Other applications of clocked regenerative comparator circuits are their usages in an analog rank-order extractor [8] or as a voltage sense amplifier in a SRAM [9–11] or for data regeneration in e.g. a fast demultiplexer [12, 13]. This work has focused on clocked regenerative comparators in a 120 nm and 65 nm CMOS technology with a nominal supply voltages of 1.5 and 1.2 V, respectively. The demands to such comparators are among many others e.g. a low power consumption, a high sampling clock frequency and sufficient sensitivity and offset.

Within this book several new implementations for clocked regenerative comparators and on-chip plus off-chip measurement setups to characterize such comparators have been proposed:

- To characterize a comparator, extensive bit-error-rate (BER) measurements were done instead of the conventional usage of counters (see Sects. 2.4.2 and 2.5).
- A delay between the enable and the reset signal was introduced to the comparator in Sects. 6.1 and 6.2 to somewhat enhance the output voltage difference and the time length of the output voltage so that a following logic gate better recognizes the comparator's logic decision.
- A voltage buffer (see Sect. 5.1) with a −3dB corner frequency of 10 GHz and with a low input capacitance, which is able to drive a 50 Ω off-chip measurement system was designed to observe the output voltages of a comparator with a fast oscilloscope.

B. Goll and H. Zimmermann, *Comparators in Nanometer CMOS Technology*,
Springer Series in Advanced Microelectronics 50, DOI 10.1007/978-3-662-44482-5_8

- An on-chip delay time measurement technique for a comparator, where only DC voltage levels have to be measured, is proposed in Sect. 5.3.
- In Sects. 6.3 and 6.5 the comparator circuits are designed so that only dynamic power is consumed by adding two cross-coupled transistors in the part with the input transistors.
- A technique to tune the sensitivity of a comparator is proposed in Sects. 6.3 and 6.5, where at lower clock frequencies than the maximum possible clock frequency the comparator is able to work, the sensitivity is enhanced.
- In Sect. 6.4 a comparator, which even works at a very low supply voltage of 0.5 V with a clock frequency of 600 MHz with 18 μW power consumption is proposed, where the active load transistors are also used to reset the comparator which saves additional parasitic capacitances at the output nodes to enhance speed.
- The possibility to fine-adjust the optimal working point, where the best BER occurs at a distinct input common-mode voltage level, and the slight improvement to the sensitivity of some comparators, when lowering the bias voltage of the n-wells of distinct p-MOS transistors are shown with BER measurements in Sects. 6.2, 6.3 and 6.4.
- Comparators with a modified latch for low supply voltages have been implemented in a 65 nm low-power CMOS technology, where the nominal supply voltage is 1.2 V and where the threshold voltages of MOS transistors are in the range of 0.4–0.45 V, are presented in Chap. 7. There is also a challenge to design fast low-power comparators , which also work at supply voltages lower than 1.2 V or which have a low standard deviation of the offset. In this technology also tunnel gate currents of MOS transistors have to be considered if the gate area is increased. The comparator in Sect. 7.1 works at a high clock frequency of 7 GHz at a nominal supply voltage of 1.2 V (1.3 mW power consumption) and when the supply voltage is reduced to 0.6 V, the circuit is able to manage a clock frequency of 700 MHz while consuming 47 μW power. In Sect. 7.2 a comparator is presented, which worked at a clock frequency of 0.6 GHz at a supply voltage of 0.65 V and consumed 128 μW power.

In Table 8.1 a comparison of the designed comparators with deep-sub-micron and nanomter clocked regenerative comparators from the literature is depicted.

There are some items to improve or continue the described work in this book:

- The on-chip delay time measurement technique, where only DC voltages are measured, may be optimized so that the offset error disappears, which can be done by achieving the same capacitive load at the input node of the first transmission gate and at each node, which is connected to a gate of a MOS transistor of the first transmission gate.
- A flash ADC may be designed, which consists of clocked regenerative comparators, which can be tuned to optimal sensitivity to enhance the resolution.
- The measurement setup may be programmed to work fully automatically, which is indeed a difficult task.

Table 8.1 Comparison of the own work with the state of the art

Technology	Topology/ application	Supply voltage	Clock	Sensitivity	Offset	Power consumption	Reference
0.18 μm CMOS	4-stage comparator: core comparator (CML, inductor loads) + 2 latches + digital flip flop in 4-bit flash ADC	1.8 V analog 2.1 V to 2.5 V digital	4 GHz	n.m.	n.m.	n.m.	[14]
0.11 μm CMOS	3-stage comparator (3rd stage consists of 2 additional amplifiers) in CML with bandwidth modulation technique in 40 GB/s DEMUX	1.2 V	10 GHz	BER < 10^{-12} for ΔU_{in} =1 V_{pp} measured	n.m.	37 mW measured with output drivers Simulated sum of CML tail currents of comparator itself: 4.2 mA	[12], [13]
0.18 μm CMOS	Investigation and comparison of a comparator with and without offset compensation, where also offset changes due to supply noise are treated	1.8 V	1.4 GHz	n.m.	$3\sigma_{OS}$ = 95 mV without offset compensation $3\sigma_{OS}$ = 16 mV with offset compensation	350 μW	[6]
0.18 μm CMOS	1-bit quantizer with rail-to-rail input range for $\Sigma\Delta$-modulators	0.8 V	10 MHz 12 MHz	10 μV at 10 MHz (simulated) 100 μV at 12 MHz (simulated)	n.m.	134 μW	[15]
0.13 μm CMOS	Yield, speed optimization of a latch-type voltage sense amplifier for SRAM	1.5 V, works down to 0.7 V	Delay time in comparison phase higher than 11 ns at 0.7 V supply	n.m.	1.5 V supply σ_{OS} = 19 mV at an input common-mode voltage of 1.5 V	n.m.	[10], [11], [16]

(continued)

Table 8.1 (continued)

Technology	Topology/ application	Supply voltage	Clock	Sensitivity	Offset	Power consumption	Reference
			Delay time in comparison phase higher than 1.5 ns at 1.5 V		1.5 V supply σ_{OS} = 8.5 mV at an input common-mode voltage of 1.05 V		
0.13 µm CMOS	Comparator with pre-amplifier in a 6-bit flash ADC	1.5 V	4 GHz	n.m.	n.m.	n.m.	[17]
0.25 µm CMOS	Comparator with pre-amplifier in a 6-bit flash ADC	1.8 V (analog part)	1.3 GHz	n.m.	n.m.	n.m.	[18]
0.13 µm CMOS	Comparator with pre-amplifier in a 6-bit flash ADC (offset compensated pre-amplifiers plus comparator latch)	1.5 V	1.6 GHz	n.m.	σ_{OS} = 80 mV of comparator latch, σ_{OS} = 2 mV with offset compensated pre-amplifiers (gain about 40)	n.m.	[5]
0.18 µm CMOS	4-bit flash ADC, pre-amplifier + 3 CML latches, ±0.4 V swing at outputs (simulated)	1.8 V	5 GHz	n.m.	σ_{OS} = 67 mV (comparator simulation)	Sum of tail currents: 1 mA × 1.8 V (simulated)	[19]
0.13 µm CMOS	pre-amplifier and comparator in 6-bit/22 MHz flash ADC	0.8 V	22 MHz	n.m.	n.m.	n.m.	[20]
90 nm CMOS	Sense amplifier	1.2 V	1 GHz, 2 GHz	measured RMS noise voltage V_{rms} = 1.5 mV	σ_{OS} = 13 mV (simulated)	113 µW@1 GHz, 225 µW@2 GHz, 92 fJ/decision	[21]
65 nm CMOS	Comparator with pre-amplifier	1.2 V	1 GHz	n.m.	Down to 20 mV (measured)	380 µW@1 GHz	[22]
40 nm CMOS	SAR-ADC with tri-level comparator	0.5 V	15–50 MHz[1]	n.m.	n.m.	1.2 µW@ 1.1 MS/s[2], 6.3 fJ/ conversion-step[2)]	[23]

(continued)

Table 8.1 (continued)

Technology	Topology/ application	Supply voltage	Clock	Sensitivity	Offset	Power consumption	Reference
0.12 μm CMOS	Pre-amplifier, comparator and transfer stage with delayed reset	1.5 V	1.5 GHz	BER of 10^{-9}: 25 mV@1.5 GHz	σ_{OS} = 20.2 mV (simulated)	660 μW@ 1.5 GHz	[24] Sect. 6.1
0.12 μm CMOS	Standard comparator with delayed reset	1.5 V	2 GHz	BER of 10^{-9}: 9.5 mV@1.5 GHz 14 mV@1.8 GHz 16 mV@2 GHz	σ_{OS} = 9.5 mV (simulated)	360 μW@ 2 GHz	[4] Sect. 6.2
0.12 μm CMOS	Clocked, regenerative comparator with tunable sensitivity and two cross-coupled transistors added above the p-MOS input transistors	1.5 V	4.2 GHz	BER of 10^{-9}: 10.8 mV@1.5 GHz 11.2 mV@2 GHz 14.5 mV@2.5 GHz 20 mV@3 GHz 26mV@3.5 GHz 118 mV@4 GHz	σ_{OS} = 32.9 mV (simulated)	812 μW@ 4 GHz	[25] Sect. 6.3
0.12 μm CMOS	Clocked, regenerative comparator with two active p-MOS transistor loads which are also used for reset	1.5 V	6 GHz	BER of 10^{-9}: 11.2 mV@3 GHz 16.5 mV@4 GHz 29.4 mV@5 GHz 38 mV@5.5 GHz	σ_{OS} = 22.1 mV (simulated)	2.65 mW@ 6 GHz	[26], [27] Sect. 6.4
		0.5 V	0.6 GHz	BER of 10^{-9}: 21.2 mV@0.4 GHz 25.8 mV@0.5 GHz 60.5 mV@0.6 GHz	σ_{OS} = 46 mV (simulated)	18 μW@ 0.6 GHz	
0.12 μm CMOS	Clocked, regenerative comparator with advanced tunable sensitivity and the complementary circuit of [25]	1.5 V	3 GHz	BER of 10^{-9}: 3.9 mV@2 GHz 9.2 mV@3 GHz	σ_{OS} = 14.5 mV (simulated)	584 μW@ 3 GHz	[28] Sect. 6.5
65 nm CMOS	Clocked, regenerative comparator with new latch for low supply voltage operation	1.2 V	7 GHz	BER of 10^{-9}: 20 mV@4 GHz 27.2 mV@5 GHz 63 mV@6 GHz	σ_{OS} = 24.5 mV (simulated) σ_{OS} = 18.3 mV (simulated, with sensitivity tuning)	1.3 mW@ 7 GHz	[29] Sect. 7.1
		0.6 V	0.7 GHz	BER of 10^{-9}: 16 mV@0.5 GHz 34 mV@0.6 GHz 90.2 mV@0.7 GHz	σ_{OS} = 47 mV (simulated)	47 μW@ 0.7 GHz	

(continued)

Table 8.1 (continued)

Technology	Topology/ application	Supply voltage	Clock	Sensitivity	Offset	Power consumption	Reference
65 nm CMOS	Clocked, regenerative comparator with new latch for low supply voltage operation	1.2 V	5 GHz	BER of 10^{-9}: 7.8 mV@3 GHz 16.5 mV@4 GHz 145 mV@5 GHz	$\sigma_{OS} = 1.9$ mV (simulated) $\sigma_{OS} = 6.1$ mV (simulated)	2.88 mW@ 5 GHz	[30] Sect. 7.2
		0.65 V	0.6 GHz	BER of 10^{-9}: 6.9 mV@0.5 GHz 12.1 mV@0.6 GHz		128 μW@ 0.6 GHz	

n.m = not mentioned, Sect. = Section, @ = at
[1] for tri-level comparator
[2] for SAR-ADC with tri-level comparator

References

1. R. van de Plasche, *CMOS Integrated Analog-to-Digital and Digital-to-Analog Converters* (Kluwer Academic Publishers, Boston, 2003)
2. R. Patzelt, H. Schweinzer, *Elektrische Messtechnik* (Springer, Wien, 1996)
3. L. Sumanen, M. Waltari, V. Hakkarainen, K. Halonen, CMOS dynamic comparators for pipeline A/D converters. IEEE Int. Symp. Circuits Syst. **1**, 157–160 (2002)
4. B. Goll, H. Zimmermann, A low-power 2-Gsample/s comparator in 120 nm CMOS technology. IEEE European Solid-State Circuits Conf. 507–510 (2005)
5. C. Sandner, M. Clara, A. Santner, T. Hartnig, F. Kuttner, A 6bit, 1.2GSps low-power flash-ADC in 0.13 μm digital CMOS. IEEE J. Solid-State Circuits **40**(7), 1499–1505 (2005)
6. K.-L.J. Wong, C.-K.K. Yang, Offset compensation in comparators with minimum input-referred supply noise. IEEE J. Solid-State Circuits **39**(5), 837–840 (2004)
7. E.L. Wong, P.A. Abshire, M.H. Cohen, Floating gate comparator with automatic offset manipulation functionality. IEEE Int. Symp. Circuits Syst. **1**, 529–532 (2004)
8. Y.-C. Hung, B.-D. Liu, 1 V CMOS comparator for programmable analog rank-order extractor. IEEE Trans. Circuits Syst. I **50**(5), 673–677 (2003)
9. B. Wicht, *Current Sense Amplifiers for Embedded SRAM in High-Performance System-on-a-Chip Designs* (Springer, Berlin, 2003)
10. B. Wicht, T. Nirschl, D. Schmitt-Landsiedel, Yield and speed optimization of a latch-type voltage sense amplifier. IEEE J. Solid-State Circuits **39**(7), 1148–1158 (2004)
11. B. Wicht, T. Nirschl, D. Schmitt-Landsiedel, A yield-optimized latch-type SRAM sense amplifier, IEEE European Solid-State Circuits Conf. 409–412 (2003)
12. Y. Okaniwa, H. Tamura, M. Kibune, D. Yamazaki, T.-S. Cheung, J. Ogawa, N. Tzartzanis, W.W. Walker, T. Kuroda, A 40 Gb/s CMOS clocked comparator with bandwidth modulation technique. IEEE J. Solid-State Circuits **40**(8), 1680–1685 (2005)
13. Y. Okaniwa, H. Tamura, M. Kibune, D. Yamazaki, T.-S. Cheung, J. Ogawa, N. Tzartzanis, W. W. Walker, T. Kuroda, A 0.11 μm CMOS clocked comparator for high-speed serial communications, IEEE Symp. VLSI Circuits. 198–201 (2004)
14. S. Park, Y. Palaskas, M. P. Flynn, A 4GS/s 4b flash ADC in 0.18 μm CMOS, IEEE Int. Solid-State Circuits Conf. 570–571 (2006)
15. M. Maymandi-Nejad, M. Sachdev, 1-Bit quantiser with rail to rail input range for sub-1 V $\Delta\Sigma$ modulators. IET Electron. Lett. **39**(12), 894–895 (2003)
16. B. Wicht, J.-Y. Larguier, D. Schmitt-Landsiedel, A 1.5 V 1.7 ns 4k × 32 SRAM with a fully-differential auto-power-down current sense amplifier, IEEE Int. Solid-State Circuits Conf. 462–463 (2003)

17. C. Paulus, H.-M. Blüthgen, M. Löw, E. Sicheneder, N. Brüls, A. Courtois, M. Tiebout, R. Thewes, A 4GS/s 6b flash ADC in 0.13 μm CMOS, IEEE Symp. VLSI circuits. 420–423 (2004)
18. K. Uyttenhove, M. Steyaert, A 1.8 V 6-Bit 1.3 GHz flash ADC in 0.25 μm CMOS. IEEE J. Solid-State Circuits **38**(7), 1115–1122 (2003)
19. S. Sheikhaei, S. Mirabbasi, A. Ivanov, A 4-Bit 5GS/s flash A/D converter in 0.18 μm CMOS. IEEE Int. Symp. Circuits Syst. **6**, 6138–6141 (2005)
20. J. H.-C. Lin, B. Haroun, An embedded 0.8 V/480 μW 6b/22 MHz flash ADC in 0.13 μm digital CMOS process using nonlinear double-interpolation technique, IEEE Int. Solid-State Circuits Conf. 308–309 (2002)
21. D. Schinkel, E. Mensink, E. Klumperink, E. van Tuijl, B. Nauta, A double-tail latch-type voltage sense amplifier with 18 ps setup-hold time, IEEE Int. Solid-State Circuits Conf. 605, 314–315, (2007)
22. X. Zhu, Y. Chen, M. Kibune, Y. Tomita, T. Hamada, A dynamic offset control technique for comparator design in scaled CMOS technology, IEEE Custom Integr. Circuits Conf. 495–498 (2008)
23. A. Shikata, R. Sekimoto, T. Kuroda, H. Ishikuro, A 0.5 V 1.1 MS/s 6.3 fJ/conversion-step SAR-ADC with tri-level comparator in 40 nm CMOSl. IEEE J. Solid-State Circuits **47**(4), 1022–1030 (2012)
24. B. Goll, H. Zimmermann, A low power 1.4 Gsample/s comparator for flash-ADCs in 120 nm CMOS technology, Austrochip 2004. 39–42 (2005)
25. B. Goll, H. Zimmermann, A low-power 4 GHz comparator in 120 nm CMOS technology with a technique to tune resolution, IEEE European Solid-State Circuits Conf. 320–323 (2006)
26. B. Goll, H. Zimmermann, A 0.12 μm CMOS comparator requiring 0.5 V at 600 MHz and 1.5 V at 6 GHz, IEEE Int. Solid-State Circuits Conf. 316–317 (2007)
27. B. Goll, H. Zimmermann, Low-power 600 MHz comparator for 0.5 V supply voltage in 0.12 μm CMOS. IET. Electron. Lett. **43**(7), 388–390 (2007)
28. B. Goll, H. Zimmermann, A clocked, regenerative comparator in 0.12 μm CMOS with tunable sensitivity, IEEE European Solid-State Circuits Conf. 408–411 (2007)
29. B. Goll, H. Zimmermann, A 65 nm CMOS comparator with modified latch to achieve 7 GHz/1.3 mW at 1.2 V and 700 MHz/47 W at 0.6 V, IEEE Int. Solid-State Circuits Conf. 328–329 (2009)
30. B. Goll, H. Zimmermann, A comparator with reduced delay time in 65 nm CMOS for supply voltages down to 0.65 V, IEEE Transactions on Circuits and Systems-II: Express. Briefs **56**(11), 810–814 (2009)

Index

B. Goll and H. Zimmermann, *Comparators in Nanometer CMOS Technology*,
Springer Series in Advanced Microelectronics 50, DOI 10.1007/978-3-662-44482-5

MIX
Papier aus verantwortungsvollen Quellen
Paper from responsible sources
FSC® C105338

If you have any concerns about our products,
you can contact us on
ProductSafety@springernature.com

In case Publisher is established outside the EU,
the EU authorized representative is:
Springer Nature Customer Service Center GmbH
Europaplatz 3, 69115 Heidelberg, Germany

Printed by Libri Plureos GmbH
in Hamburg, Germany